T0315124

THE SELECTION PROCESS OF BIOMASS MATERIALS FOR THE PRODUCTION OF BIO-FUELS AND CO-FIRING

IEEE Press
445 Hoes Lane
Piscataway, NJ 08854

THE SELECTION PROCESS OF BIOMASS MATERIALS FOR THE PRODUCTION OF BIO-FUELS AND CO-FIRING

Najib Altawell

Mohamed E. El-Hawary, *Series Editor*

Published simultaneously in Canada

Library of Congress Cataloging-in-Publication Data:

Altawell, Najib.
The selection process of biomass materials for the production of bio-fuels and co-firing / Najib Altawell.
pages cm
Includes index.
Includes bibliographical references and index.
ISBN 978-1-118-54266-8 (hardback)
1. Biomass chemicals. 2. Renewable energy sources. I. Title.
TP248.B55A48 2014
662'.88–dc23

2013047869

Printed in the United States of America

10 9 8 7 6 5 4 3 2 1

For those who are dedicated to look after
the environment
and
life as a whole

CONTENTS

PREFACE

More than 15 years ago, the following statement was made: "In Sweden today, power production with bioenergy systems is more costly than with fossil energy system" (Gustavsson and Börjesson, 1998).Today, although the cost of solid biomass still does not match that of coal, the legislative framework is established through mechanisms, such as the renewable obligation (RO) in the United Kingdom, to encourage biomass utilization for power generation. However, the factors surrounding biomass utilization in co-firing together with dedicated biomass power plant and combined heat and power (CHP) are complex, as they depend on business, scientific, and technical factors. In this book, a methodology is developed to assist in the selection of the most suitable biomass materials for co-firing, as well as for the production of bio-fuels.

At least nine scientific and technical (S&T) factors, including calorific value, ash content, and combustion performance, are evaluated. Similarly, more than 30 business factors (BF) concerning the overall viability, including environmental and human health risks, have been considered. Weightings have been applied to each of the factors based on expert input. Of the biomass samples considered, rapeseed was the highest rated, followed by black sunflower seeds, niger seed, apple tree pruning, and sunflower striped seeds.

The scenario of using a mixed biomass blend based on these samples (super fuel sample, SFS) is explored as a means to reduce the cost in relation to performance. Although the methodology is designed in the first instance for comparing different biomass samples for co-firing, it can be applied to any scenario involving biomass utilization. Examples of this would be pyrolysis and gasification, along with the sole production of a new bio-fuel.

This book has been designed and compiled for the widest possible range of readers, researchers, businessmen, and economists who are connected in one way or another to the renewable energy field in general, and biomass energy in particular. Most important, this book has been compiled for the general reader who may or may not have a technical or scientific background. This means the use of the technical language has been avoided wherever possible. As a result, the style of the writing is simple; that is, this book should be accessible to the majority of people with little or no higher education.

With regard to methodology, the approach and mechanism is also simple and flexible. This means that the design and application of the methodology has not been written specifically for any particular energy business or, for that matter, a "specific" power station. In essence, the aspects of the methodology can be applied to a variety of biomass energy projects and businesses. In fact, with minor adjustments, the methodology itself can be applied to any type of commercial renewable energy enterprise. Therefore, the methodology is "universal" in its approach and application; that is, a commercial energy business has the option and the flexibility to fit the methodology to its own type of functionalities, business dealings, and calculations. In a way, the methodology resembles a "skeleton" or "backbone" that can fit into any situation related to any commercial energy scheme. It gives the opportunity for those who are using it to "tailor" it for their own particular use in order to help in the production of their fuel. Methodology factors, percentage values, scoring values, and other related variables or constants are left open so the users can insert their own values during the application of the methodology. To achieve this kind of flexibility, the methodology is designed so that the weighting percentage factors, boundary level scoring, and the addition/removal (or change) of both BF and S&T factors can all be manipulated. A power generating company, or any energy business, can create their own default values in a form they find more beneficial to their own business.

In some sections of the book, such as in Chapter 4, topics related to ash obtained from the samples used in this book have been briefly mentioned; that is, only the main elements for each sample have been listed in a graph. The reason for this is that another book will be forthcoming. This book will deal with ash for the aforementioned samples as well as other aspects related to the field of bio-energy.

Finally, whether the need arises from the original objectives of the research in the form of selecting suitable biomass materials for the purpose of generating electricity or from the principle aim of reducing carbon dioxide from the atmosphere, the flexibility of the methodology has provided a platform for future development and regular updates.

NAJIB ALTAWELL

REFERENCE

Gustavsson L, Börjesson P (1998) CO_2 mitigation cost: bioenergy systems and natural gas systems with decabonization. Energy Policy 26(9):699–713.

ACKNOWLEDGMENTS

I would like to thank the staff at the Centre for Energy, Petroleum and Mineral Law and Policy (CEPMLP) (Dundee University), the Centre for Water Law, Policy and Science (CWLPS) (Dundee University), the Institute of Energy and Sustainable Development (IESD), the Faculty of Technology (De Montfort University), and the rest of the staff at De Montfort University (DMU). In particular, I would like to thank Dr. Rafael Macatangay (CEPMLP), Professor Chris Spray (CWLPS), Dr. Neil Brown (IESD), Dr. Nicole Archer (CWLPS), Dr. Leticia Ozawa-Meida (IESD), and Mr. Jim Boulton (DMU).

N.A.

ABBREVIATIONS

AD	Absolute density
AF	Ash factor
APF	Applicability factor
Apple P	Apple pruning
BF	Business factors
BFB	Bubbling fluid bed
BMF	Baseline methodology factor
BP	By-products
Btu	British thermal unit
BVF	Business viability factor
CCS	Carbon capture and storage
CFB	Circulating fluid bed
CHP	Combined heat and power
CIF	Combustion index factor
CV	Calorific value
DEC	Decarbonized electricity certificate
Defra	Department for Environment Food and Rural Affairs
DF	Density factor
DTF	Drop tube furnace
DTI	Department of Trade and Industry
ECS	Energy crop scheme
EF	Energy Factor
EJ	Exajoule (10^{18} joules)
EOR	Enhanced oil recovery
EPA	Environmental Protection Agency
ERDP	England Rural Development Programme
ESF	Emerging systems factor
EU-ETS	European Union—Emission Trading Scheme
EWP	Energy White Paper
FAOSTAT	Food and Agriculture Organization Corporate Statistical Database
FGD	Flue Gas Desulphurisation
GGETS (GETS)	Greenhouse Gas Emissions Trading Scheme

GHG	Greenhouse gases
GRO	Government's renewable obligation
HHV	Higher heating value
IAFRE	International Association for Renewable Energy
IMF	International Monetary Fund
IPCC	Intergovernmental Panel on Climate Change
ISBF	International Standard-Biomass Fuel
ha	Hectare
KP	Kyoto Protocol
kW_e	One thousand watts of electric capacity
LCI	Life cycle inventory
LECs	Levy exemption certificates
LHV	Lower heating value
LWIF	Land and water issues factor
MC	Moisture content
MF	Moisture factor
MJ	Megajoule (1000 j)
MT	Million tons
MUV	Manufactures unit value (World Bank)
MW	Megawatt
Mwe	Megawatts of electrical output
MWh	Megawatt-hour (1,000,000 watts for 1 hour)
MWt	Megawatts of thermal output
NEF	Nitrogen emission factor
NI	National Insurance
NNFO	Nonfossil fuel obligation
Odt	Oven dry ton
PC	Pulverized coal
PD	Packing density
PF	Pulverized fuel
PV	Photovoltaic
QA	Quality assurance
QC	Quality control
QF	Quality factor
RD&D	Research, development, and demonstration
REA1	Renewable Energy Analyser One
RET	Renewable energy technology/technologies
RO	Renewable obligation
ROC	Renewable obligation certificate
SB	Suspension burner
SF	System factor
SFS	Super fuel sample
SG	Stoker grate
S&T	Scientific and technical (factors)
SUF	Supply factor

Sunflower BS	Sunflower black seed(s)
Sunflower SS	Sunflower striped seed(s)
TI	Technology issues
TMT	Thousand metric tons
VED	Volumetric energy density
VM	Volatile materials
VMF	Volatile materials factor

1

INTRODUCTION

1.1 WHY THIS BOOK?

The motivation to find an efficient and economical way of obtaining energy from renewable sources, such as from biomass materials, is vital for the present and future generations of this world. There is an increasing demand for energy and an urgent need to protect the world's climate and environment, as a whole. Governments around the world are positively encouraging research and application in this field. There has been a degree of competition in recent years among participating countries, and among international power generating companies, in an attempt to be the first to find a more economical source of energy, mainly from renewable sources. Local and national government laws and emerging international regulations give the same indication, that is, the urgent need for a new type of energy, mainly for the reasons mentioned above. As a result, power generating companies, particularly those in Europe, are facing increasing demands from their local and central governments to reduce their dependency on fossil fuels.* In consequence, these power generating companies, alongside their own internal research, now allocate external

* There are two issues facing power generating companies. The first issue is how to reduce emission at a lower cost. The second issue is how to increase the plant efficiency, that is, during electricity generation, distribution, and up to the end-user applications.

The Selection Process of Biomass Materials for the Production of Bio-fuels and Co-firing, First Edition. Najib Altawell.

budgets to sponsored projects in this field at various educational and research institutes across the globe.

Regarding this book, the message is very clear, stating that biomass materials are a good source of energy and are relatively economical. Furthermore, as natural materials, biomass should not affect the environment and/or increase the emission of CO_2 in the atmosphere. In fact, using energy crops as a source for sustainable energy is the only practical renewable energy system that can actually decrease CO_2 in the atmosphere over a long period of time. Biomass energy is the only source of renewable energy that can be produced in three different states, that is, solid, liquid and gas, similar to the fuel production obtained from fossil fuels. This means that there is no need to reinvent the combustion engine or even to replace it with an electric motor. However, despite a large number of past and present research projects, no economically affordable (at lower cost than fossil fuel) and efficient biomass fuel has yet been found. In other words, a fuel produced without affecting the market in relation to the price of biomass materials used, such as energy crops. This means that the prices of energy crops are kept low even when some of these resources are diverted toward the production of biofuels. For this reason, the main part of this work is to achieve what has not been achieved so far within the field of renewable energy. Fortunately, all the indications and results suggest there is a method that can be used to obtain and harness the chosen biomass materials for the purpose of generating electricity, as well as for use as a fuel for transportation and heating/cooling systems.

The emphasis in this work is given to four/five biomass samples. These samples carry with them the main research justification, as they are the makeup of the fuel which will turn the turbines to generate electricity in the not too distant future. Therefore, the aim of this book is to apply a systematic approach to biomass materials for the purpose of finding the most economical and efficient type, capable of producing sufficient energy on a commercial scale. This should help in reducing harmful gasses and stabilize CO_2 in the atmosphere.

1.2 THE BOOK STRUCTURE

1.2.1 Introduction

This book begins with a literature review and basic revision in areas related to renewable energy in general and biomass energy in particular.

A period of 7 months was spent in the preparation and examination of various projects and research within the above field of energy, that is, biomass energy. The literature review produced a vast amount of material with multiple answers to different types of questions. This in turn produced a number of new ideas on how to proceed with the work during the following steps and stages.

At the beginning of the second stage of writing this book, for example, Chapters 2–4, the laboratory work for selected samples began and lasted for about 8 months. The data obtained from the laboratory tests produced interesting results in that certain biomass samples have similar amount of energy as those obtained from fossil fuels, such as coal. These results were the input for the methodology system REA1.

When the first part of the research approached the final stages, the principal idea of the book was about how to design and implement a new methodology in dealing with the selection of biomass materials. This new methodology should be able to examine various selected samples from different aspects related to commercial, legal, and business to scientific and technical factors. As the research progressed, it was decided that the work itself should be divided into four different parts. "Part 1" deals with investigation of the samples. "Part 2" deals with the building of the methodology. "Part 3" deals with the implementation of the methodology. "Part 4" deals with the biomass commercial aspects scenario during the initial period for the introduction of a new type of bio-fuel, together with an economic analysis report. The fact that the methodology was researched, compiled, and written within a reasonably short period of time indicates the huge amount of effort and time spent in achieving initially unexpected goals. The aim was in part to investigate aspects of this new technology and its commercial use in the early part of the twenty-first century. Biomass in general, among other renewable sources of energy, is the science and technology for a new type of energy, which many predicted would be the challenge facing this century. However, biomass energy, as we know it today, is both engineering and a branch of science. Here we research and investigate various biomass samples for long-term commercial use. This kind of investigation will help in producing new materials and technology to replace the fossil fuels being used at the present time.

1.2.2 Structure

There are four different subjects (Fig. 1.1), which are, nevertheless, integrated into each other, as all of them work in order to achieve certain function(s) for the purpose of obtaining certain result(s). In "Part 1," investigations took place into the characteristics, composition, and suitability of biomass materials in general, and energy crops in particular, for the production of bio-fuels. "Part 2" builds a new methodology specifically to deal with biomass samples and their final selection. The two main factors on which the methodology relies, that is, scientific and technical (S&T) and business factors (BF) are divided into other factors, such as systems, approach, business viability, applicability, biomass supply, emission, technical and technical risks, commercial and commercial risks, and environmental and human health risks. These factors themselves are then divided into further subfactors. REA1 methodology, therefore, looks at various factors and possibilities in the field of technology and science,

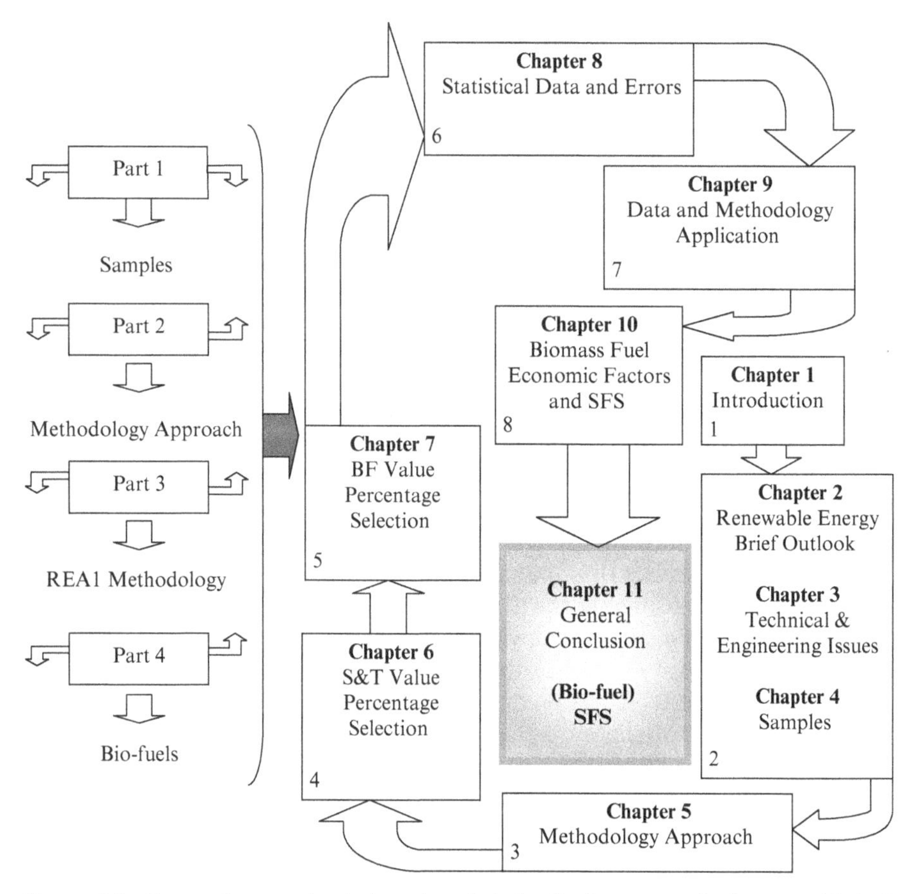

Figure 1.1 Four-part research strategy (concluded with the prospect of a new bio-fuel. *Source*: Author.

in the field of business, and commerce, as well as at government regulations, together with local and international laws.

In "Part 3," a deciding factor was made as to how to proceed to fast and accurate calculation in applying the principles of "Part Two" on each sample. The methodology, as it stands, can be used to do the proposed calculations for each sample. The completion of REA1 methodology for biomass selection, and other possible future applications and development leads to "Part 4."

Finally, "Part 4" is mainly concerned with the scenario of a final biomass sample creation to be used on a commercial scale, as well as the economic analysis for what has been achieved so far. To briefly explain the idea behind this, it would be better to look at the point where each stage during the research itself formed the next part of the work, starting with the four/five final selected samples. These four/five final samples have already been checked from the market point of view and their business viability, in general. The removal (or addition) of any unwanted elements in a single sample (or samples) would be applied commercially on a large-scale plantation. The economic cost can easily be known in advance, depending on the market fluctuations and stability, by using the actual data that "energy businesses" already have (for both BF and S&T factors). The new sample would be ready to produce from the chosen four or five samples. However, this stage would take the book beyond its present scope. The new sample should be higher in quality but cost almost the same (if not less) in comparison to any of the original four or five samples used before the final production.

1.3 ENERGY UTILIZATION

The aim for all types of renewable sources of energy is very similar, in that they all aim for the same target. This target can be summarized as the production of energy that is affordable/economical, sustainable, and environmentally friendly. In comparison with other types of renewable sources of energy, biomass research, development, and applications have been historically the dominant source of energy for thousands of years. The present development within the biomass energy commercial sector is also a market leader in a number of countries across the globe.

Figures related to the percentage usage of energy indicate that around 15% of the world's primary energy is from plant materials. Developing countries use around 38% from the same source as a fuel, while for Sub-Saharan African and as well as a number of Asian countries, plant materials provide between 60% and 90%. In rural areas within the developing countries, the use of biomass accounts for more than 90% of total daily use (EIA, 2013; IPCC, 2012). When biomass energy, as well as other sources of renewable energy, is discussed in the media, daily conversation, or within projects and schemes, together they create a strong momentum that contributes to the creation of new and useful ideas, which in turn implement the use of environmentally friendly energy, anywhere in the world.

In addition to this, the momentum highlights the local, national, and international stage concerning environmental and energy issues and, therefore, can help to provide resources urgently needed for positive actions concerning the environment and safer and more sustainable sources of energy.

When it comes to sources of energy from fossil fuels, sustainability and environmental aspects have been ignored for decades by the majority of policy makers and international commercial energy companies across the globe. In 1949, M. King Hubbert predicted that fossil fuel would be short lived, and in 1956, he predicted that peak oil production in the United States would occur during the year of 1970 (Hubbert's peak). These predictions only recently started to make sense to politicians and world leaders. Consequently, the action taken so far is too late and too little. For this reason, a book such as this can play an important role, regardless of the size of contribution it may provide

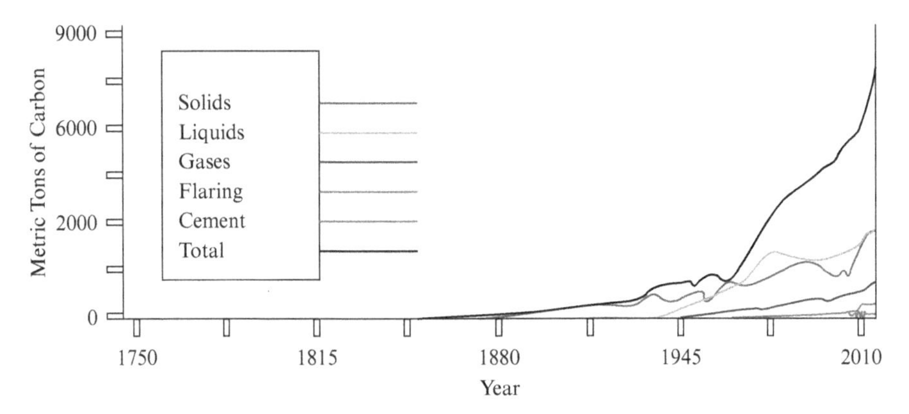

Figure 1.2 Global carbon emissions since 1880. *Source*: Adapted from U.S. Department of Energy.

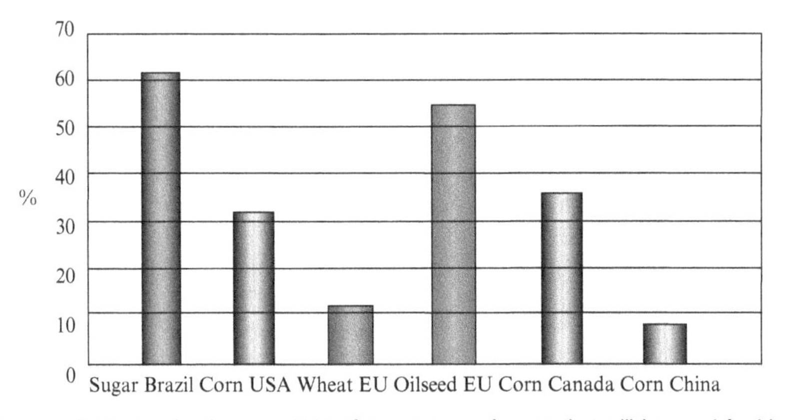

Figure 1.3 Projection for the year 2016 of percentage of crops that will be used for bio-fuel production in various countries. *Source*: Data from OECD (2013).

later on. Supporting and publicizing projects about renewable sources of energy can result in solutions for some of the problems facing everyone on this planet, with regard to environmental issues and energy needs. Increasing global emissions of CO_2 (Fig. 1.2) and the rise of energy crops prices, as a result of their usage in the bio-fuel sector (Fig. 1.3), all make an urgent case for further research, as do new ideas and different approaches in the usage and application of energy sources. As mentioned previously and repeated once more here, the need for energy can be summarized as *sustainable, environmentally friendly, economical, efficient, and adaptable.* The question is if biomass energy can fulfill these criteria.

1.4 THE NEED FOR EFFECTIVE BIOMASS UTILIZATION

The need for an alternative source of energy rises with each passing day. When environmental issues, long-term supply/availability, and economic reasons are taken into account, the design and implementation of a new biomass energy system can provide benefits in many ways. The following points are examples of why the need for a new energy system exists:

a. Developing systems to economically produce fuels and chemicals from biomass will help power generating companies to create their own resources, while simultaneously helping rural economic development.
b. Adding value to agricultural products will economically enhance many local industries.
c. Demonstrating full-scale biomass conversion systems promotes increased adoption of these technologies.
d. Biomass materials for energy stimulate the development of new products and technologies, as well as create a new market (with new jobs) that has export potential.
e. Development of a new biomass fuel (e.g., SFS—see Chapters 9 and 10).
f. Implementing technological and behavioral intervention can stop or reduce GHG before it is too late "Without technological and/or behavioral intervention, atmospheric concentration of GHGs will continue to increase . . ." (DTI Project Report, 2005).

1.5 RENEWABLE ENERGY IMPACT ON BIOMASS ECONOMY

There are several barriers to the adoption of renewable energy technologies; however, opportunities do exist to overcome them. These barriers include:

- Financial constraints, which limit greater deployment of renewable technologies. This barrier lies in the perceived risk associated with investing in this field, which is generally higher than competing in conventional technologies, and the effects of this higher perceived risk on the market.
- The RE technologies are relatively new to the capital markets and as such there is more risk than in using established technologies. The higher the perceived risk, the higher the required rate of return demanded on capital.
- The perceived length and difficulty of permitting process in this field is an additional determinant of risk.
- The high financing requirements of many renewable energy technologies often present additional cost-recovery risks for which capital markets demand a premium.

Possible recommendations could be the following:

- Low interest loans or loan guarantees might serve to reduce perceived investor risk.
- Tax credits for renewable energy technologies production through the early high-risk years of a project may provide another mechanism for further development.
- Regulatory cost recovery mechanisms, which today often favor low initial cost fuel-based technologies, can be modified to recognize life-cycle cost as an appropriate determinant of cost effectiveness.
- Effective redistribution of government spending on research and development that directly reflects the potential of RE technologies.

In the United Kingdom, the government has set up a renewable obligations certificate (ROC) (Biomass Task Force, 2005) in relation to the use of energy. This certificate details:

A. 15% of electricity should be generated from renewable sources by 2015.
B. 20 p/l tax rebate for biodiesel.
C. Direct support for renewable energy.
D. 20% GHG reduction target by 2020.
E. Climate change levy/Carbon Trust.
F. Emissions Trading Scheme (2002).
G. Set aside payments for nonfood crop production.

According to Ofgem (Ofgem, 2009): "A Renewables Obligation Certificate (ROC) is a green certificate issued to an accredited generator for eligible renewable electricity generated within the United Kingdom and supplied to customers within the United Kingdom by a licensed electricity supplier. One ROC is issued for each megawatt hour (MWh) of eligible renewable output generated." The ROC became law in 2005 when the government issued the Renewable Obligation Order 2005.* The ROC obliged the power generating providers to obtain a percentage of their electricity produced from renewable sources. According to previous government legislation as early as 2002, every year the percentage of electricity from renewable sources should be increased, for example, 2006, 6% and 2007, 7%, reaching 14% by 2014. Power generating companies, who cannot provide proof (certificates) related to the above, can be fined. As a digital certificate, the ROC holds information concerning the production of renewable electricity per unit. The certificates can be traded as they are guaranteed by the government (Box 1.1).

* In the United Kingdom, "The Renewable Obligation Order 2005" that updated the previous orders of 2002 and 2004, obliged power generating companies to supply their customers with 6.7% of its energy derived from renewable sources.

Box 1.1

UK RENEWABLE ENERGY SUPPORT SCHEMES

Renewable Obligation Certificates (ROCs)
Each MWh of green electricity produced 1 ROC is delivered.

An option to use the generated electricity either for the energy business or fed into the grid. By fulfilling ROC quota obligation, an option is available for the ROC to be sold to energy suppliers.

The ROC selling price is between £30 and £50, that is, 0.034–0.056 €/kWh.

Tax: VAT reduced to 5%.

Government Grants: Under Phase I and II Low Carbon Buildings Programme—residential buildings: 50% of project cost or €2276/kW (€2841 maximum). The public sector as well as nonprofit organizations buildings: 50% of project cost or €2841 maximum.

EPIA (2009)

1.6 SUMMARY

Two important topics related to energy supply and climate issues were discussed in the DTI white paper "Meeting the Energy Challenge" (2007):

> The International Energy Agency (IEA) forecasts that $20 trillion of investment will be needed to meet these challenges by 2030. The investment decisions that will be taken over the next two decades will be critical in determining the world's climate and the security of its energy supplies. At home it is likely that the UK will need around 30–35 GW of new electricity generation capacity over the next two decades and around two thirds of this capacity by 2020. This is because many of our coal and most of our existing nuclear power stations are set to close. And energy demand will grow over time, despite increased energy efficiency, as the economy expands.

The IEA figure of $20 trillion of investment may or may not result in what the IEA is forecasting for the next two decades. This kind of forecasting, even if it is built on solid and accurate data, is unreliable. The reason(s) for this usually lie within various constantly changing factors. These factors can range from the degree in which our climate is changing and/or the increase in the earth's population, to a change in politics worldwide, in particular when individual countries are concerned with their own interests rather than the world as a whole. This can contribute to a very difficult situation and can produce disunity, rather than the unity which is very much needed to solve the global warming and energy crises. In the case of internationally vital decisions (concerning everyone on this planet), no country should consider its own interests alone, but rather how its decisions and laws/regulations may influence the present and future global environment.

Box 1.2

BIOMASS MATERIALS

Writing a book to investigate the best way to produce energy from biomass materials can be a long process, especially if creativity prompts new ideas which were not considered seriously at the beginning.

Biomass materials will be one of the main sources of energy in the twenty-first century and beyond it, just as they were in ancient times seen as the main source of energy since man discovered how to make fire from these materials.

Regarding the United Kingdom's future energy need, there will certainly be an increase in demand in this sector, and possibly higher than the figure of 30–35 GW mentioned in the white paper. The reason for this is simply because the UK population is increasing at a higher rate than at any other time in history, according to the latest population survey and home office prediction. One of the reasons attributed to this is the number of East European countries that have joined the European Union. The possibility is that (for the next three decades) the United Kingdom will import most of its energy needs at a higher rate than ever before. Of course, this will depend mostly on the developments taking place in the energy field within the United Kingdom.

As mentioned previously, this book is divided into four different stages in order to allow each stage its own productive environment within that particular field, keeping in mind the connection with the other parts and the final outcome for the book, as a whole. The question about energy and a new system are important issues that should be taken into consideration, that is, relevant issues in the field of energy and viability can be fitted into and lead to the aims and objectives of the book.

Looking at basic factors such as climate change, economic, and political factors to unlock the main discussion, as well as understanding public views and concerns, will open the door in examining more closely the situation of all renewable energies and their possible future impacts. Particular attention is given to the economic and social aspects, which apart from the environmental factor, can affect everyone, both directly and indirectly (Box 1.2).

REFERENCES

Biomass Task Force (2005) Biomass task force report to the government. Department of environment, food and rural affairs (Defra) publications, London. http://archive.defra.gov.uk/foodfarm/growing/crops/industrial/energy/biomass-taskforce/pdf/btf-final-execsumm.pdf (last accessed December 3, 2013).

DTI (2007) Meeting the energy challenge. A White Paper on energy presented to Parliament by the Secretary of State for Trade and Industry, May 2007. http://www.berr.gov.uk/files/file39387.pdf (last accessed December 3, 2013).

DTI Project Report (2005) Best practice brochure: co-firing of biomass (main report). Report No. COAL R287 DTI/PUB URN. http://www.berr.gov.uk/files/file20737.pdf (last accessed December 3, 2013).

EIA (2013) World energy demand and economic outlook. International Energy Outlook 2013. http://www.eia.gov/forecasts/ieo/world.cfm (last accessed December 3, 2013).

EPIA (2009) Overview of European PV support schemes (last update on 8/05/2009). http://www.epia.org/index.php?id=463 (last accessed July 7, 2009).

IFPRI (International Food Policy Research Institute) (2008) Testimony biofuels and grain prices impacts and policy responses. Rosegrant M. W., testimony before the U.S. Senate Committee on Homeland Security and Governmental Affairs http://www.ifpri.org/pubs/testimony/rosegrant20080507.asp (last accessed December 3, 2013).

IPCC (Intergovernmental Panel on Climate Change) (2012) Managing the risks of extreme events and disasters to advance climate change adaptation. http://www.ipcc.ch/pdf/special-reports/srex/SREX_Full_Report.pdf (last accessed December 3, 2013).

OECD (2013) Biofuels and agriculture: projections to 2016. http://www.oecd.org/agriculture/biofuelsandagricultureprojectionsto2016.htm (last accessed December 3, 2013).

Ofgem (2009) Renewables obligation—what is the renewables obligation (RO)? Sustainability, environment. http://www.ofgem.gov.uk/Sustainability/Environment/RenewablObl/Pages/RenewablObl.aspx (last accessed December 3, 2013).

U.S. Department of Energy (2006) Biomass energy data book. Energy Efficiency and Renewable Energy, USA.

2

BACKGROUND

2.1 RENEWABLE ENERGY: A BRIEF OUTLOOK

This chapter provides a brief generalized overview concerning renewable energy, biomass applications, and co-firing. Further detail and discussion, regarding the same and/or similar topics, is provided in the following chapters.

2.1.1 Introduction

Renewable energy comes under different headings and can cover a wide array of energy sources. Natural sustainable resources are the origin of this kind of energy. The main sources are the following:

a. Wind
b. Water (waves, underwater currents, flowing water from higher ground and osmosis or "osmotic power")
c. Geothermal
d. Solar (sunlight and sun's heat)
e. Biomass.

These sources of energy are called "renewable" as they are naturally replenished. Concerning development and growth in this field, the current fast development occurred mainly due to the recently increased level of investments

The Selection Process of Biomass Materials for the Production of Bio-fuels and Co-firing, First Edition. Najib Altawell.

from a number of countries across the world. For example, around $71 billion was invested worldwide during the year 2007, in comparison to $55 billion in 2006 and $40 billion in 2005 (REN21, 2008). Financial investments in this field are still rapidly rising, not just in the West. China, India, and Japan are just a few examples of where heavy investments in this field are continuously experiencing growth.

By utilizing a range of technologies, renewable energy can be obtained from a variety of sources, mainly to replace the present limited resources associated with fossil fuels, but also to find an economical and sustainable source of energy. However, there are other urgent reasons to search for alternative sources of energy connected to the issue of "global warming," and protection of the environment.

By using certain types of renewable sources of energy, it is possible to reduce the emission of known gaseous substances associated with fossil fuels, such as CO_2, which among other gases, is accepted to be one of the main causes of global warming. In most cases, fossil fuels are used for the production of hardware, and possibly during transportation, installation, operation, and decommissioning of the renewable energy systems. This means that these systems are not zero-carbon technology, at least not in the earlier-mentioned procedures currently taking place worldwide.

Some of the costs associated with renewable energy are in decline, and may continue to decline in the foreseeable future. For example, the future cost for biomass energy is predicted to be around $6–10 G/J (or less), compared to $8–25 G/J during 2007 (Table 2.1).

The capacity factor* (CF) is an important aspect of energy generation, in general and renewable energy in particular (Table 2.2). The CF is the average electricity generated during 1 year divided by the amount of electricity generated continuously during 1 year at full capacity, as in the following equation:

$$\mathrm{CF} = \frac{\text{Average electricity generated/Year}}{\text{Electricty generated at full capacity continuously/Year}}.$$

To illustrate: the capacity factor of technology such as wind power in the United Kingdom has been estimated to be 0.3, that is, the average energy generated from wind turbines (maximum-rated capacity) is around 30%, as wind speeds vary considerably. The output (kWh) therefore, is equal to the capacity factor multiplied by the number of working hours/year.

According to Lester Brown, the following average values of CF (for generating electricity) are related to fossil fuels as well as renewable energy in the United States. These have been calculated as shown in Table 2.2 (Brown, 2009).

* The capacity factor should not be confused with the load factor, and neither with the declared net capacity. The load factor is the ratio of the average load to the peak load. The declared net capacity is the maximum total capacity a power station can operate for during a sustained period of time minus the amount of electricity consumed by the plant.

Table 2.1 The status of RET during 2007

Technology	Capacity Factor (%)	Turnkey Investment Costs (US$/Kw)	Current Energy Cost of New Systems	Potential Future Energy Cost
Biomass energy				
Electricity	1	900–3000	5–150/kWh	4–100/kWh
Heat		250–750	1–50/kWh	I–50/kWh
Ethanol			8–25 $/GJ	6–10 $/GJ
Wind electricity	20–30	1100–1700	5–130/kWh	3–100/kWh
Solar photovoltaic Electricity	8–20	5000–10,000	25–1250/kWh	5 or 6–250/ kWh
Solar thermal Electricity	20–35	3000–4000	12–180/kWh	4–100/kWh
Low-temperature Solar heat	8–20	500–1700	3–200/kWh	2 or 3–100/ kWh
Hydroelectricity				
Large	35–60	1000–3500	2–80/kWh	2–80/kWh
Small	20–70	1200–3000	4–100/kWh	3–100/kWh
Geothermal energy				
Electricity	45–90	800–3000	2–100/kWh	I or 2–80/kWh
Heat	20–70	200–2000	0.5–50/kWh	0.5–50/kWh

Source: Adapted from Shrestha (2007).

Table 2.2 CF average value for generating electricity from fossil fuels and renewable energy sources in the United States (Brown, 2009)

Energy Source	Capacity Factor (%)
Coal	72.2
Oil	18.9
Natural gas	37.3
Nuclear	89.0
Wind	36.0
Solar-PV	22
Solar thermal	24.4
Geothermal	90.0
Biomass	80.0
Hydropower	44.2

2.1.2 Old Graphs

Old graphs from the National Renewable Energy Laboratory (2002) have been intentionally used in the following sections. The reason for this is that these graphs are estimating and predicting the costs of energy generated from wind power, hydropower, geothermal, solar and biomass from the year 1980 and up to the year 2020. Examining these graphs, it is true that up to the year 2012/2013 the cost related to the above sources of energy has already declined, as these graphs predicted (compared with the available data at the time of compiling this book). However, the cost so far did not decline as steeply as predicted, not just for biomass energy but for all the other renewable sources of energy in various parts of the world. Nevertheless, these graphs can be applied today and can be used progressively until the year 2020 for the sake of past/present cost, prediction,* and comparison.

2.2 WIND

By using energy from moving air, large blades on a wind turbine rotate (to turn the turbine/alternator) in order to generate electricity. Energy from wind turbines for the purpose of generating electricity require large open spaces with a high occurrence of wind throughout the year in order to be commercially viable. The cost regarding hardware technologies, and the cost of generating electricity is steadily declining. By the year 2015/2016, the prediction is that the wind energy cost will start to level off, as illustrated in Figure 2.1. The positive aspect of wind energy is that the source is free. It generates no air or water pollution; however, the present cost is still high as without a generous subsidy many of the present turbines may not be commercially viable. Wind farms are not labor intensive, and therefore creating jobs on a large scale may not be possible either. Also, turbines do not generate electricity when the wind speed is too low or too high. Noise and light fluctuation from the sun caused by the wind turbine blades movement can be a negative side effect, especially if turbines are constructed near residential areas. In addition, the limited

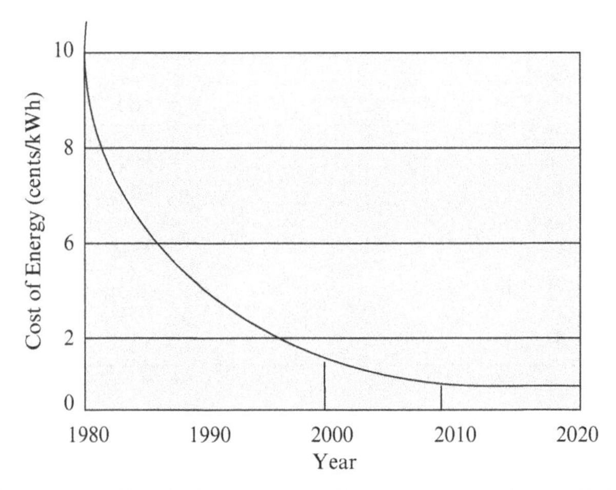

Figure 2.1 The decline of wind energy cost. *Source*: Adapted from NREL (2002.)

* The prediction is that renewable power could surpass natural gas as a source for generating electricity within a short period of time, possibly within the next 3 years according to the International Energy Agency (IEA).

availability of space with a regular wind pattern can be a problem. Finally, the possibility that large wind turbines may harm wildlife, such as birds, should be investigated alongside the design and construction of the hardware system.

There are a number of obligations related to the construction of wind turbines, that is, in the form of local government regulations, such as planning exclusion zones and noise controls. However, these kinds of regulations have been issued only in certain countries.

According to Solarbuzz, today's guideline electricity generation cost of wind energy is around 4–7 cents/kWh (Solarbuzz, 2009). The total wind power capacity installed globally by the end of 2011 = 237 GWH (GWEC, 2012).

2.3 WATER

Water movement, in general, can be utilized to generate energy. This may range from ocean waves to falling water and underwater currents, as well as osmosis.* The hydropower system is used to produce electricity by spinning a turbine generator or simply for mechanical purposes. Currently, hydropower is the most common way to generate electricity worldwide in the form of renewable energy. A hydro dam built for the purpose of generating electricity can help with flood control and irrigation. Depending on the type of hydro project and the state of the land before the project, the lake produced can be a positive aspect of an aquatic ecosystem.

The structures of a hydropower plant, such as canals, tunnels, dams, reservoirs, access roads, and so on, are useful in relation to the area's development. On the other hand, negative aspects can be: siltation, soil erosion, soil and water salinity, obstruction of the free passage between oceans and rivers, weed growth, floods due to dam failures, the possibility of diseases spread by organisms that live in stagnant water, and damage to natural resources, such as fish. Some dams may result in the loss of farmland, habitats, and housing.

In many parts of the world, development within this sector of hydropower is still expanding, mostly for generating electricity. Hydropower in the USA (Fig. 2.2) peaked within the last two decades of the last century (ESA21, 2009). The Total hydropower capacity installed globally by the end of 2011 = 3500 GWH (EPI, 2013).

2.4 GEOTHERMAL

With appropriate hardware technology and relevant data related to geological mapping, geothermal energy can be obtained anywhere in the world. However, parts of the globe where volcanic eruptions happen more often may be

* Employing seawater and fresh water at the same time, power can be generated. The process starts with seawater in the chamber under pressure. By pumping freshwater (via a membrane) into the chamber, the difference in the pressure results in generating power used to turn the turbine.

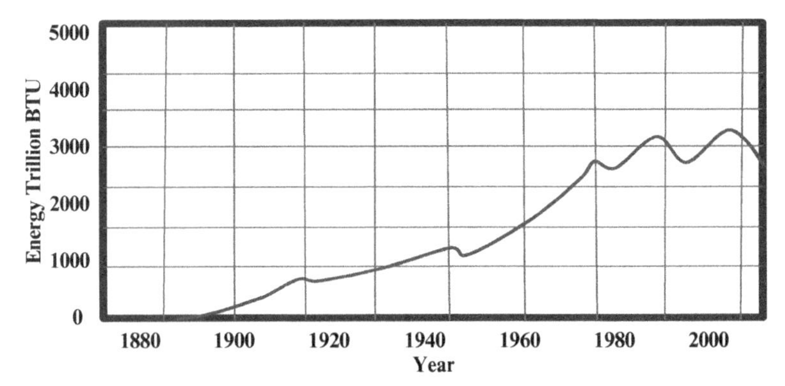

Figure 2.2 Hydroelectric energy generation development and expansion for more than 100 years in the United States. *Source*: Adapted from ESA21.

considered for drilling at a lower cost. A high temperature emitted from the earth's core heats water deep in the ground. The hot water can be used to heat (or cool) homes and commercial buildings and to generate electricity. This kind of natural energy is usually referred to as a "geothermal energy system."

A number of variations of geothermal energy use can be found across the world. For example, ocean thermal energy is currently being researched and used in some countries. There are three main classes of geothermal energy sources; these are (Ostridge, 1998):

1. *Direct Usage.* Water is heated by hot rocks (via magma heat) beneath the ground
2. Steam that comes from superheated water (when there is pressure, this can be used to turn a turbine)
3. *Dry Steam.* An external water source is pumped to fracture very hot rock and the steam resulting from this process is used to turn turbines.

The positive aspect of geothermal energy is that it has very high efficiency, while on the negative side,* it may not be considered to be a renewable energy source since more energy is taken out than nature can put in. Heating using this method can be more costly than conventional systems. Another negative aspect is that geothermal energy is a local resource only. As geothermal energy technology improves, the cost gradually decreases (Fig. 2.3). This predicted reduction in cost† will be more obvious in the near future as development in

* Negative aspects may also include toxic air emissions as geothermal materials (e.g., fluids) contain dissolved gases (e.g., H_sS and CO_2) that can easily escape into the atmosphere. Also, by removing these fluids, the pressure for underground reservoirs will be reduced, and in many cases, the surface of the earth will sink. Geothermal fluids may contain toxic elements, such as lithium, boron, mercury, and arsenic.

† Well drilling cost has been estimated to be around two-third of the overall total cost.

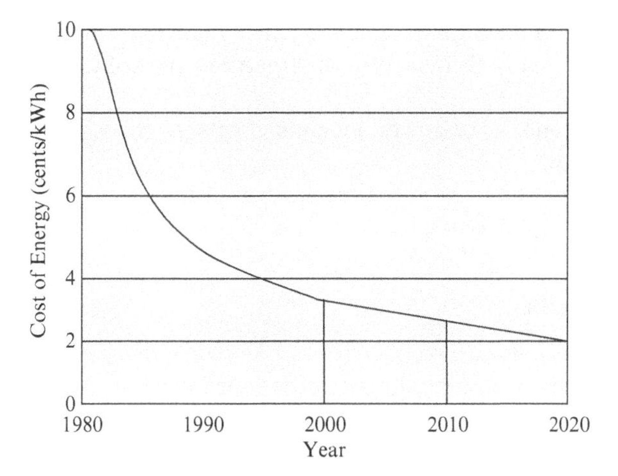

Figure 2.3 The decline of geothermal energy cost. *Source*: Adapted from NREL (2002).

> **Box 2.1**
>
> **GEOTHERMAL ENERGY: SOURCE AND VALUE**
>
> The main sources for geothermal energy are the heat flow from the earth's core and mantle (~40%), and that generated by the gradual decay of radioactive isotopes in the earth's continental crust (~60%). Together, these result in an average terrestrial heat flow rate of 44 TWth (1400 EJ/yr), nearly 2.8 times the 2009 worldwide total primary energy supply, 509 EJth/yr, (IEA, 2011), which is about 1% less than the 2008 value (514 EJth/yr).
>
> IEA (2012)

this field is already at an advanced stage. Lessons can be learned from crude oil technology drilling approaches, mechanism, and applications.

The total geothermal capacity installed globally by 2011 = ~11 MWH. However, a prediction has been made that the capacity will be increased to around 14.3 GW by 2020 (Pike Research, 2011) (Box 2.1).

2.5 SOLAR

Using sunlight and/or sun heat to generate energy in the form of electricity (or for heating systems) is one of the growing ways in which solar energy is used and applied. From a small calculator to the satellites orbiting the earth, solar cells can provide energy in different ways and for different processes.

The geographical areas of the world with high insolation are located as follows: The Middle East; part of northern and southern Africa; Australia; south eastern and part of western Asia (e.g., India and China); the southern part of Europe; Mexico and the southern part of the United States; part of South America, that is, Bolivia, part of Argentina, part of Chile, and the southern part of Peru.

As solar systems have fewer mechanical parts, the cost of maintenance predictably is low.

Sunlight or heat can be used directly or indirectly, as in the following brief examplesin the next sections.

2.5.1 Solar Cells

Solar cells are usually referred to as *photovoltaic* or *photoelectric cells*. These cells convert sunlight directly into electricity using semi conducting materials, such as silicon. Solar cells depend on what has been termed as *the photoelectric effect*,* which is a natural phenomenon of matter (e.g., metal) emitting electrons whenever light shone on its surface. For example, on a sunny day, sunlight falling on 1 m^2 of solar panel can produce enough power to run a 100 W light bulb (MNRE, 2008), that is, 1 m^2 of solar panel could produce 100 W.

2.5.2 Solar Water Heating

In some domestic settings, by pumping water through a pipe painted black, the heat from the sun is used to heat water in glass panels placed on the house roof. In order to trap and keep the absorbed heat from the sun under the glass, where the black painted pipe is located, the system uses the principles of the greenhouse effect. For this reason, the glass type used is an important factor in retaining the absorbed heat. Reportedly, a prism glass is recommended in order to improve efficiency (Cariou, 2010). There are two different types of solar water heating system, direct (through collectors, the potable water circulates) and indirect (using a heat exchanger) systems. Solar water heating systems are most suitable in places where large amount of heat regularly emanates from the sun.

2.5.3 Solar Furnaces

By collecting sun rays (e.g., by the use of curved or flat mirrors) and concentrating them in a small space (focal point), a high temperature can be achieved in a specially designed furnace. Using this design, it was reported that a temperature of 3000–3500°C was reached at one of Odellio's laboratories in France (Tseng, 2007).

There is no pollution from the energy obtained and it can be used in various ways, including the generation of electricity, in nanomaterials, hydrogen fuel production, and high temperature applications, including steel melting (foundry

* Discovered by Heinrich Hertz in 1887.

applications). On a smaller scale, the solar furnace principles have been applied to manufacture solar cookers/grills.

These are very brief examples of how sunlight and sun heat are currently being used as another type of renewable source of energy. The most positive aspect of using sunlight and sun heat is that there is no waste or pollution.* It can be used in sunny countries to get electricity in remote places when a connection to the main electricity grid is not possible. Also, it can be used for small domestic appliances (e.g., heating water) as well as for charging batteries. However, it is costly to convert solar energy to electricity for large power stations (Khosla, 2008). Finally, solar power can be unreliable in an area where there is no regular sunlight; however, the system may function normally using a diffused sunlight.

According to Solarbuzz, today's guideline electricity generation cost from Solar PV central station energy is 20–30 cents/kWh and from Solar PV distributed is 20–50 cents/kWh (Solarbuzz, 2009). Total solar capacity installed globally by 2011 = 63.4 GWH (BP, 2013). Figure 2.4 and Figure 2.5 illustrate past and possible future costs for both thermal and sunlight energy.

2.6 BIOMASS

Various types of plentiful biomass material can be located and used anywhere in the world, such as agricultural crops, wood, animal/human waste, different species of grasses, and fast-growing bushes. All can be used to produce heat and electricity and fuels for transportation/cooling systems. As these materials are widely available, on occasion they can be free to obtain or at a very low cost to purchase. To produce biomass materials (e.g., energy crops), there is a need for land and a regular supply of water. Also, in the process of obtaining energy from biomass, various by-products will be produced, which if not dealt

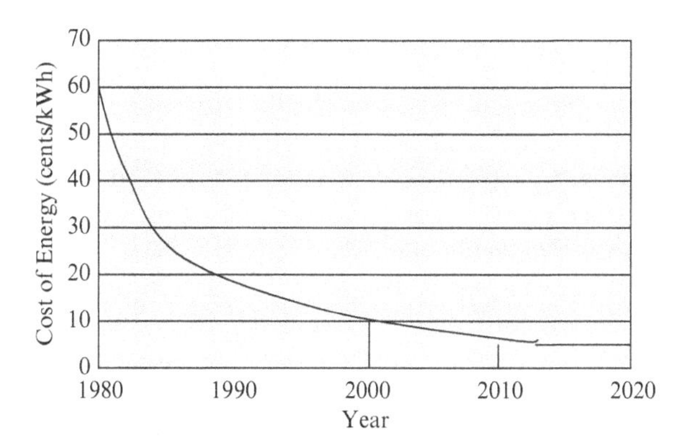

Figure 2.4 The decline of solar energy cost. *Source*: Adapted from NREL (2002).

* If the hardware manufacturing process for the system using fossil fuels has been excluded.

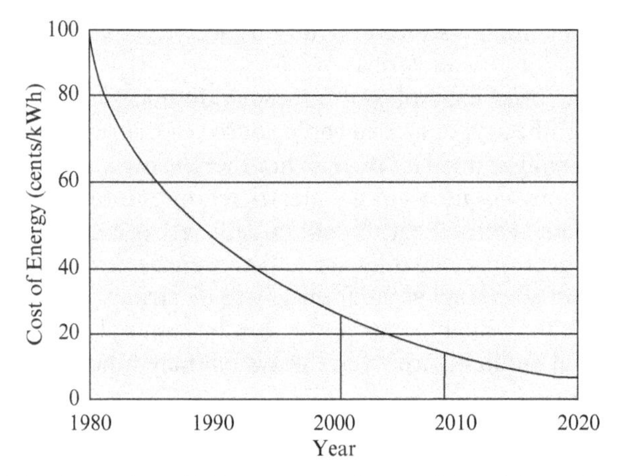

Figure 2.5 The decline of PV energy cost. *Source*: Adapted from NREL (2002).

with, may harm the environment. Deforestation may also occur if strict laws and regulations are not followed. According to NASA, deforestation has accounted for approximately 12% of all human carbon dioxide emissions (NASA, 2013).

The positive aspect of biomass materials is that they are part of the bio-cycle of the environment, that is, natural raw materials, and therefore not as toxic as crude oil. This means they can break down relatively quickly into their natural elements (Fig. 2.6).

However, biomass fuels are still relatively expensive to produce on a large commercial scale. In comparison to fossil fuel usage, the biomass industry is still small (U.S. Department of Energy, 2009). Having said that, the current technical and commercial development in this field is growing fast, even to the extent that a number of predictions have already been made suggesting that biomass energy will be one of the dominant modern sources of fuel in the not-too distant future (European Climate Foundation, 2010). One of the reasons attributed to this growth* is that biomass energy is the only renewable energy that can be obtained in three different forms, that is, solid, liquid and gas. This means that there is no need to reinvent the combustion engine, as bio-fuel can be used to replace the current fuels derived from crude oil and presently used worldwide. Other reasons supporting this development, in common with other renewable energy sources, are the instability of crude oil prices, locally and internationally. It is also an accepted fact that crude oil is a limited source of energy, and therefore an alternative sustainable source is the only answer to provide energy security and stable competitive prices.

Cost for biomass is steadily decreasing, as illustrated in Figure 2.7, in the shape of a downward straight line, that is, present and future prediction. The

* Biomass energy from landfill gas is the cheapest source of generating energy compared with other sources of renewable energy, unless, of course, there is a regular wind occurrence throughout the year, then energy from wind turbines can be slightly cheaper.

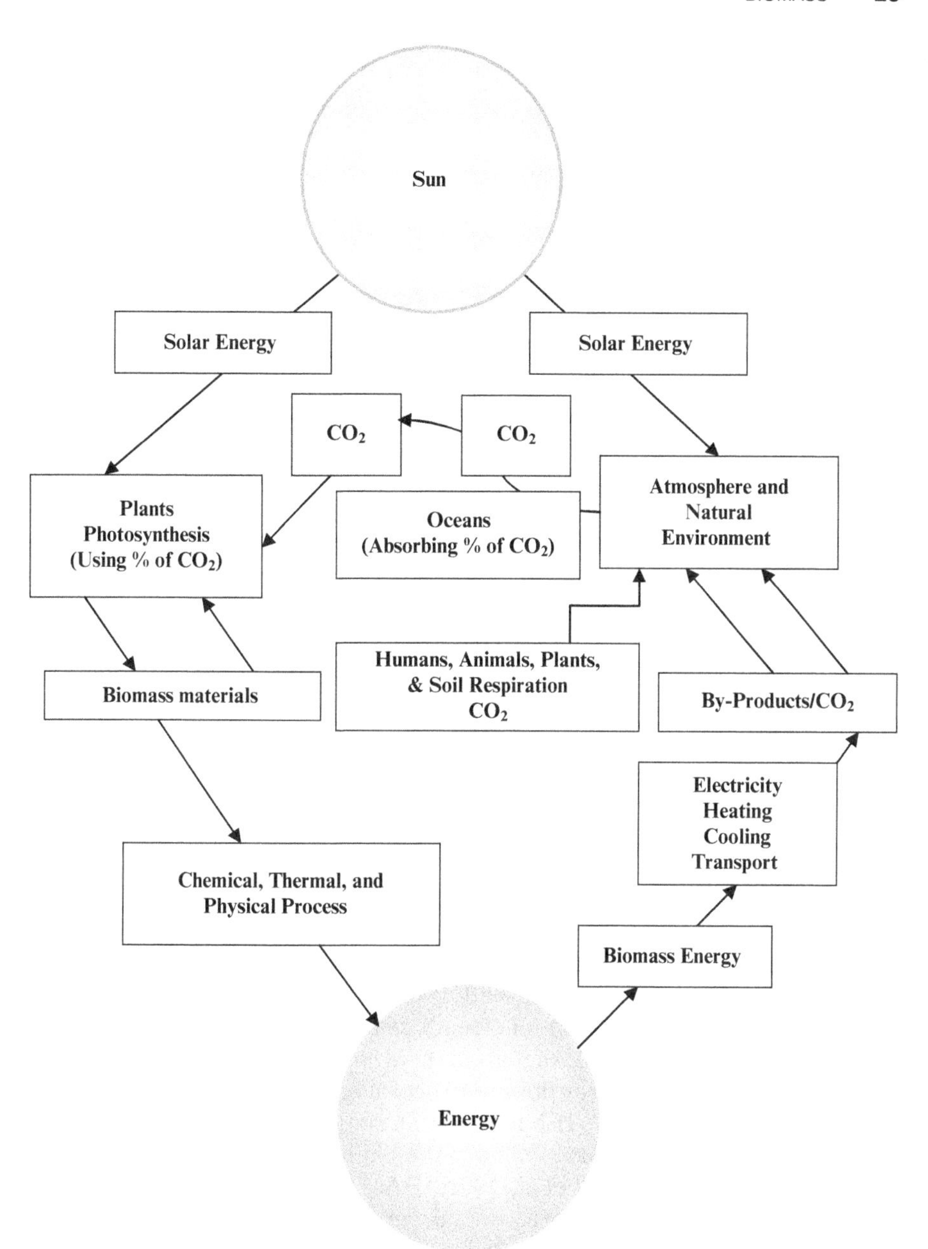

Figure 2.6 The cycle of biomass and carbon dioxide in nature. *Source*: Author.

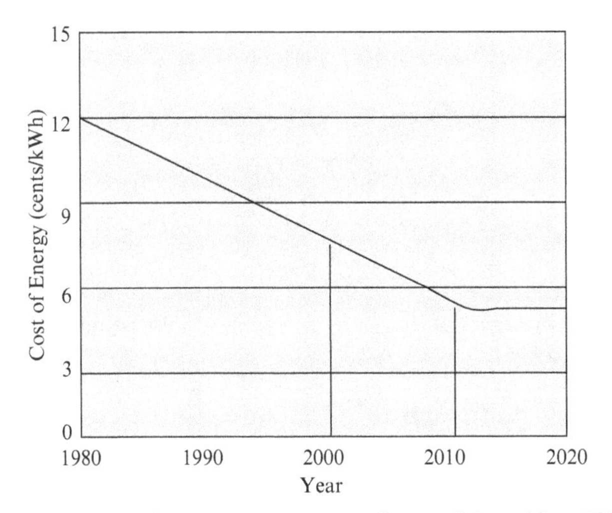

Figure 2.7 The decline of biomass energy costs. *Source*: Adapted from NREL (2002).

prediction is that the biomass energy cost will be stabilized around 2016–2020 (NREL, 2002).

According to Solarbuzz, today's guideline electricity generation cost from biomass gasification energy is 7–9 cents/kWh and from remote diesel generation 20–40 cents/kWh (Solarbuzz, 2009).

For electricity generation, the total biomass capacity installed globally by 2011 = 72 GWH (REN21, 2012) (Box 2.2) (Table 2.3).

2.7 BIOMASS AS A SOURCE OF ENERGY

From ancient time to the present day, biomass was and remains a reliable source of energy in our daily lives. With around 70% of the world using biomass materials as a source of energy in the nineteenth century, biomass dominated world energy supply until the middle of that century (Grubler and Nakicenovic, 1988).

From the data available at present concerning biomass, it supplies 14% of the world's energy sources (IENICA, 2007). In developing countries 40% of energy is derived from biomass, while in the United States, it is 4%, in Sweden 14%, and Austria 10% (Hall et al., 1992; IEA, 2012).

The term "biomass" refers to organic matter that can be converted to energy and was introduced in 1975 to describe natural materials used as energy sources. It is well known that biomass energy is the oldest source of energy in human history and dates back to man's first source of energy: fire.

In biomass, there is a process called photosynthesis, whereby plants capture sunlight and transform it into chemical energy, as shown in the photosynthesis equation (Eq. 2.1) (see Section 4.3.5 for further details):

$$6CO_2 + 12H_2O + \text{light} \rightarrow C_6H_{12}O_6 + 6O_2 + 6H_2O. \qquad (2.1)$$

Box 2.2

RENEWABLE ENERGY HYBRID SYSTEMS

- Applying the strong points of multiple technologies
- Hybrid system should be close to optimal in both performance and cost
- The capital cost is high (e.g., cost of solar, cost of batteries, and cost of maintenance)
- Better reliability of power supply
- The need for technical and engineering know-how
- Ideal as a backup for power generating systems.

From this chemical equation, biomass energy is obtained by reversing the photosynthesis process. Organic resources generating energy employing the earlier-mentioned processes are collectively called biomass. The main elements in biomass are carbon and hydrogen, and this stored energy is released when the chemical compounds within the biomass materials are broken. Biomass is a good source of renewable energy, but it is not a good source of fuel as it typically contains more than 70% air and void space. Consequently, this low volumetric energy density makes it difficult to collect, ship, store, and use. Various types of organic material can be burned to produce energy or converted into a gas to be used as fuel. Research in this area has shown that the net energy available in biomass materials when combusted ranges from about 8 MJ/kg for green wood to 20 MJ/kg for oven dry plant matter, to 55 MJ/kg for methane, compared with about 23–30 MJ/kg for coal (Fletcher et al., 2005).

There are obvious differences between the properties of biomass fuel in comparison with coal fuel. These differences can be summarized in the following points:

1. Biomass is more volatile than coal.
2. Coal has a lower content of oxygen than biomass.
3. Coal produces more ash than most biomass materials.
4. The percentage of sulfur, in general, is less in biomass than coal.
5. Biomass has more potassium and chlorine than coal.

The main benefit of using co-firing to generate electricity is the reduction of gas emissions, which can range from CO_2 to other gases, such as NO_x and SO_x (DTI, 2007). In addition to the environmental benefits, there is a reduction of ash treatment, as well as a cost saving.

Biomass energy can be divided into two categories:

1. Modern biomass
2. Traditional biomass.

Table 2.3 Summary of positive and negative aspects of renewable energy sources

Wind Power Total wind Power Capacity Installed Globally by 2011 = 237 GWH[a]		Hydropower Total Hydropower Capacity Installed Globally by 2011 = 3,500 GWH[b]		Geothermal Total Geothermal Capacity Installed Globally by 2011 = ~11 MWH[b]		Solar Total Solar Capacity Installed Globally by 2011 = 63.4 GWH		Biomass Total Biomass Capacity Installed Globally by 2011 = 72 GWH[c] (for Electricity Generation Only)	
Positive	Negative	Positive	Negative	Positive	Negative	Positive	Negative	Positive	Negative
No pollution during the time generating electricity	Noise and light fluctuation	Flood control	Soil problems	High efficiency	It may not be renewable	No waste or pollution during the time generating electricity	The initial cost can be high	Not toxic	Need land and water
The source of energy is free	May harm wildlife, that is, birds	Ecosystem	Possible dam failure	Minimal land and freshwater requirements	Costly in comparison to conventional systems	The source of energy is free Generates electricity in remote places when connection to the grid is not possible	Does not work at night	Widely available Unlimited resource	By-products Can be harmful to the environment if not controlled
Cost declining	Limited space With regular wind pattern	Area development	Effect on natural resources	Highly scalable	Local resource only	Domestic use for example, heating water and power supply	Unreliable where there is no regular sunlight	Economical (low price) to purchase	May cause degradation of soil quality and may result in deforestation if not strictly regulated

[a]Data source: Global Wind Energy Council (GWEC), 2012.
[b]Data source: Earth Policy Institute (EPI, 2013) Data source: BP (British Petroleum) (2013).
[c]Data source: REN21 (2012).

Source: Author.

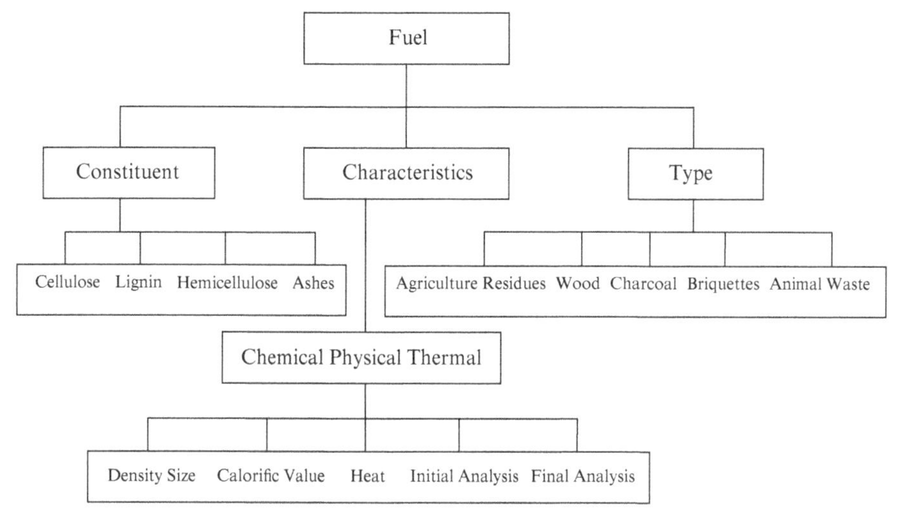

Figure 2.8 The process of obtaining fuel from biomass materials is to examine the biomass properties. *Source*: Author, based on FAO diagram.

In most cases, the proposed modern uses of biomass involve large-scale applications as it tries to replace conventional fossil fuels. These large biomass energy production facilities may use forest wood and agricultural residues, urban wastes, biogas, and energy crops. Generally, most traditional biomass is found in developing countries, and biomass materials in this case may include various types of wood, charcoal, rice husks, animal dung, and other plant residues (Fletcher et al., 2005).

To obtain fuel from biomass materials, three basic initial steps have to be considered. These are related to "type" of biomass (selection process), "characteristics," and "constituents" (Fig. 2.8). From these main three steps other subdivided steps will be considered in the same manner (see Chapter 4).

2.7.1 Energy Crops

Energy crops are becoming an important source of energy as well as gradually forming the basis for future development in this field. The four basic reasons for this are the following:

A. Climate change
B. Economic factors
C. Political factors
D. White paper drivers.

A. Climate Change

Scientific studies and research have shown that there are changes taking place in our climate (DTI, 2003). These changes are attributed mainly to the use of fossil fuels. This means that large quantities of CO_2 are constantly being added to the atmosphere—producing what is known as the "greenhouse effect" (Farabee, 2006). This phenomenon can lead to a rise in the earth's temperature with the possibility of climate change in the form of drought, frequent heat waves, water shortages, and decreasing soil moisture. Also, it is believed that the greenhouse effect will melt the ice in the Earth's poles, raising the sea level, and as a result, submerging areas of land in many parts of the world. This is the prediction and the official line behind the causes of global warming. However, not all scientists accept the above approach. While regulations and the new zero CO_2 emission systems have been introduced recently to combat the greenhouse effect, climate issues will continue to be the dominant factor for many years to come. Energy crops can help avoid this danger by simply reducing the input of CO_2 into the atmosphere, and in the long term, may stabilize the climate. This means that by planting the same amount of energy crops as those being burned, the amount of CO_2 will stay the same, if not reduce, as it is believed that plants* can absorb three times more CO_2 than they produce (Reich, 2006).

B. Economic Factors

It is a fact that present oil reserves can last only for a limited period of time (Rempel, 2000). Expanding economies (e.g., China and India) are increasing the demand for oil (and other sources of energy). In consequence, depending on the health of the world economy, the price of a barrel of oil may continue to rise as and when the world economy develops. Unstable oil prices can affect the economy worldwide, especially in those countries that depend mainly on imported oil. Energy crops can change the high demand for traditional forms of energy and in turn may strengthen the economies of many countries. In this respect, it may also help the worldwide economy without the possibility of causing inflation.

C. Political Factors

Politics and the economy are connected in various ways. This means that for the economy and the political situation of any country to be safeguarded, an uninterrupted supply of energy should be present. Depending on other countries for these supplies can be a problem, especially when relying on regions that have the highest oil reserves in the world, but with potentially

* In the future, in certain parts of the world, climate change may make it harder to regrow harvested forests.

unstable political situations (e.g., the Middle East). When it comes to vital energy supplies, a country's own renewable source of energy can make a difference to the daily lives of its citizens. Politics and politicians can play an important role in speeding up or delaying the path of progression in the field of renewable energies, depending mostly on issues of current affairs, at least in countries where a genuine democracy is in place.

D. White Paper Drivers

The present government legislation, according to the white paper of May 2007 "Meeting the Energy Challenge" and the ROC (Renewable Obligation Certificate), requires that the power generating companies produce ~13% of their 2013 output from renewable sources (DTI, 2007). As mentioned earlier, by the year 2014, the required percentage will be ~14%, and by the year 2015, it will be ~15%, and so on. The incentive is that for every MWh of renewable energy produced, the power generating company can claim from the government £34.30, as illustrated in Table 2.4.

2.7.2 Examples of Energy Crops

Any type of energy crops can be used as a "biomass material" to generate energy. However, as the aim of many biomass projects is to generate energy economically on a commercial scale, the selection process in choosing the most suitable biomass materials needs to undergo strict testing. This testing procedure is related to scientific and technical as well as business and environmental aspects.

There are a number of popular examples of energy crops with possible future commercial use. One of these examples is the "short rotation coppice" (Defra, 2002). Eight illustrations of a possible source for commercial energy are the following:

Table 2.4 Support for various energy technologies

Band	Technologies	Level of Support ROCs/MWh
Established	Sewage gas; landfill gas; co-firing of nonenergy crop (regular) biomass.	0.25
Reference	Onshore wind; hydroelectric; co-firing of energy crops; energy from waste with combined heat and power; other not specified.	1.0
Postdemonstration	Offshore wind; dedicated regular biomass.	1.5
Emerging technologies	Dedicated biomass burning energy crops, with or without CHP (combined heat and power); dedicated regular biomass with CHP; solar photovoltaic; geothermal.	2.0

1. Alfalfa (fixes nitrogen in the soil).
2. Willow.
3. Switch grass (protects the soil and the water in the watershed).
4. A wide range of crops that can be used for biodiesel production, such as canola (rapeseed), palm oil, sunflower oil, soya bean oil, animal fat (tallow), as well as recycled oil (e.g., frying oil).
5. Common crop residues (waste matter).
6. Sorghum.
7. Forestry crops, that is, fast growing trees, which should be suitable for coppicing. Coppicing involves harvesting the tree after a few years and then allowing the tree to sprout again from the stump, followed by subsequent harvesting (usually between 2- and 5-year periods).
8. Forestry residues, for example, generated by operations such as thinning of plantations, natural attrition, extracting stemwood for pulp, and clearing for logging roads. Various types of work on wood can also generate large volumes of residue, such as sawdust, bark and woodchip rejects, and off-cuts. There are plenty of these types of by-product materials around, but they are usually not utilized on a large commercial scale.

2.7.3 Biomass Utilization

There are two main options available for utilizing biomass:

1. *Construction of Stand-Alone or Dedicated Biomass.* Defined in the Renewable Obligation as those which have been commissioned since 1st January 1990 and are "fueled wholly by biomass in any month" (DTI, 2005).
2. *Co-firing of Biomass with Other Fuels.* Biomass energy and the utilization of biomass sources are both still lagging behind. The main sources of energy in many parts of the world are largely fossil fuels. In fact, 79% of world energy use depends on fossil fuels (REN21, 2008). Concerning Europe's use of biomass as a source of energy, more than 40% of the European Union's energy supply depends on oil imported from OPEC countries (Ignaciuk et al., 2004).

   Research predicts that in the coming years, there will be an increasing dependence on oil and gas imports. This will result in the share of imports in the European Union (EU) rising by 70% by 2030 (Ignaciuk et al., 2004). The energy obtained from biomass materials can be through combustion (direct/indirect) or via a thermochemical or biological conversion process (Fig. 2.9). In both instances, the energy obtained is in the form of heat or electricity. Co-firing with biomass is the preferred method when it comes to the reduction of harmful emissions and lowering the costs of energy production (see Section 2.9 for further details).

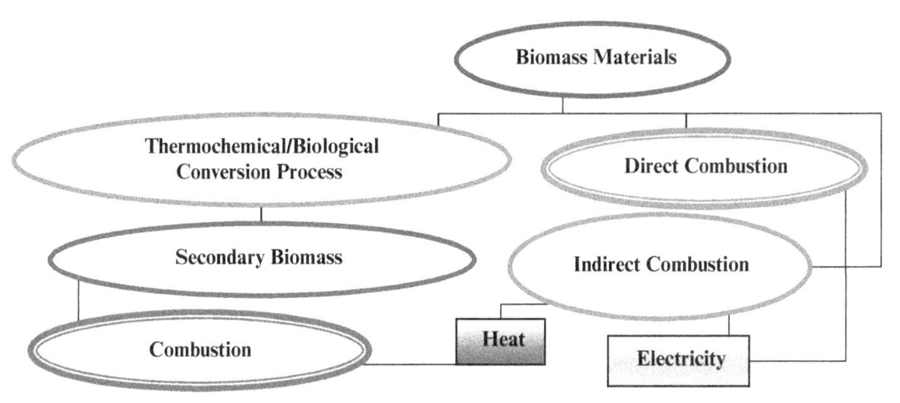

Figure 2.9 Bio-energy conversion from biomass materials providing heat and electricity. *Source*: Author.

2.7.4 Biomass and Coal Components

A. Biomass Components

Biomass is mainly made up of the elements carbon and hydrogen. The use of technology (e.g., dedicated biomass energy systems) can free the energy bound up in these chemical compounds. The other main components of biomass materials are the following:

1. *Cellulose* $[C_6(H_2O)_5]_n$. A carbohydrate usually found in the cell wall of any plant.
2. *Hemicellulose* $[C_5(H_2O)_4]_n$. Short-chain carbohydrates that differ from cellulose because it is built up from five different types of sugar.
3. *Lignin* $[C_{10}H_{12}O_3]_n$. The agent in wood that binds the cells together.
4. *Extractives.* A deposit of chemical compounds in wood, which takes place during the transition from sapwood to heartwood. Usually, the extractives give heartwood its dark color and can provide insect and decay resistance.
5. *Ash.* The inorganic mineral elements of plants.

B. Coal Components

There are three main types of coal:

1. Anthracite (hardest and high carbon, i.e., higher energy content)
2. Bituminous (soft middle-rank coal)
3. Lignite (the softest and lowest in carbon, high in hydrogen and oxygen).

Table 2.5 The percentage of various elements related to anthracite, bituminous, lignite, and wood

Constituent	Anthracite (%)	Bituminous (%)	Lignite (%)	Wood (%)
Ash	2.4	1.8	5.0	0.1
Carbon	94.88	89.47	72.38	47.4
Hydrogen	1.83	4.93	5.30	5.2
Nitrogen	0.67	1.66	1.12	0.01
Oxygen	1.78	3.49	20.53	37.9

Source: Adapted from Elliot (1981) and Vuthaluru (2003).

Coal contains the following main elements:

1. Carbon
2. Hydrogen
3. Oxygen
4. Nitrogen
5. Sulfur (varying amounts).

These elements are part of the makeup of wood as well. However, an important component of wood is the volatile matter with a varying percentage, depending on the type of wood, age, and location. Table 2.5 provides a comparison of the elements (percentage) makeup of anthracite, bituminous, lignite, and wood.

2.7.5 Types of Energy Crop Needed

High energy output and availability at a low production cost are the two most important factors with regard to the selection process of biomass materials. The following general and basic requirements should be taken into account when considering the types of biomass materials required for generating energy:

1. Easy to grow and available most of the year
2. Secure and in regular supply
3. High energy content
4. Simple to process/sort
5. Quality and/or quantity of by-products
6. Density
7. Low market value
8. Transportation
9. Storage
10. Moisture content
11. Nitrogen emission

12. Ignition temperature
13. Burning period
14. Percentage of volatile materials.

2.7.6 Biomass Energy Influencing Factors

Having mentioned the 14 factors in Section 2.7.5, there are vital but basic starting factors which need to be taken into consideration when biomass energy is considered for wider use. These factors are significant to the success of any commercial project dealing with biomass materials and in particular a project researching the viability of biomass energy. These factors are the following:

A. Distance
B. Transportation
C. Processing
D. Sorting
E. Storage (temporary and permanent)
F. By-products.

A. Distance

> Transports longer than 50–100 km are not economically or environmentally viable. (Rupar and Sanati, 2005)

There is no point in trying to reduce the CO_2 in the atmosphere if the method of transportation of biomass materials employs the use of fossil fuels for this purpose, that is, pumping the same amount (or more) of CO_2 back into the atmosphere. For this reason, when it comes to the transportation of biomass materials, consideration of distance is vital.

Emissions, from whatever source, should be tackled purely for the purpose of achieving the balance needed.

> The results show that emissions from long range transportation, 1200 km, performed with ships, is of minor importance compared to emissions from local bio-energy systems in a local market. (Forsberg, 2000)

B. Transportation

There are four basic points which should be considered when dealing with all aspects related to a biomass transporting system:

1. Method of transport and cost
2. Density
3. Moisture
4. Seasonal availability.

The method of transportation can be looked at in two different ways. The first is to make sure that the emissions generated are low and the second is the cost of the transportation itself. Transportation cost is one of the main components in the overall total biomass power cost. A study in western Canada has estimated that for a truck operating at optimum power to transport one completely chipped tree, forest residues, and agricultural residues, such as straw and chaff, it would cost respectively US$10 per dry ton, $38 per dry ton, and $20 per dry ton (Kumar et al., 2003). This makes up 14%, 38%, and 25% of the biomass power cost. Usually, it costs more for forest residues; however, the maximum distance for transportation is 330 km (137 MW). A 450 MW biomass power plant would require 10–15 trucks per hour. The same study also suggests the idea of a pipeline to transport biomass materials, which was evaluated by using water or oil as the carrying liquid.

Transportation systems are a condition for development of an effective bioenergy system, and therefore an important factor in the present stage of energy crop development. Density, moisture, and seasonal availability are other factors that should be considered when transporting biomass materials. Depending on the type of materials being transported, and how these materials are processed before the transportation takes place, density will vary accordingly.

Moisture should be at a minimum level in the majority of biomass materials during transportation, and/or when waiting in temporary storage for the purpose of transportation later on. Seasonal availability can also be linked to the transportation process.

C. Processing

Processing can include crushing, shredding, and chipping preparation. The use of biomass energy can be classified according to generation method technology, such as direct burning, physical conversion, biological conversion, and liquification. Physical pretreatment techniques, grinding, milling, irradiation and steaming, are all part of the biomass process (additional details are in Chapter 3).

D. Storage (Temporary and Permanent)

When storing biomass materials a number of factors should be taken into account, especially if these materials will be stored over a long period of time. Temporary and permanent stores may have different requirements. Factors to be considered are the following:

1. Moisture content
2. Type of material
3. Age of the material
4. Size of the material.

Not taking the above factors into consideration can be a danger to the environment and human health (e.g., hydrocarbons and fungal spores).

E. By-products

There are numerous by-products, but ash is the cause of various problems related to processing equipment (for details, see Chapter 3, Section 3.6, and Chapter 6, Section 6.3.5). Best biomass processing practice is the reduction of by-products, such as the ash volume (if considered as unwanted by-product), the improvement of ash quality, reducing harmful gaseous emissions, and preventing/or reducing by-product hardware corrosion.

2.7.7 Characteristics/Co-firing Properties and Testing Method

Characteristics and co-firing properties can be classified under the following areas (Biomass Task Force, 2005):

1. Heating value
2. Reactivity
3. Moisture content
4. Ash level
5. Composition
6. Milling properties.

Standard biomass analytical procedures as developed by the National Renewable Energy Laboratory (NREL) and the American Society for Testing and Materials (ASTM) have been summarized in the following list. Prior to laboratory testing, some of the testing methods were investigated in order to understand more about the biomass samples:

1. Preparation of samples for compositional analysis
2. Determination of structural carbohydrates and lignin biomass
3. Determination of ash in biomass
4. Measurement of cellulose activities
5. Determination of ethanol concentration in biomass
6. Test methods for moisture
7. Determination of sugars and by-products
8. Determination of starch in biomass
9. Determination of protein in biomass.

There are other laboratory analytical testing methods and procedures in addition to these tests. These laboratory tests will be discussed in detail in Chapter 6, Sections 6.3.1 to 6.3.7.

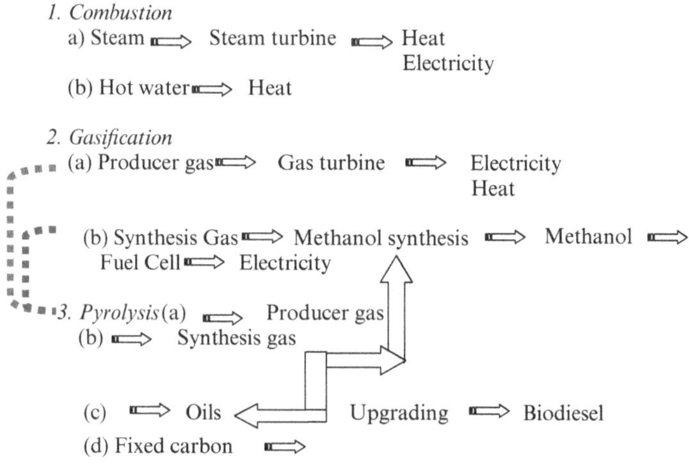

Figure 2.10 Combustion, gasification, and pyrolysis.

2.8 BIOMASS APPLICATIONS

Typical biomass applications can be categorized under combustion, gasification, and pyrolysis, as illustrated in Figure 2.10. Details of the applications in the figure are explained briefly in the following sections.

2.8.1 Bio-fuels

Past and present research shows that we can change and transform many transport systems by adapting them to use bio-fuels, such as ethanol and biodiesel. In this way, the large quantity of fossil fuels presently in use can be replaced by environmentally friendly fuels from renewable sources of energy (Biomass Task Force, 2005). The use of ethanol in Brazil (e.g., combustion engines within the transport system) demonstrates that bio-fuels can be used commercially and on a large scale. Also, some progress is being made in the use of bio-fuel in Europe, the United States, and Australia. In these parts of the world, fuel products exist where the bio-fuel is mixed with petrol, such as E20, where 20% is ethanol and the remaining 80% is petrol. There is another fuel product called E10, which is 10% ethanol and 90% petrol. At present in Australia, there are a number of car manufacturers (e.g., Ford, Toyota, and Mitsubishi) who have supported the use of E10, and this interest shown by car manufacturers in bio-fuel will help develop bio-fuel use further. In fact, these mixed fuels can be used with most car engines (Wyman, 1994).

With regard to biodiesel, the fuel is derived from vegetable oils as a result of alcoholysis, a transesterification reaction. It is characterized as light, lubricating and less viscous, and suitable for a variety of diesel engines. As bio-fuel is still a new product, and therefore not widely used, cost can be a problem. This is why government incentives in this area are important. In the future,

Table 2.6 The top five countries producing electricity from renewable sources in TWh/Year during 2011

Country	~Total	Hydro	Wind	Biomass	Photovoltaic	Geothermal
China	797	687	73	34	3	
Brazil	459	372	270	32		
USA	520	325	120	57	2	17
Canada	399	373	20	6		
Japan	116	83	4	23	4	3
Germany	127	18	47	44	19	

Source: Data from various sources.

this may change as biomass energy gradually establishes itself in the international market. In consequence, bio-fuel commercial companies will start to compete against each other, prices may fall, and government support will probably no longer be needed.

2.8.2 Electricity Generation

Currently, the main focus worldwide is to convert from the use of fossil fuels to renewable sources, in particular when it comes to the generation of *electricity* (Table 2.6). As mentioned previously, the production of electricity from renewable sources does not contribute to the greenhouse effect. In the case of biomass sources, the amount of carbon dioxide released by the biomass during combustion (either directly or after a bio-fuel is produced) is equal (or less) than the carbon dioxide absorbed by the biomass material during its growth (Johansson et al., 1993).

The electricity market has changed considerably in the last 15 years, leading to a reform of the energy sector and increased competition, therefore reducing electricity prices. The European Emissions Trading Scheme (EU ETS) adopted and created new CO_2 emission allowances. These are expected to give greater incentives for bioelectricity production and to offer further opportunities for bio-energy development in addition to national implementation schemes.

The upper limit for a bioelectricity plant generally ranges from 30 to 100 MW, mostly depending on the geographical context and biomass fuel sources used. There are however, larger plants already in operation. The largest biomass-fired plant in the world began operating in Finland in 2002: Alholmens Kraft has an electrical power output of 240 MWe and uses wood-based bio-fuels (45%), peat (45%), and coal as a reserve fuel (10%).

2.8.3 Heat, Steam, and CHP

Burning or the combustion of biomass materials (or biogas) can be used for the purpose of generating heat and steam. Most of the steam generated by power stations in various parts of the world is via the use of fossil fuels. In fact, a large portion of the electricity generated worldwide (approximately 80%) is still via steam-driving rotary generators (Electropaedia, 2013).

Heat can be the main product (direct use: e.g., within a domestic environment, such as heating and cooking) or generated as a "by-product," usually within a commercial/industrial environment. By-product heat (or waste heat) generated at conventional power stations can be from biomass materials or fossil fuels (CHP). CHP has three stages: power generation, heat recovery, and heat use. There are three types of CHP: large-scale and small-scale CHP and micro-CHP. A CHP unit of around 40 MW is typically a steam turbine when coal/oil is used, or a gas turbine if the main fuel is natural gas. If plants have a higher output than 50 MWe, then a CCGT will be used (Carbon Trust, 2010). The use of micro-CHP is mainly for small businesses and for domestic applications. CHP provides a secure and reliable energy supply and can reduce costs as well as benefit the environment. This is because it can make a better use of the primary fuel, that is, an efficient application of energy while generating power and heat. In order to achieve an efficient CHP, high capital investments as well as investment in resources will be required.

2.8.4 Combustible Gas

A number of uses can be made of biogas produced from anaerobic digestion or pyrolysis. These are the following:

1. Fuel for internal combustion engines
2. To produce heat (and/or a cooling system) for commercial and domestic needs
3. Transportation fuel.

The following are three methods of obtaining gases as an energy source from biomass materials:

A. Gasification

Gasification is described as the process of converting the organic fraction of biomass at higher temperatures (with the presence of air) into a gas mixture with a fuel value. A greater variety of fuels can be obtained than from the original solid biomass. This gas can be combusted to produce heat and steam. It can be used in internal combustion engines or gas turbines to produce electricity, as well as for the purpose of obtaining mechanical energy. Biomass gasification can also be integrated with fuel cells.

The production of electricity via gas turbines combined with steam cycles is the most effective and economical use of the gaseous product. Several biomass gasification processes have been developed (or are under development) for electricity generation that offer advantages over direct burning, such as higher efficiency and cleaner emissions.

Some of the gasification systems are currently at the demonstration stage, but the development of these efficient systems for electricity production is essential: BIGCC (Biomass Integrated Gasification and Combined Cycle) and BIG-STIG (Biogas Integrated Gasification Steam Injected Gas Turbine) plants can achieve efficiencies of 42–47% (EUREC Agency, 2002).

Significant developments have been made over the past 20 years in the field of biomass gasification, especially in the area of medium to large-scale electricity production. Gas cleaning to improve the quality of gas is a crucial issue in both combustion and gasification systems, and requires measures such as reduction of emissions and removal of particulates and tars.

Air gasification net product can be expressed by summing up the partial reactions, as illustrated in Equation (2.2) (Susta et al., 2003):

$$\text{Carbohydrate matter } (C_6H_{10}O_5) + O_2 \Rightarrow C_XH_Y + C_LH_MO_N + CO + H_2 + \text{Heat.} \tag{2.2}$$

B. Anaerobic Digestion

Anaerobic digestion is the decomposition of wet and green biomass through bacterial action in the absence of air (Gunaseelan, 1997). The anaerobic digestion process is made up of four main biological and chemical stages:

1. Hydrolysis
2. Acidogenesis
3. Acetogenesis
4. Methanogenesis.

The output is a mixed gas of methane (CH_4) and carbon dioxide (CO_2), called biogas. Landfill gas is the result of the anaerobic digestion of municipal solid waste buried in landfill sites. The methane gas produced in landfill sites eventually escapes into the atmosphere. This gas can be extracted by inserting perforated pipes into the landfill.

There are a number of benefits related to anaerobic digestion, which can be described as environmental benefits rather than technical or commercial benefits (see Box 2.3). Anaerobic digestion decreases methane emissions and can provide a good treatment system for organic waste. Consequently, this prevents groundwater contamination and reduces odor from the local environment.

The anaerobic digestion process is illustrated in Figure 2.11.

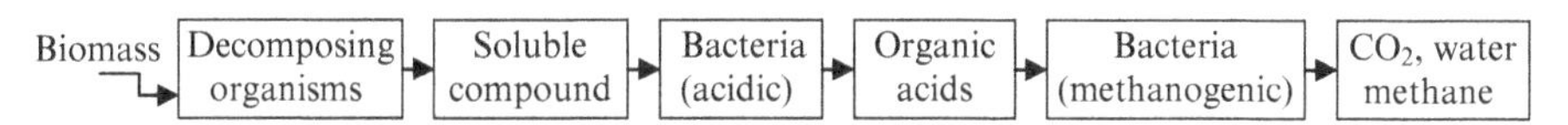

Figure 2.11 The anaerobic digestion process.

Box 2.3

ANAEROBIC DIGESTION

Potentially by 2020 with a medium expansion curve of AD plants and allowing for Government policy on reducing the amount of foodwaste in general, a reasonable expectation for England could be in the range of 5 Mt of foodwaste realistically available for AD and somewhere in the region of 20–60 Mt of animal waste. If this 5 Mt of food waste was digested this would replace 47,500 tonnes of nitrogen (N), 14,720 tonnes of diphosphorus pentoxide (P2O5) and 20,400 tonnes of potassium oxide (K2O), saving a total of 386,000 tonnes of CO2-equivalent in GHG emissions. Combined with 40 Mt of manures this gives the potential to generate approximately 3.5 TWh of electricity, enough to supply 913,000 households and saving 1.8 Mt of CO2-equivalent GHG**13** from grid-based electricity production.

Defra (2011)

C. Pyrolysis

Pyrolysis is an old method used many centuries ago for the purpose of producing charcoal (fixed beds). This old method is associated with low temperatures and slow chemical reaction producing a small amount of liquid. Today, high quality commercial pyrolysis usually takes place at a temperature ranging from 300 to 700°C with the absence of oxygen. In most cases, the presence of oxygen cannot be eliminated completely. The chemical decomposition of organic materials via heating (usually lasting for a few seconds and under normal atmospheric pressure) yields various products. The final outcome of the pyrolysis process is that the organic materials are transformed into gases/vapor and leave a solid residue (coke) made up from carbon and ash.*

Using pyrolysis, a solid biomass can be liquefied; this is called "direct hydrothermal liquefaction." Flash pyrolysis systems produce bio-oils (condensed vapor) from biomass materials, that is, up to 75% of dry feedstock is converted into bio-oil. The bio-oil can be upgraded into fuels, although these fuels may contain a high percentage of oxygen, that is, can be corrosive and unstable in nature. Further processing will therefore be required. In addition to the above, synthetic diesel fuel can be produced using the pyrolysis method (Biomass Engineering, 2013).

Some flash pyrolysis systems are able to safely dispose of waste, avoiding the contamination of the atmosphere (U.S. Department of Energy, 2005, Biomass).

One of the main benefits of flash pyrolysis is that fuel production is separated from power generation.

* Char output from biomass during pyrolysis process is far less than coal but higher when it comes to volatile matter.

Table 2.7 The biomass pyrolysis technology process and its products

Technology	Residence time	Heating rate	Temperature (°C)	Products
Carbonation	Days	Very low	400	Charcoal
Conventional	5–30 minutes	Low	600	Oil, gas, char
Fast	0.5–5 seconds	Very high	650	Bio-oil
Flash-liquid	<1 seconds	High	<650	Bio-oil
Flash gas	<1 seconds	High	<650	Chemicals, gas
Ultra	<0.5 seconds	Very high	1000	Chemicals, gas
Vacuum	2–30 seconds	Medium	400	Bio-oil
Hydro-pyro	<10 seconds	High	<500	Bio-oil
Methanopyro	<10 seconds	High	>700	Chemicals

Source: Adapted from Livingston (2007).

According to Livingston (2007), general types of biomass pyrolysis technology processes and their products require considerable variations in heating rate, temperature, and residence time (Table 2.7).

As the present pyrolysis technologies are in relatively early stages of development, similar to the rest of the present bio-oil upgrading processes, solutions to negative aspects should be found, such as corrosivity and low heating value. Fortunately, there are many projects in the United Kingdom and abroad (some of them already in advanced stages) searching for solutions to the above challenges. Finally, in conjunction with the existing systems, pyrolysis can be used for large-scale electricity production and consequently a reduction in the usage of fossil fuels can be successfully achieved.

2.8.5 Additional Bio-energy Technologies

There are other new types of technologies by which bio-energy can be produced. Some are at the development and experimental stages, and may not be economically viable (e.g., organic Rankine cycles [ORC], bio-synthetic natural gas [SNG], dimethyl ether [DME]) (Box 2.4).

Box 2.4

VERTICAL BIO-REFINING

Vertical bio-refining is the mechanism and approach related to the recycling and processing of biomass materials via various steps and stages.

HORIZONTAL BIO-REFINING

Horizontal bio-refining is the production of bio-fuels, heat, power, and material products from a single feedstock.

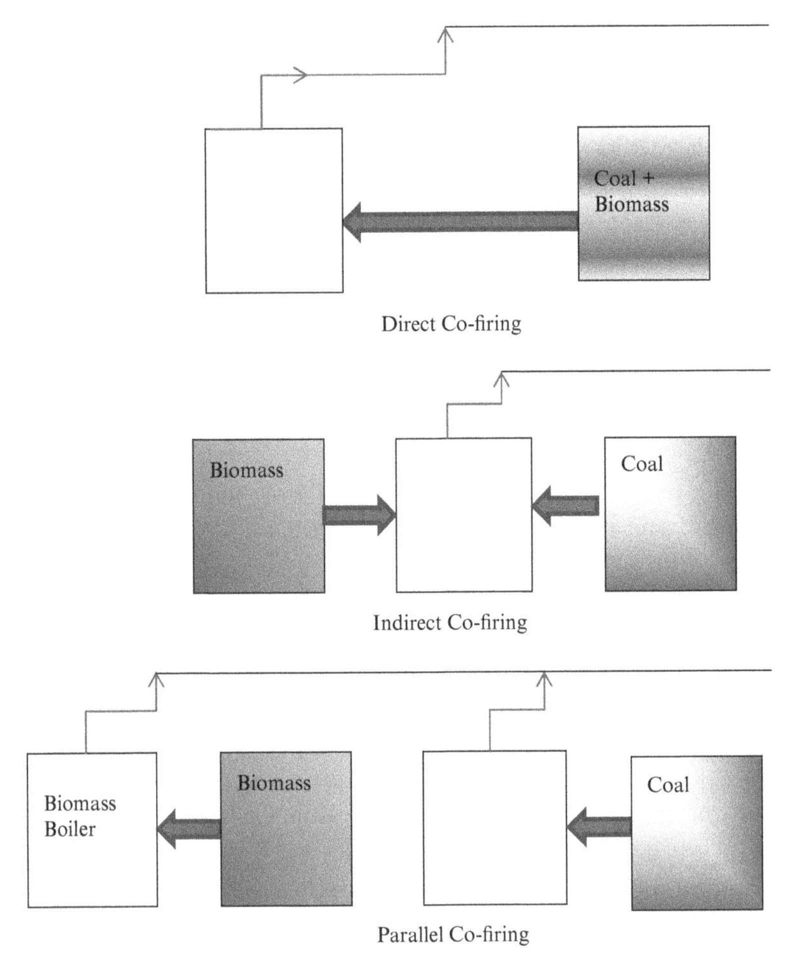

Figure 2.12 Illustration of three types of co-firing methods. *Source*: Author.

2.9 CO-FIRING

Co-firing of biomass with coal is the preferred route in the United Kingdom toward electricity generation from biomass. In simple terms, co-firing is a method of supplementing coal in a coal fired boiler with a different type of fuel, such as biomass materials.* Presently, there are three different types of co-firing: direct co-firing, indirect co-firing and parallel co-firing (Fig. 2.12). For direct co-firing, two different methods have been developed. The first method involves blending the biomass and coal in a fuel handling system and

* Co-firing should not be confused with multiple fuel boilers. Multiple fuel boilers are designed to burn a wide range of fuels. Co-firing on the other hand is done in a boiler specifically designed to burn only a specific type of coal and biomass materials.

feeding the blend to the boiler. The second method involves the separation of the fuels during handling, that is, biomass injected into a separate special burner, thus there is no impact on the conventional coal delivery system. Indirect co-firing is based on the thermal conversion of biomass or waste to gases or liquid fuel, and the co-firing of these converted fuels together with the main fuel.

Three different types of indirect co-firing exist. These are "indirect co-combustion with pre-gasification", "indirect co-combustion in gas-fired power plants," and "parallel co-combustion (steam side coupling)".

Co-firing can reduce the emission of a number of gases. It has already been established that these gases pollute the environment and can cause global warming. Co-firing, therefore, can be beneficial in a number of ways. The most important benefit obtained from the co-firing system is CO_2 reduction. In addition to this, the benefit extends to a form of NOx reduction and a reduction in flame temperature. Co-firing with biomass can reduce the emission of SOx due to the lower sulfur content in biomass materials. Other important beneficial factors are cost saving, as a variety of biomass materials are much cheaper than fossil fuels and that there is no threat of exhausting reserves, unlike fossil fuels. For this reason, co-firing can increase sustainability of energy supplies from power production and producing fewer by-products than the burning of coal alone (clearly this depends on the type of biomass being used in the co-firing system).

Co-firing improves combustion due to the higher volatile content in many of the biomass materials. By increasing the use of biomass materials as part of the co-firing process new jobs can be created within the above field, in general, and within the production of energy crops in particular.

2.9.1 Barriers for Biomass Co-firing

Grasses contain potassium and sodium compounds, which include various crops such as corn and wheat. These alkali compounds exist in all annual crops and crop residues.

During combustion, the alkali combines with silica. This reaction causes slagging and fouling problems in conventional combustion equipment designed for burning wood at higher temperatures. Additionally, volatile alkali lowers the fusion temperature of ash. In conventional combustion equipment, which has a furnace gas exit temperature above 788°C, the combustion of agricultural residue causes slagging and deposits on heat transfer surfaces. Specially designed boilers with lower furnace exit temperatures could reduce slagging and fouling from the combustion of these fuels. It is possible to reduce slagging and fouling caused by the alkalis by adding additives (e.g., dolomite, kaolin, bauxite, and limestone). Low temperature gasification may be another method of using these fuels for efficient energy production, while avoiding the slagging and fouling problems encountered in direct combustion. Regarding cost, according to US figures from 2010, the cost of using biomass to generate electricity is still relatively high. However, only nuclear and certain solar systems cost more than biomass prepared for co-firing.

2.9.2 Additional Challenges for Co-firing

There are a number of factors which can affect co-firing. Some of these are purely technical, while others are related to commercial, business, and law/regulation aspects.

Many of these factors do not apply to fossil fuels, and this is what makes co-firing with biomass a more challenging area. To establish biomass energy in a more competitive environment, a number of systems and fuels must be tested and marketed at the same time. Co-firing will need more support and time to smooth out its various problems before it can level out "commercially" with fossil fuels.

Factors that may give rise to problems are usually related to fuel preparation, storage, delivery, and fuel flexibility, that is, "quality and quantity." There are also problems related to ash deposition,* that is, an increased need for soot blowing and a more intensive cleaning of heat transfer surfaces may be required. In addition to this, there are problems of pollutant formation, increased corrosion rates of high temperature components, an increased number of bed material changes per day (in fluidized bed combustion), fly ash utilization (unburned carbon and contamination), higher in-house power consumption, difficulty in achieving complete combustion along with difficulties in mixing coal, and biomass in the boiler. Fouling and corrosion of the boiler (alkalis and chlorine) are other negative aspects of co-firing.

As part of the technical challenges, there is a negative impact on flue gas cleaning (SCR DeNOx) as the condensation of flue gasses is a major cause of corrosion. Nontechnical factors may range from the economic aspects (lack of financial incentives, uncertain fuel prices, and the open market) to legislative aspects (utilization of fly ash in cement, determining green share, and emission legislation), and public perception of co-firing of biomass/waste (further details are in Chapters 3, 6, and 9).

2.9.3 Further Advancement in Co-firing Engineering

The engineering part will play a vital role in helping to efficiently obtain energy from hybrid fuel in a less costly system and with fewer unwanted by-products. To achieve this, a number of technical and engineering steps will be required:

1. Improving the present type of boilers that can separately inject the biomass from coal in order to achieve a higher biomass co-firing

* Elements such as Si, Ca, K, and P can have direct influence on ash fusion and deposition in furnaces.

percentage, as well as reducing problems related to blending biomass with coal

2. Developing a network database for biomass fuels to be used by prospective installations in order to evaluate the possibilities available in the new co-firing hardware system
3. Building biomass fuel profiles, including their chemical structure. This can greatly help when taking into consideration the engineering aspects for the entire system and may reduce slagging, fouling, and corrosion.

Finally, installing a device to properly prepare the size of the biomass fuel may help to avoid the need for additional modifications to the existing boiler.

2.9.4 Promoting Co-firing

There is a great need for a system engineering approach to promote co-firing. This can be considered from different angles, keeping in mind all the factors that can affect the outcome directly and/or indirectly from the use of co-firing.

Some of the technical and nontechnical problems related to biomass co-firing have been already discussed. A basic general approach for solutions to these challenges can be seen in the following section.

2.9.4.1 General Approach Solutions. The solutions for co-firing challenges can be found and tackled in any of the following areas:

a. Engineering
b. Regulations
c. Environmental factors
d. Economical factors
e. Public attitude.

Engineering can be a short- or a long-term solution, depending on the engineering complexity required, together with other factors such as human skills and the natural resources required. The "regulations" solution, as the name indicates, is dependent upon the local, national, and international law and policies, that is, supporting and/or speeding up the process related to co-firing, with incentives and positive regulations for this kind of energy system.

The environmental factors relate to regulations that aim to protect the environment, but this in itself can *sometimes* be an obstacle when it comes to co-firing as an industry. This is because, as in the fossil fuel case, emissions and environmental hazards generated from biomass materials are important issues that need to be tackled in order to make co-firing more environmentally friendly. By following and applying environmental regulations, co-firing can benefit the environment in the form of a net reduction of harmful emissions.

The economic* approach is one of the most important factors in finding some of the solutions, in particular those solutions that can be applied within a short period of time, if finance can be provided immediately. A number of co-firing schemes have been neglected or overlooked simply because of a lack of finances.

The public attitude to co-firing and the engineering process, along with the prices associated with it, can be another important factor where part of the solution can be found, that is, readiness to accept any new concept related to co-firing.

The earlier points are areas in which possible solutions can be found and implemented, depending on the factor itself and the circumstances surrounding it. A new design for the systems engineering approach solutions is what many of the co-firing systems presently require. This is because co-firing is a new concept of hardware and software engineering, backed up with scientific know-how and a sensible commercial approach (Box 2.5).

2.10 SYSTEM ENGINEERING

When it comes to the efficient production of energy, the technical and engineering factors are vital. Without it, progress for sustainable fuels would be

Box 2.5

CO-FIRING CHALLENGES AND SOLUTIONS

Engineering
Present and future hardware issues, challenges, and development

Regulations
Governments support and policy—local, national, and international approach

Environmental Factors
Environmental protection and health issues—short- and long-term implications

Economical Factors
Economical return to businesses (power generation companies and farmers)

Public Attitude
Understanding the benefits related to the environment and job creation issues

* Existing system and location are two important factors in relation to co-firing overall cost.

difficult. For this reason, additional examinations have been added to enhance and complete various sections of this book.

The basic engineering factors may include fuels sourced from biomass materials in the form of gas, liquid and solid fuels. As this book is mainly concerned with the selection process of biomass materials for power generating companies and establishing the ground for the production of a new hybrid bio-fuel the engineering factors will focus on this area.

The first step is to look at the scientific factors and see their connection with the other two factors, that is, the technical/engineering and the business factor. Using the scientific factor, a model can be constructed to create an improvised co-firing scenario that makes sense within a commercial environment. The technical/engineering side will run in parallel with indications of the scientific results and demonstrate what may be possible to achieve within an acceptable market scenario. If these two factors work hand-in-hand, then the business part can take over, according to the pyramid representation (Chapter 5, Section 5.2). A systematic engineering approach has been outlined in Figure 2.13.

Out of the 15 selected samples, eight of them have already been tested commercially in a co-firing setting. These were exhaustively characterized on a technical, scientific, and business basis. A model was established to correlate the characteristics of the fifteen samples in order to ascertain the suitability of each type of biomass for co-firing. This model, as a possible approach to improving a co-firing scenario, is an important part of this book as it works in the form of a "guideline" for the construction of the methodology (REA1) and within the application of the energy crops selection process that should lead to the production of a hybrid bio-fuel (Box 2.6).

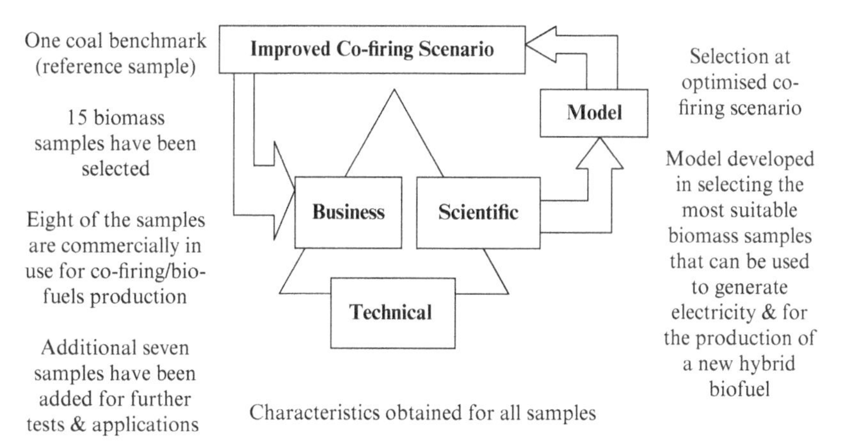

Figure 2.13 A schematic model illustrating a co-firing scenario leading to optimized hybrid bio-fuel production, *Source*: Author.

Box 2.6

BIOMASS AND LAND

Increasing the supply and use of biomass has implications for land use, biodiversity and other environmental factors and landscape. If properly planned, biomass development, in addition to helping climate change objectives, may have positive environmental effects, consistent with enhancing the overall health of ecosystems.

Defra (2007)

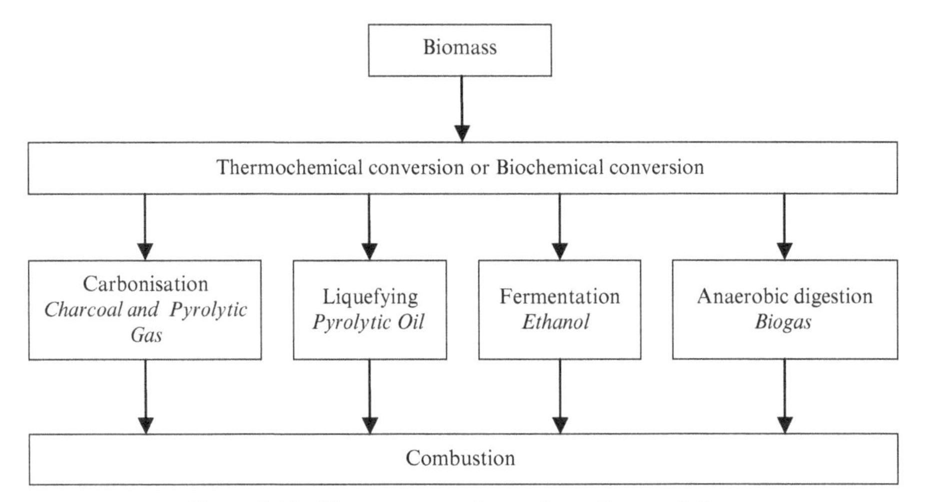

Figure 2.14 Biomass conversion systems. *Source*: Author.

2.11 BIOMASS CONVERSION SYSTEMS

There are two main methods which cover a wide area of biomass conversion technologies, thermochemical conversion, and biochemical conversion (Fig. 2.14). Indeed, to obtain the energy, the combustion factor is the key to both technologies. Hardware biomass conversion systems can be stationary or mobile The mobile hardware systems are usually used in rural areas supplying power for a small number of homes, such as in a village, or powering small- to medium-size countryside businesses. However, the principle for both stationary and mobile hardware combustion systems (depending on the method used) is similar.

The combustion can be made either using a furnace or a boiler. A furnace (direct combustion) is one of the simplest methods used to obtain energy by burning the biomass materials in a chamber to obtain heat in the form of released hot gases.

Table 2.8 Comparison between furnace and boiler

Furnace	Most of the fuel energy is in the hot gasses
Boiler	Most of the fuel potential biomass energy is in the steam

Source: Author.

A boiler for biomass can be used to transform the heat into steam. This steam is then used to turn the turbine in order to generate electricity.

There are three different types of boilers:

1. Pile burners
2. Stationary or travelling grate combustors
3. Fluidized-bed combustors.

Table 2.8 illustrates the basic difference between a furnace and a boiler, but what is important here is the percentage of efficiency for each system.

Direct firing can be divided into four different methods. These methods come under the titles of Pile burner, spreader stoker, fluidized bed, and suspension. The other method is Gasification, which can be divided into five different subbranches, that is, biological gasification, landfill gas, pyrolysis, thermal gasification, and microscale biomass.

Biomass technology conversions have been listed in Table 2.9, where technologies, conversion process type, major biomass feedstock and energy/fuel produced have been compared. Direct combustion, gasification, pyrolysis, and methanol production all come under the "thermochemical" conversion process. Anaerobic digestion and ethanol production come under "biochemical" conversion process type. Biodiesel production is classed as a "chemical" conversion process.

Emission rates for various technologies are illustrated in Table 2.10, covering CO_2, NO_x, CO, SO_2, and particulate matter (PM). The data in this table are more than 10 years old yet the present emission rates today from these technologies are relatively similar, despite technical development in this field over the past 12–13 years.

2.12 ENERGY CROPS SCHEME (U.K.)

The outline of the Energy Crops Scheme can be summarized in the following points.

- Grants to establish energy crops
- Funding to set up producer groups
- Establishment grants of £1000 per hectare for a short rotation coppice (willow or poplar) and £920 per hectare for Miscanthus.

Table 2.9 Listing of biomass technologies

Technology	Conversion	Biomass Materials	Fuel/Energy
Anaerobic digestion	Biochemical (anaerobic)	Animal manure Agriculture waste Landfills Wastewater	Medium Btu gas (methane)
Biodiesel production	Chemical	Rapeseed Soy beans Waste vegetable oil Animal fats	Biodiesel
Direct combustion	Thermochemical	Wood Agricultural waste, municipal solid waste, residential fuels	Heat Steam Electricity
Ethanol production	Biochemical (aerobic)	Sugar or starch crops Wood waste Pulp sludge Grass straw	Ethanol
Gasification	Thermochemical	Wood Agricultural waste, municipal solid waste	Low- or medium-Btu producer gas
Methanol production	Thermochemical	Wood Agricultural waste, municipal solid waste	Methanol

Source: Adapted from Oregon.gov (2002).

Applicants need to demonstrate that they have, or will have, an energy end use for the crops. This could be:

A. A biomass power plant
B. A community energy scheme using combined heat and power (CHP)
C. Heat for a small business or home, including own use (10 mi, small installations, 25 mi for power plants, 3 ha, and subject to environmental checks)
D. Producer groups must be legally formed by and consist of members who are growing short rotation coppice for an energy end use
E. Up to 50% of the costs of setting up the group are available.

Eligible expenditure includes:

- Purchase of specialist equipment
- Staff costs
- Specialist fees
- Office accommodation
- Publicity and promotion.

Table 2.10 Emission rates for various energy technologies

	Natural Gas	Coal	Oil	Direct Combustion Biomass	Gasification[a]	Landfill Gas/Biogas[b]	
Technology Type	Generic Natural Gas Fired Turbine Combined Cycle Plant (lbs/MWh)	Generic Utility Boiler Firing Bituminous Coal Plant (lbs/MWh)	Generic Utility Boiler Firing Only No. 6 Fuel (lbs/MWh)	Typical Stoker Grate Wood-Fired Biomass Plant (lbs/MWh)	Combined Cycle Biomass Gasification Plant (lbs/MWh)	Typical Gas Turbine (>3 MW) (lbs/MWh)	Fuel Cell (PAFC) (lbs/MWh)
NO_x	0.06	3.1	2.5	1.5	1.08	0.44–2.2	0.003
CO	0.03	0.21	0.35	3.556	0.001	0.6	0.015

[a]Mann M, Spath P (December 1997) Life cycle assessment of a biomass gasification combined-cycle system.
[b]Arthur D. Little (October 10, 1998) Profile of leading renewable energy technologies for the Massachusetts Renewable Energy Trust Fund.

Source: Adapted from Xenergy (2003).

Box 2.7

CLEAN ENERGY

> While clean energy has enormous potential, outcomes are not predetermined. The success of individual technologies that are not yet commercial hinges on developers overcoming current technical, social and cost barriers.
>
> Australian Government (2012)

2.13 RENEWABLE OBLIGATION CERTIFICATE (ROC) (U.K.)

The renewable obligation certificate (ROC) target:

- Electricity companies to source an increasing proportion of their supply from a renewable source of energy.
- Technologies to reach the 14% goal by 2014 (the Obligation will extend to 2027).
- Affording long-term security for the renewable market.
- Renewable obligation certificates (ROCs) will be awarded to accredited generators of eligible renewable electricity produced within the United Kingdom (including energy crops).
- Companies that are unable, or unwilling, to source the required amounts have the option of "buying out" their obligation.
- The "buy out" price is initially set at £30/MWh (US$43/MWh), and will be adjusted annually in line with the retail price index.
- Monies raised from companies "buying out" in this way will be redistributed to companies that have met their obligation, in proportion to the number of ROCs they presented in that year, so acting as a further means of market stimulation (Box 2.7).

2.14 CLIMATE CHANGE LEVY EXEMPTION CERTIFICATE (LEC) (U.K.)

The summary of the LEC is outlined in the following points:

1. The Fossil Fuel Levy, a dormant levy on the electricity bill, was originally created to support the nuclear industry. It will be reintroduced (or is possibly already introduced) to provide long-term funding for zero-carbon technologies.
2. In 2001, the government introduced a Climate Change Levy, a tax on energy used in the nondomestic sector, in order to encourage energy efficiency and reduce emissions of GHG. The levy was repaid to businesses in the form of lower National Insurance.

3. The 2003 report "Towards a Noncarbon Fuel Economy, Development and Demonstration" introduced a carbon tax, a direct payment to the government based on the carbon content of the fuel being used. This was a replacement for the Climate Change Levy.
4. Preference is shown for the purchase of goods from countries that have taken demonstrable steps to reduce their greenhouse emissions.
5. The 2003 report raises questions, such as "Does the UK lead in biomass technology?" and "What are the benefits?"
6. Export opportunities.
7. Before investment occurs, there is a need for a long-term price signal, regulation and ownership.
8. Decommissioning of the United Kingdom's electricity generator fleet, which will reach 30% by 2020.
9. Decommission of North Sea oil/gas.
10. Plant assessments, laboratory and rig testing (for a particular application).
11. Furnace, types and problems.
12. Fuel compatibility: If not mixed with coal, what other alternatives are available?
13. Mills (single mill): How much energy is used just in crushing the biomass?
14. Operational problems.
15. Fuel logistics: type of furnace and maximum load.
16. Energy crop and co-firing post-2009.
17. The technology used for biomass (co-firing) (Box 2.8).

2.15 CONCLUSION

The use and application of renewable energy in general is still in the development process despite a number of factors pushing hard for an alternative

Box 2.8

SOME ASPECTS OF BIOMASS

No limitation
Three different forms of energy sources (solid, liquid, and gas)
Energy from landfill gas
Zero GHG emission
Possible reduction/stablization of CO_2 in the atmosphere
Ash as a fertilizer and/or for cement/road-building materials
Direct energy—thermal, mechanical, and electrical—competitive
Liquid fuel for combustion engines

source of energy. Currently, the world is experiencing an evaluation process for all types of renewable energy development before one type or another can take the lead and possibly dominate the commercial energy market, as fossil fuels have been doing since the combustion engine was invented.

Concerning biomass energy in general, the methods and factors related to this field are vitally important. The following points should be taken into consideration as they can help the prospect of developing biomass energy further:

1. Knowing which types of biomass materials (e.g., energy crops) are produced locally in terms of higher energy output and regularly availability.
2. Many options present benefits in terms of saving nonrenewable energy sources, reducing GHG emissions, and providing income diversification for farmers. Support for farmers is vital in encouraging them to produce the required energy crops for this purpose.
3. Finding efficient ways to transport, process, and store biomaterials.
4. Biomass will be an important sustainable energy source, but only if we are able to supply the energy vectors demanded by modern energy services, based on economically and environmentally sound fuel chains.
5. Reducing by-products to improve the efficiency of the devices used to convert biomaterials into energy.
6. Biomass is intrinsically linked to energy, environmental, and agricultural policies, and these will shape the biomass energy markets.
7. Whenever possible, it is important to learn from other systems presently in use around the world.
8. Farmers should be supported by the government to encourage them to allocate part of their farming land to renewable energy plantation on a regular basis.
9. Government legislation should give more support to businesses and companies dealing with/working within the renewable energy sector.
10. The fossil fuel levy, a dormant levy on the electricity bill, should be investigated as a means of providing long-term funding for zero-carbon technologies.

Using biomass energy should be considered to be as part of protecting the environment by helping in the reduction (or balancing) of the emission of carbon dioxide. Regulations and various governmental incentives have resulted in some reduction of CO_2 emissions in the energy production field. This in turn resulted in the development of different co-firing technologies. At present, there are a number of new and advanced co-firing technologies, some of which have been passed from the experimental and testing stage into large commercial operations. On the other hand, many new co-firing technologies are still in the process of being developed or are close to the final testing phase. In addition to the above, PC and fluidized bed boilers used in direct co-firing are good examples of how the hardware industry in this field is establishing itself widely on the commercial market. There are other mature methods, such as co-firing of the product gas in PC and gasification. Alternatively, hot gas cleaning is still being developed and may take some time before being used on a commercial scale. Examples of indirect co-firing, such as using parallel boilers in a steam integrated cycle, is another good working engineering design that may soon be used. In some parts of the world, this is already in use on a commercial scale.

New co-firing technologies should be supported by direct contact and discussion with the power generating companies and governmental organizations—at various levels. This kind of regular high level contact will speed up the development of various co-firing industries to the level reached by the fossil fuels hardware systems, presently in operation today (Box 2.9).

Box 2.9

STRAW PELLETS (SP)

Long before the idea of using biomass materials to generate electricity, pellets in general, and straw pellets in particular, were (and are still) part of the industry for the production of animals food. They are also used as fuel for heating systems.

The basic process for making pellets has been summarized in the following points:

(1) Sorting (removing heavy contaminant);
(2) grinding;
(3) drying (generally moisture reduced to about 10%);
(4) binding agent added (this can be organic binding agent or using steam, depending on the final use of the pellets);
(5) shaping/moulding;
(6) cutting to the required length;
(7) cooling.

The difference in the makeup of wood and straw is that wood has less hemi-cellulose than straw, while straw has less cellulose and less lignin. The makeup of straw can be summarized in the following points:

1. Straw cell walls make up 80–90% (dry weight)
2. Silica (5–10%), the remaining are extractives (5–15%)
3. Cellulose, hemicellulose, and lignin (carbohydrates).

The chemical elements of straw pellets:

Carbon	43.34%
Hydrogen	5.90%
Nitrogen	3.08%
Oxygen	47.60%
Sulfur	~0.08%

Source: Author.

REFERENCES

Australian Government (2012) Energy white paper 2012: Australia's energy transformation. http://www.ret.gov.au/energy/Documents/ewp/2012/Energy_%20White_Paper_2012.pdf (last accessed December 3, 2013).

Biomass Engineering (2013) Pyrolysis. http://www.biomass.uk.com/pyrolysis.php (last accessed December 3, 2013).

Biomass Task Force (2005) Biomass task force report to the government. Department of Environment, Food and Rural Affairs (Defra) publications, London. CVBP interim test burn emissions report (2003).

BP (British Petroleum) (2013) Solar capacity. http://www.bp.com/extendedsection genericarticle.do?categoryId=9041560&contentId=7075261 (last accessed December 3, 2013).

Brown L (2009) Plan B 4.0, Mobilization to Save Civilization. WW Norton and Company.

Carbon Trust (2010) Introducing combined heat and power. http://www.carbontrust.com/media/19529/ctv044_introducing_combined_heat_and_power.pdf (last accessed December 3, 2013).

Cariou J (2010) Solar water heater. Global Energy Network Institute. http://www.geni.org/globalenergy/research/solar-water-heaters/solar-water-heater.pdf (last accessed December 3, 2013).

Defra (2002) Growing short rotation coppice: best practice guidelines. Department of environment, food and rural affairs (Defra) publications, London.

Defra (2007) UK Biomass Strategy. http://www.biomassenergycentre.org.uk/pls/portal/docs/PAGE/RESOURCES/REF_LIB_RES/PUBLICATIONS/UKBIOMASS STRATEGY.PDF (last accessed December 3, 2013).

Defra (2011) Anaerobic digestion strategy and action plan. http://www.defra.gov.uk/publications/files/anaerobic-digestion-strat-action-plan.pdf (last accessed December 3, 2013).

DTI (2003) Our energy future: creating a low carbon economy. Department of Trade and Industry (DTI) Publications Cm 5761, HMSO, London.

DTI (2005) UK Carbon Abatement Technologies Programme. http://www.dti.gov.uk/energy/coal/cfft/ (last accessed January 26, 2014).

DTI (2007) Engineering report: guidance document on biomass co-firing on coal fired power stations. DTI Project 324-2.

Electropaedia (2013) Steam turbine electricity generation plants. http://www.mpoweruk.com/steam_turbines.htm (last accessed December 3, 2013).

Elliot MA (1981) Chemistry of Coal Utilization. Second supplementary volume. New York: Wiley-Interscience.

EPI (Earth Policy Institute) (2013) Website data from various articles. http://www.earth-policy.org/about_epi/C122 (last accessed January 26, 2014).

ESA21 (Environmental science activities for the 21st Century) (2009) Renewable energy: Hydropower. http://esa21.kennesaw.edu/activities/hydroelectric/hydroactivity.pdf (last accessed December 3, 2013).

European Climate Foundation (2010) Biomass for heat and power: opportunity and economics. http://www.europeanclimate.org/documents/Biomass_report_-_Final.pdf (last accessed December 3, 2013).

Farabee JM (2006) On-line biology book. Estrella Mountain Community College, Avondale, Arizona, USA. http://www.emc.maricopa.edu/faculty/farabee/BIOBK/BioBookTOC.html (last accessed December 3, 2013).

Fletcher S, Lyon K, Rayner M with assistance from John Todd and Philip Jennings of Murdoch University. (2005) Biomass energy systems. Edited and updated by Mark McHenry. http://reslab.com.au/resfiles/biomass/text.html (last accessed December 8, 2009).

Forsberg G (2000) Biomass energy transport: analysis of bioenergy transport chains using life cycle inventory method. Biomass and Bioenergy 19:17–30.

Grubler A, Nakicenovic N (1988) The dynamic evolution of methane technologies. In TH Lee, HR Linden, DA Dryefus and T Vasko, eds., The Methane Age. Dordrecht: Kluwer Academic.

Gunaseelan VN (1997) Anaerobic digestion of biomass for methane production: a review. Elsevier Science, Biomass and Bioenergy 13(1):83–114, 32.

GWEC (Global Wind Energy Council) (2012) Global Wind Report Annual Market Update 2012.

Hall DO, Hemstock SL, House J, Rosillo-Calle F (1992) Second World Renewable Energy Congress, Reading. September 13–18.

IEA (2011) World Energy Outlook 2011.

IEA (2012) World Energy Outlook 2012.

IENICA (2007) Reed canary grass. Interactive European network for industrial crops and their applications. http://www.ienica.net/crops/reedcanarygrass.htm (last accessed June 25, 2008).

Ignaciuk A, Vöhringer F, Ruijs A, van Ierland EC (2004) Competition between biomass and food production in the presence of energy policies: a partial equilibrium analysis. Environmental Economics and Natural Resources Group, Wageningen University.

Johansson TB, Kelly H, Reddy AKN, Williams RH (1993) Renewable energy: sources for fuels and electricity. Washington, DC: Island Press.

Khosla V (2008) Scalable electric power from solar energy. Partner, Khosla Ventures. Breaking the Climate Deadlock, Briefing Paper. The Climate Group. http://www.khoslaventures.com/wp-content/uploads/2012/02/Scalable-Electric-Power-from-Solar-Energy.pdf (last accessed December 3, 2013).

Kumar A, Cameron JB, Flynn PC (2003) Pipeline transport of biomass. Oral Presentation 1A-04 Department of Mechanical Engineering, University of Alberta.

Livingston WL (2007) Biomass ash characteristics and behaviour in combustion, gasification and pyrolysis systems. Technology & Engineering, Doosan Babcock Energy.

MNRE (Ministry of New and Renewable Energy) (2008) Solar photovoltaic programme. Government of India. http://mnes.nic.in/ (last accessed December 3, 2013).

National Renewable Energy Laboratory (2002) Renewable energy cost trends. NREL Energy Analysis Office. http://texas.sierraclub.org/press/images/RenewablesCostsFalling.pdf (last accessed January 2, 2009).

NASA (2013) Effects of changing the carbon cycle. Earth Observatory. http://earthobservatory.nasa.gov/Features/CarbonCycle/page5.php (last accessed December 3, 2013).

Oregon.gov (2002) Biomass energy home page: biomass energy. Oregon Department of Energy, USA. http://www.oregon.gov/ENERGY/RENEW/Biomass/Pages/BiomassHome.aspx (last accessed December 3, 2013).

Ostridge R, (1998) Written for Physics 261, University of Prince Edward Island.

Pike Research (2011) Geothermal power capacity could more than double by 2020. http://www.pikeresearch.com/newsroom/geothermal-power-capacity-could-more-than-double-by-2020 (last accessed December 3, 2013).

Reich P (2006) Hold your breath; plants may absorb less carbon dioxide than we thought. Morrison D., University of Minnesota, EurekAlert. http://www.eurekalert.org/pub_releases/2006-04/uom-hyb041206.php (last accessed December 3, 2013).

Rempel H (2000) Will the hydrocarbon era finish soon? Federal Institute for Geoscience and Natural Resources, Stilleweg 2, 30655 Hannover. http://www.hubbertpeak.com/Rempel/ (last accessed December 3, 2013).

REN21 (2008) Renewables 2007 global status report. http://www.ren21.net/Portals/0/documents/activities/gsr/RE2007_Global_Status_Report.pdf (last accessed May 4, 2009).

REN21 (2012) Renewables 2012 global status report. http://www.map.ren21.net/GSR/GSR2012_low.pdf (last accessed December 3, 2013).

Rupar K, Sanati M (2005) The release of terpenes during the storage of biomass. Biomass and Bioenergy 28:29–34.

Shrestha MR (2007) Financial analysis of renewable energy projects. e-learning course Bio-energy for achieving MDGs, Lecture 7, School of Environment, Resources and Development, Asian Institute of Technology, Thailand.

Solarbuzz (2009) Photovoltaic industry statistics: costs (website). http://www.solarbuzz.com/StatsCosts.htm (last accessed December 3, 2013).

Susta MR, Luby P, Mat SB (2003) Biomass energy utilization & environment protection: commercial reality and outlook. http://www.powergeneration.siemens.com/NR/rdonlyres/FDD06929-8B80-49FA-B20B-A80294CDDBFC/0/4_Biomasse_Energy.pdf (last accessed December 3, 2013).

Tseng T (2007) Solar power: an alternative energy source. http://cosmos.ucdavis.edu/archives/2007/cluster2/tseng_tiffany.pdf (last accessed December 3, 2013).

U.S. Department of Energy (2005) Energy efficiency and renewable energy: biomass. http://www1.eere.energy.gov/biomass/pyrolysis.html#thermal (last accessed December 4, 2007).

U.S. Department of Energy (2009) Biomass program: information resources: ABC's of biofuels. http://www1.eere.energy.gov/biomass/abcs_biofuels.html (last accessed April 20, 2009).

Vuthaluru HB (2003) Thermal behaviour of coal/biomass blends during co-pyrolsis. Fuel Processing Technology 85:141–155.

Wyman EC (1994) Alternative transportation fuels from biomass. Alternative Fuels Division, National Renewable Energy Laboratory, Golden, CO 80401.

Xenergy (2003) Securing a place for biomass in the northeast United States: a review of renewable energy and related policies. http://www.nrbp.org/pdfs/nrbp_final_report.pdf (last accessed December 3, 2013).

3

CO-FIRING ISSUES

3.1 TECHNICAL AND ENGINEERING ISSUES

3.1.1 Introduction

Before considering the biomass and fossil fuel samples obtained for the application of the methodology (Chapter 4), there is a need for an investigation into biomass energy technical and engineering issues. By examining factors and basic hardware from an engineering point of view, prior to the testing of biomass samples, a clearer picture helps us to understand some of the complexities that power generating companies face. For these companies, the main challenge is the high level of technical risk. In order to find a suitable solution, while at the same time reducing costs and creating better efficiency, particularly concerning the technical aspects, biomass properties have to be analyzed objectively and adjusted accordingly. Problems that should be considered concerning biomass materials used as a viable fuel can range from hardware engineering issues (milling, fuels mixing, combustion, and by-products) to access to unwanted chemicals in the fuels themselves.

When the biomass materials selected for usage have been already examined and tested thoroughly, an environmentally friendly and economically sound co-firing method can be achieved from both a scientific/technical and commercial standpoint. This kind of consideration is usually made long before the implementation of the final engineering hardware.

The Selection Process of Biomass Materials for the Production of Bio-fuels and Co-firing, First Edition. Najib Altawell.

Looking at the biomass energy utilization in general and co-firing with coal in particular, barriers and related technical and engineering issues are easily recognizable. This is simply because biomass energy industries on a large commercial scale are still relatively new, in some countries still at the planning stage. Therefore, collecting biomass materials on a large scale on a regular basis without sufficient knowledge in biomass management and without technical and engineering know-how makes the whole task of trying to produce energy from biomass materials costly, challenging, and uneconomical. Many of the boilers at power generating companies have been designed for burning coal, that is, not designed for burning biomass materials (Perry and Rosillo-Calle, 2006). Even the basic adjustment to these boilers in order to improve the co-firing mechanism will incur a huge cost for the business itself. This is one of the reasons there is still widespread reluctance toward new investment to upgrade and/or develop hardware systems (DTI, 2007).

3.1.2 Hardware and Biomass Materials

Different biomass materials may possess different chemical qualities and therefore behave differently with each other during combustion, as well as behaving differently with coal during the co-firing process (Ireland et al., 2004) (Fig. 3.1). This means that whether or not the process involves co-firing with coal, biomass materials should be treated *individually* and certainly not in a generalized fashion.

Related problems can be traced to the way in which biomass materials are processed to a specified size, quality, and moisture content. Mechanical and/or feeding problems prior to the fuel being injected should be examined. Second, a blockage during the feeding of biomass materials into the combustion chamber may occur from time to time. This can be due to the uneven size of the biomass materials and/or due to the mechanical design itself.

Another important technical and engineering issue is the impact of biomass materials on the hardware. This can cause corrosion, fouling and slagging within the boiler itself. The effect on the hardware from alkali chloride, which is one of the chemical components of various types of biomass materials, has a damaging effect on the hardware, with or without regular maintenance (DTI, 2007). The conditions of hardware systems and the required maintenance they should receive, along with the types of biomass and coal used in co-firing, are all important factors in achieving a better operational and higher energy output. This is notwithstanding the variations between the combustion systems, the type of boilers, as well as the fuel injection method. These factors all may have an additional impact on the outcome (Box 3.1).

3.2 TECHNICAL AND HARDWARE ISSUES

Part of the problem associated with co-firing (biomass with coal) is related to fuel preparation, storage, delivery, ash deposition, fly ash utilization, higher

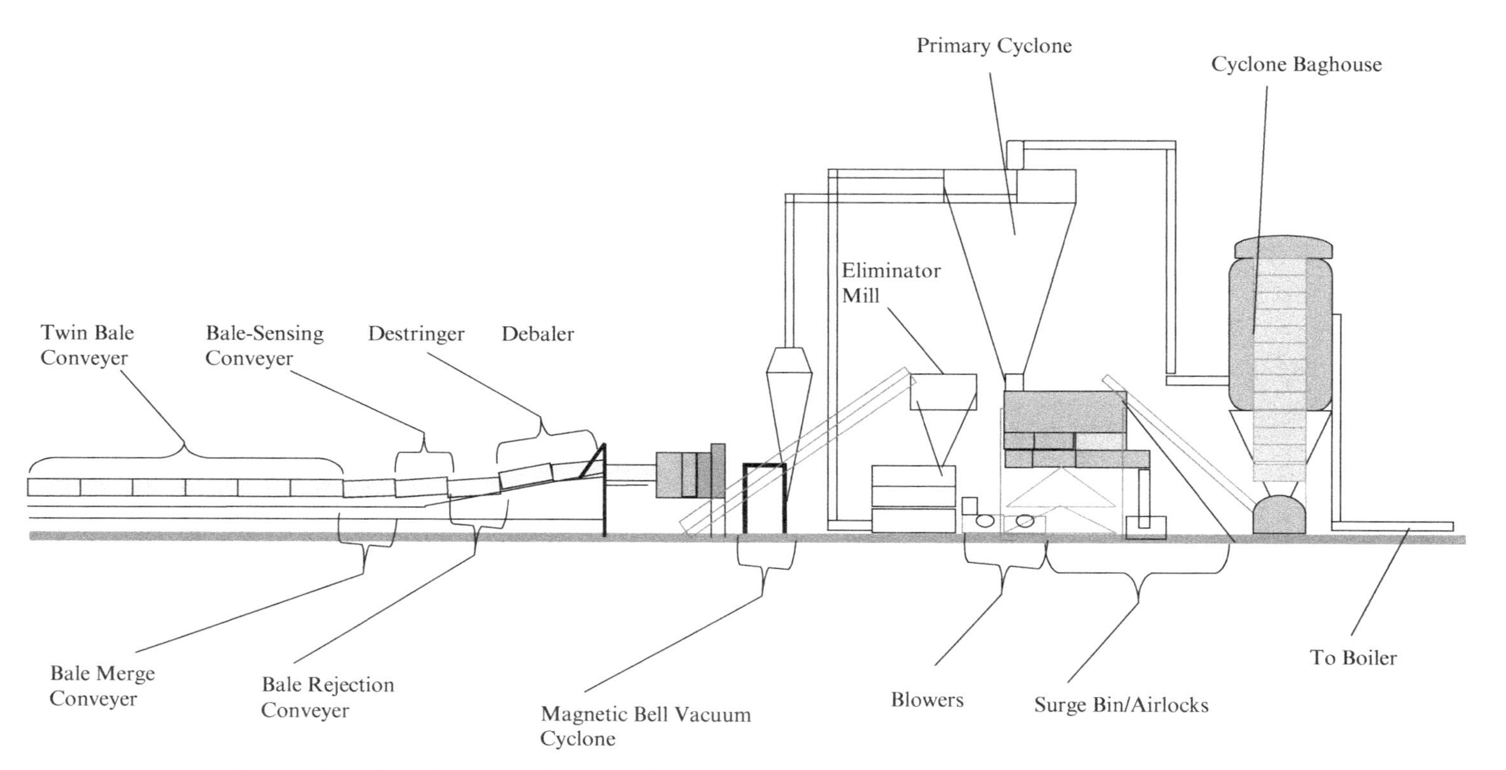

Figure 3.1 Schematic diagram illustrating the co-firing process. *Source*: Adapted from Prairie Lands (2013).

Box 3.1

TUBELESS BOILERS

"Tubeless" boilers use tubing coils instead of rigid tubes. "Direct contact" water heaters have no tubes, tubing or coils; they have heat transfer media such as spheres or cylinders and allow flue gases to come in direct contact with the water.

Boiler Burner (2013)

in-house power consumption, pollutant formation, higher corrosion within high temperature components and, in fluidized bed combustion, an increase in the number of bed material changes (Mitchell et al., 2004). Some of the main issues related to the co-firing technical and engineering field are the following:

1. Degradation can be prevented by lengthening the storage period for biomass materials, either by building storage specifically for biomass materials (a biomass silo) and/or adding a preservative agent to them during or after the grinding process.
2. Biomass materials should be milled to a fine powder or pressed and shaped into pellets or briquettes (whenever this is possible) to help in reducing the space requirement for storage, and for transportation purposes.
3. Biomass fuel can be injected into the boiler only when the level of moisture is reduced to a minimum. In this way, corrosion of boilers can be reduced and energy efficiency can be kept at a higher level.
4. By mixing different types of biomass material to create the final single sample, the high volatility* can be reduced by simply adding a small percentage of ash to the total volume of the new sample.
5. This kind of blend, that is, ash and dry biomass materials (powder), not only lengthens the burning period, but also adds additional useful properties to the new product and maintains low market costs through the use of different biomass materials.
6. If a by-product, such as ash, is part of the economy of production at the power station when sold for use in cement, then finding alternative ways to produce ash for cement will be required.
7. The best way to achieve efficiency when using biomass materials (e.g., during combustion) is to treat each different type of material separately.

* There are advantages and disadvantages regarding high volatility. The ignition can be aided with high volatility, while the burning time scale (burning period) will be reduced, which is a disadvantage when it comes to fuel applications.

In this way, the by-product and its effects on the boiler are reduced, as it is well known that by-products produced from similar types of materials are much simpler than those produced from mixed different types of combusting materials (National Renewable Energy Laboratory, 2006). Following this method, biomass blending with coal will not be needed, and this in itself reduces some of the cost for the power generating company, especially as some of the blending can be at sites remote from the power stations (DTI, 2007).

8. Avoiding and/or reducing chloride and alkali content in biomass materials, used as fuels. This may help to reduce corrosion, and create higher operational efficiency, as well as lengthen the life cycle of the hardware.
9. An increase in the level of submicron aerosols and fumes in the flue gases may degrade the collective efficiency.
10. Increasing the co-firing ratio of biomass as well as increasing the different variations of biomass materials can in turn increase the technical risk, and as a consequence, the cost of co-firing with biomass also increases.
11. If required, the establishment of an international standard for biomass fuels (in order to provide outlined acceptable boundaries related to technical and nontechnical issues possibly via the United Nations).

3.3 MILLING

The milling process should always be performed separately, that is, coal milling is separate from biomass milling (DTI, 2007).

The following points provide some of the characteristics and behavior of milling coal and biomass materials used by power generating companies:

1. Coal mills generally pulverize coals, depending on the level of brittle fracture of the coal particles.
2. It is difficult to mill biomass materials, in comparison with the milling of coal. This is because, unlike coal, there is no brittle fracture in them.
3. If biomass and coal are milled together or a coal milling machine is used to mill biomass materials, then as mentioned in point two above, the difficulty and the outcome of milling biomass can have an effect on the co-firing percentage level of biomass with coal.
4. All biomass materials should be dried before use in the mill. Without this, the heat balance in the mill will be affected and consequently the end result will be poor.

5. Some biomass materials accumulate within the mill. This may affect the milling quality, and affect the co-firing ratio.
6. The majority of power stations co-fire by preblending the biomass with coal, before it enters the mill.
7. Co-firing ratios are typically 5% on a heat input basis.
8. There is a minimal effect on boiler performance and the environment.
9. Most of the technical problems are associated with reception, storage, and handling.
10. The constraints on co-firing ratio are the following:
    a. Fuel availability
    b. Handling/blending system capacity
    c. Limitations of coal milling equipment.
11. Present methods of biomass co-milling are the following:
    a. Ball and tube mills (Fig. 3.2)
    b. Vertical spindle ball
    c. Ring and roller mills.
12. Mill performance depends on the brittle fracture of coal into particles.
13. Biomass can accumulate in the mill so it takes longer to clear the mill during shutdown.
14. Mill differential pressure and power intake increases on vertical spindle mills.
15. Mill product top-size increases as larger biomass particles exit the classifier.
16. Biomass moisture affects mill heat balance, which can be a limiting factor (e.g., inadequate drying).
17. Safety issues when co-milling biomass, in accordance with optimization of mill operating procedures.

Most current designs of coal mills can handle between 10% and 15% biomass (Perry and Rosillo-Calle, 2006). In order to increase the percentage of biomass in co-firing, a new design will be required that takes into consideration the design of the boiler itself, that is, the possibility of two (or more) separate boilers, one for the biomass and the other for the coal. This may provide a better solution and consequently overcome problems related to the present design of a single boiler for the co-firing method.

3.4 FUEL MIXING

There are three areas of fuel mixing for the purpose of generating electricity: (1) fuel mixing from different sources of energy; (2) fuel mixing for the purpose

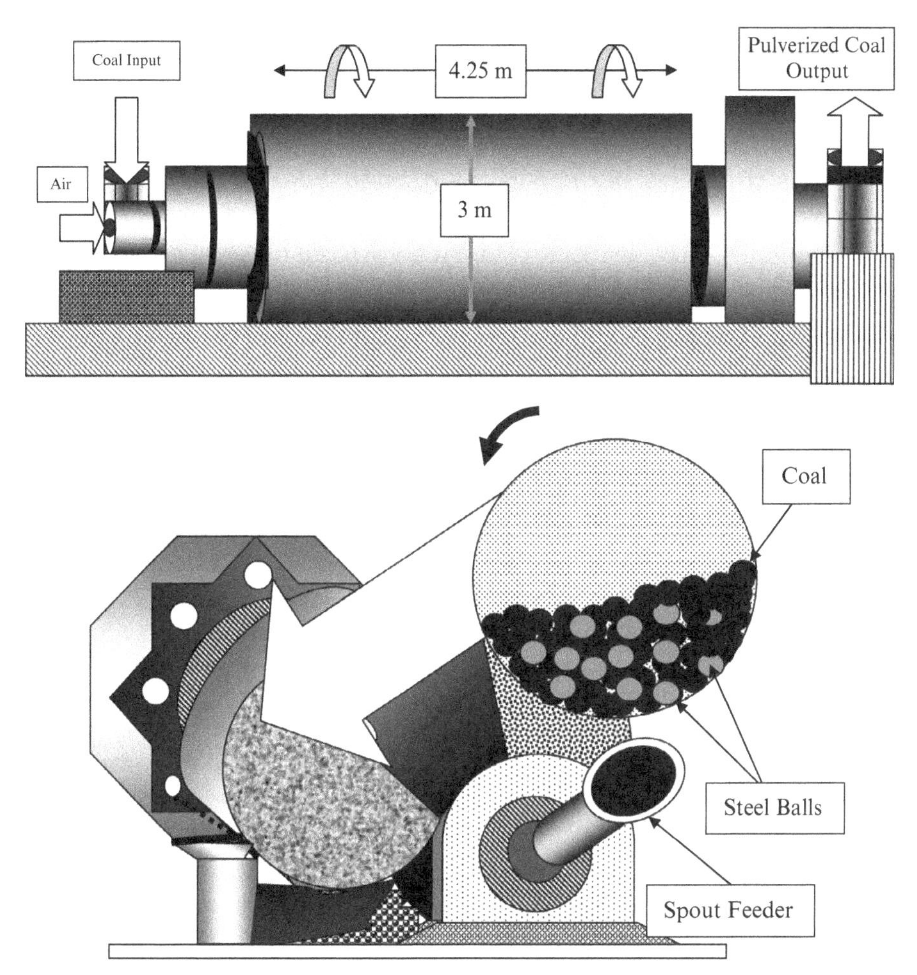

Figure 3.2 Tube ball mill schematic diagram: general view (above) and cross section of the cylinder containing coal and steel balls (below). *Source*: Author.

of co-firing (e.g., biomass with coal); (3) fuel mixing from different types of biomass materials.

1. *Fuel Mixing from Different Sources of Energy.* Fuel mixing from different sources of energy, that is, fuel mixing information, as it is sometime referred to, is a term used to explain how the electricity supplied to customers is generated. The electricity supplied is mostly generated from varieties of energy sources. These sources are coal, natural gas, renewable, and/or nuclear. The sources of electricity with proportional percentages from each source give the total electricity output production for a particular power generating company. The UK electricity is regulated by Ofgem, which enforces the condition that all power generating

companies must inform their end user about the sources of their electricity. Table 3.1 is an example of fuel mixing information for some of the power generating companies in the United Kingdom (Electricity guide.org.uk, 2009).

2. *Fuel Mixing for the Purpose of Co-firing (e.g., Biomass with Coal).* Biomass co-firing with coal requires an appropriate mixture in order to find a balanced fuel which is both efficient and economical as well as having a lesser impact on the environment. The present legislation in the United Kingdom states that by 2014, a minimum of 14% of biomass should be part of the fuel used to generate electricity (ROC) (DTI, 2003) of which 14% is obtained from renewable sources.

   Scientific and technical (S&T) and business factors (BF) in the REA1 methodology can solve the issue of using one type of biomass material with coal. This process is much simpler than the use of a variety of biomass samples in a co-firing method with coal. Regardless of the number of biomass materials, that is, the number of biomass samples needed to be used as part of the mixture or to be used in a co-firing method, the selection process using REA1 methodology would speed up the operation. This is presuming that data are already available for each individual type of biomass material. The co-firing of solid biomass is by premixing with coal and processing the mixed fuel through the installed coal handling systems (e.g., milling and firing) are the common methods used at power generating companies. In many cases, power generating companies use the mixing of fuels by co-milling when they embark on the co-firing method for the first time (Livingston, undated).

3. *Fuel Mixing: From Different Types of Biomass Materials.* The mixing process for the production of fuels of various biomass materials is an important area in which a number of factors are considered first prior to the mixing method. The basic principle is to increase the standard fuel characteristics required in relation to legal obligations (human health and environment), to reduce costs, and at the same time raise the fuel quality. The new approach is to apply a researched recipe in order to produce a quality fuel close to the required specifications. This can be achieved by following strict conditions within a workable methodology, such as REA methodologies (see Chapter 10, Section 10.6, as an example). Apart from the selection procedure, the optimizing process for a new hybrid fuel (mixed bio-sources of fuels) is completed by knowing the required percentage for each biomass sample in the overall mixture (Fig. 3.3). The energy business should be able to provide the actual required percentage value for each biomass sample, according to the required quality and standard needed. The selection process of biomass materials and the mixing values (percentage) of these samples are the main body of this work. REA1 methodology explains the first part, that is, the selection process of biomass materials.

Table 3.1 Electricity sources: fuel mix information

Supplier	Coal %	Natural Gas %	Nuclear %	Renewable %	Other %	CO_2 Emissions per kg/kWh	Radioactive Waste per g/kWh
Atlantic Electric and Gas	29.5	57.3	4.1	7.5	1.6		
British Gas	14	62	16	5	3	0.368	0.00187
Countrywide Energy	26.7	36.4	29.6	0.9	6.4	0.406	0.00355
EDF Energy	46	14	33	4	3	0.555	0.0017
Equipower (EBICo)	29.5	57.3	4.1	7.5	1.6		
Ecotricity	22.3	30.3	24.7	17.4	5.3	0.338	0.0030
Good Energy	0	0	0	100	0	0	0
London Energy	46	14	33	4	3	0.555	0.0017
Lloyds TSB	47.7	41.1	4.1	5.9	1.2	0.590	0.0005
Manweb	47.7	41.1	4.1	5.9	1.2	0.590	0.0005
npower	46	35	13	3	3	0.558	0.00159
Powergen	56.2	33.3	8	0.5	2	0.642	0.001
Sainsburys Energy	47.7	41.1	4.1	5.9	1.2	0.590	0.0005
Scottish Hydro-Electric	29.5	57.3	4.1	7.5	1.6		
Scottish Power	47.7	41.1	4.1	5.9	1.2	0.590	0.0005
Seeboard Energy	46	14	33	4	3	0.555	0.0017
Southern Electric	29.5	57.3	4.1	7.5	1.6		
Swalec	29.5	57.3	4.1	7.5	1.6		
SWEB Energy	46	14	33	4	3	0.555	0.0017

Source: Adapted from Electricity guide.org.uk (2009).

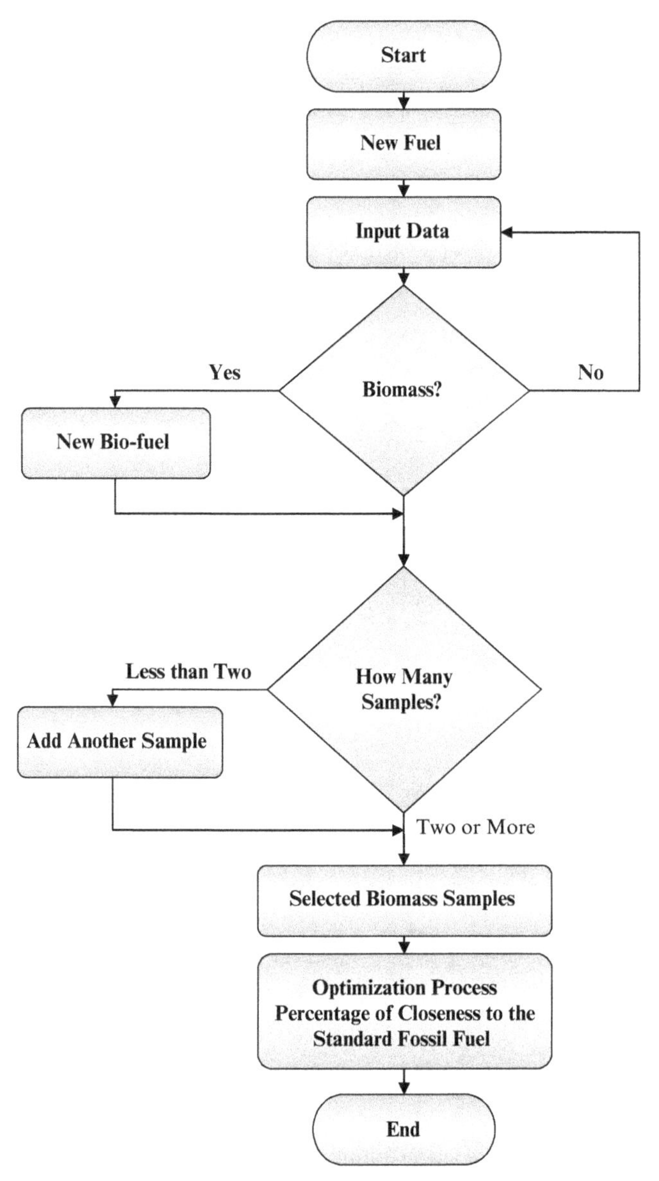

Figure 3.3 Flow chart to explain the basic points of an optimization program for a new bio-fuel (percentage mixing). *Source*: Author.

3.5 THE COMBUSTION SYSTEM

In coal-fired boilers, recent development related to biomass material and coal co-utilization, has produced tangible progress in co-firing methods. Worldwide, there are more than 150 power generating companies, using mostly coal to apply the co-firing method and/or have already experimented with using mainly biomass or waste alongside coal (Koppejan and Baxter, undated). Employing a co-firing method, an average power generating company generates power in the range of 50–700 MWe (Baxter and Koppejan, undated). Therefore, utilization of hardware in co-firing is vital. This is because the technical and cost considerations for energy businesses, considering the use of a co-firing method, are paramount. Concerning combustion, the following statement summarizes this phenomenon:

> Combustion is a complex phenomenon involving simultaneous coupled heat and mass transfer with chemical reaction and fluid flow. (Jenkins et al., 1998)

When it comes to the combustion of biomass materials, incomplete burn could leave large amounts of carbon in the ash deposited in areas of the boiler. This kind of deposit can drastically reduce the efficiency and reliability of the hardware. Other hardware and technical issues resulting from the use of biomass materials without the application of S&T factors are the increased emissions. These mainly consist of volatile organic compounds (VOCs) and particulate matter (PM), which can have other effects, such as exceeding permit levels or possibly overloading the bag house flue gas cleaning equipment. Also, a premature system failure may occur as a result of the handling of biomass materials as it speeds up corrosion and hardware wear. Regarding the efficiency of combustion, the size of particles is an important factor, that is, smaller particles combust easily and consequently can achieve complete combustion. For this reason, the particle size for pulverized coal should be smaller than 75 μm (Quaak et al., 1999). Finally, a brief look at different types of boilers can help in providing some answers to a variety of technical and engineering hardware issues.

3.5.1 Boilers

The characteristics of a boiler are reflected in the way it produces combustion and the results, that is, flue gases and the transfer of heat. Examples of these methods can be firetube, watertube, thermal oil transfer, and hybrid, such as fire tube/water tube (Applachian Hardwood Centre, 1998). Using a co-firing method, the efficiency of the boiler, apart from the quality of the hardware design, depends mostly on the moisture content and the type of biomass materials being used, as well as for the co-firing ratio. With an approximate value of 3–5% related to a mass basis, there is only a small loss of boiler efficiency (Van Loo and Koppejan, 2008). Table 3.2 illustrates the efficiency of a boiler

Table 3.2 Heat loss from boiler

Loss	Contents
Heat loss in dry flue gas	Sensible heat loss of flue gas Largest heat loss
Water heat loss in air	Latent heat loss to vaporize water in air
Radiation heat loss	Heat loss from outer surface of furnace to atmosphere
Unburned fuel	Flammable material unburned and remained in residue
Unburned carbon in flue gas	CO or carbon
Other heat loss	Blowing or atomizing steam loss

Source: Adapted from Hirayama (2006).

in the form of heat loss and their causes (Hirayama, 2006). According to Soo (2006), the percentage of efficiency for a boiler is defined as:

$$\%\ \text{Efficiency} = \frac{\text{Heat Out of Boiler}}{\text{Heat Supplied to Bioler}} \times 100\%$$

There are currently two different commercial systems for boilers: (1) fixed bed system; and (2) fluidized bed system.

The fixed bed type is an old method which goes back to the first steam systems, while the second type is around three decades old (Quaak et al., 1999).

3.5.1.1 Fixed Bed Combustion Systems. Fixed bed systems are characterized by the type of grates and how the fuel is transported and/or supplied. Ash is usually removed manually, although some new designs have an automatic ash removal mechanism. The system's temperature can range from 850 to 1400°C. Examples of fixed bed systems include: manual-fed, spreader stoker, static grates, under screw and inclined grate (Quaak et al., 1999). Examples of this system are briefly discussed below:

A. Pile Burner (PB)
   The PB is an old industrial method for combusting wood. Ash needs to be removed from time to time. The PB operates in an erratic cyclic way that makes it less efficient than other types of combustion systems. PBs can handle biomass material such as wood with a content of up to 65% moisture (Applachian Hardwood Centre, 1998). Reportedly, the system has a low efficiency estimated to be 50–60%. Benefits arising from this system are that it can handle dirty and wet fuel (Bain et al., 1998).
B. Stoker Grate (SG)
   The next step from the PB is the SG, which has a mechanism to allow continuous ash collection through the use of a moving grate. Another

notable advantage is that the fuel in the combustion zone is organized in a thin layer and spread evenly. Additional development in the SG is the lowering of the furnace temperature by increasing the air access over its fire, which can result in lowering NO_x emission. Usually, the temperature is maintained at about 980°C, which means that the furnace temperature is below the ash deformation temperature of most biomass fuels (Bain et al., 1998).

C. Suspension Burner (SB)
   There is an interest in this type of boiler, not just in relation to coal firing, but also the possibility of its use in a co-firing process.

In comparison with the previous types of boilers, the SB is smaller in size and more efficient. However, cost and power usage can be higher as it needs a specially designed burner such as a vertical–cylindrical or a scroll cyclonic burner. To make this type of boiler more efficient, the biomass materials should contain less than 15% moisture and the particle size be no more than 1.5 mm (Bain et al., 1998).

3.5.1.2 Fluidized Bed Combustion Systems. There are two main groups: atmospheric systems (operating under normal atmospheric pressure) and pressurized systems (which require elevated pressures). These are divided into two subgroups: bubbling and circulating fluidized bed. Examples of bubbling and circulating FB systems have been considered briefly below.

1. *Bubbling Fluid Bed* (BFB). This kind of combustion (Fig. 3.4) is characterized by a better quality of mixing, heat transfer, and with low NO_x emissions as a result of combustion occurring at a relatively low temperature, that is, around 980°C. The combustion is very efficient, especially in regard to carbon, with approximately 99% burnout in most cases (Hansen, 1992).

   Particles circulate within the bed as gas flows upward—since the gas has a high speed. As a result, the particles are separated and start moving in and outside the bed (Perry and Chilton, 1973).
2. *Circulating Fluid (Fluidized) Bed* (CFB). Turbulent bed solids are collected as part of the system in the CFB, that is, separated from gas first and then brought back to the bed, completing a cycle. CFB differs from BFB in that the separation in the dilute solid zone and the dense solid zone is, unlike BFB, not clear.

   The temperature in the CFB is kept at around 870°C, which helps in reducing gaseous emissions as well as increasing the limestone–sulfur reactions (Bain et al., 1998). Densities between CFB and BFB differ in that BFB is 720 kg/m^3 while CFB is around 560 kg/m^3 (Babcock and Wilcox, 1992). In circulating fluidized beds (Fig. 3.5), coal can provide stability during the combustion process, as it produces large amounts of char, which is the main cause of its stability. Biomass fuel does not

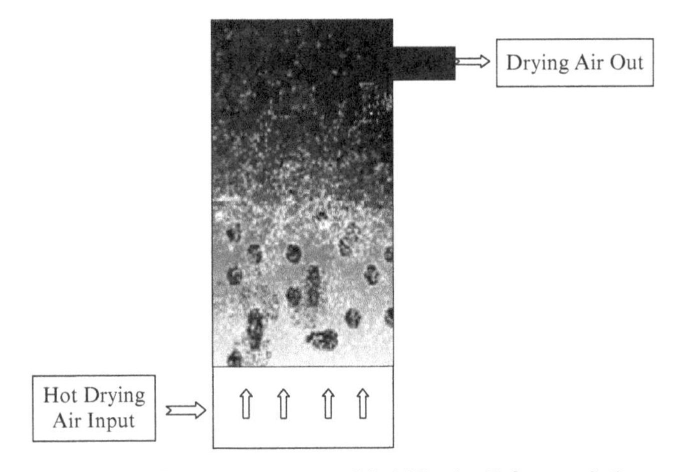

Figure 3.4 Schematic diagram of "bubbling bed." *Source*: Author.

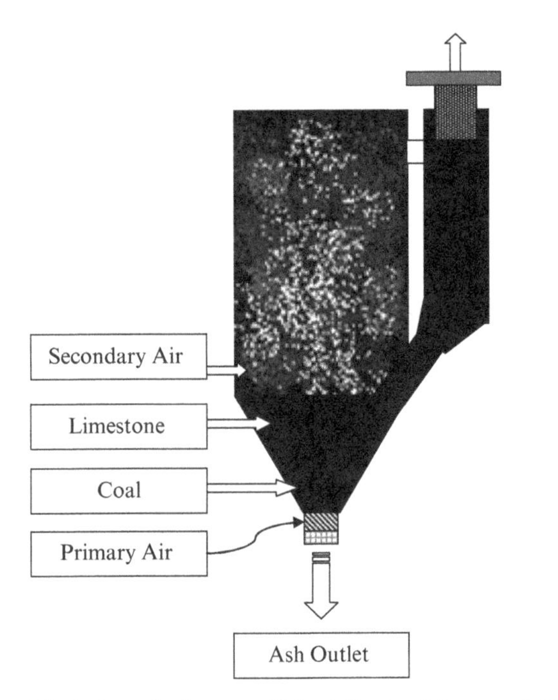

Figure 3.5 Schematic diagram of "circulating bed." *Source*: Author.

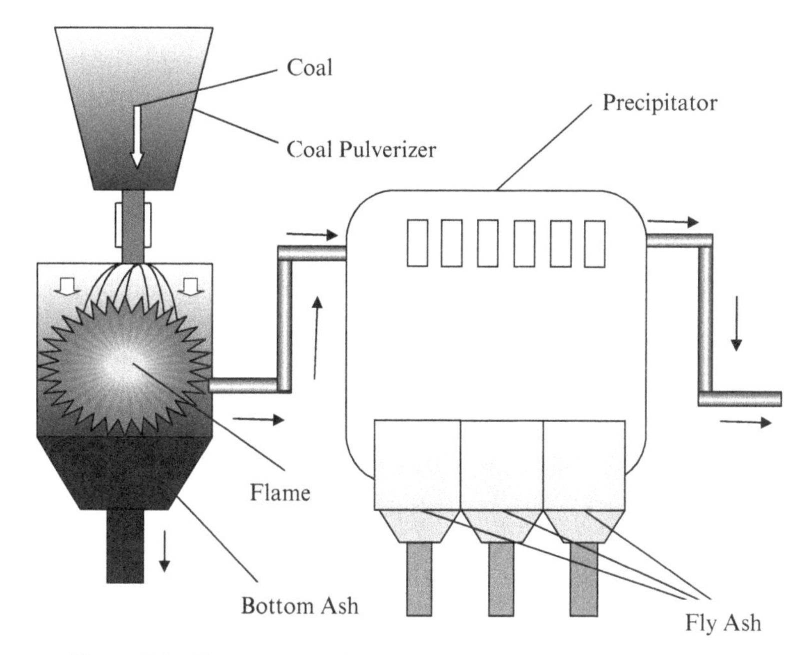

Figure 3.6 The formation of bottom ash and flying ash. *Source*: Author.

provide a similar stability as char buffer stock is not stored in the bed (Nevalainen et al., 2007).

3.6 BY-PRODUCTS

3.6.1 Ash Formation and Deposition

There are two types of ash,* bottom ash and flying ash,† both of which are sources of a variety of hardware and energy output problems (Fig. 3.6). The accumulation of ash in a boiler (e.g., stoker fuel beds and fluidized bed-fired boilers) usually leads to a lower quality of ignition and combustion as well as difficulties in ash removal and de-fluidization of fluidized beds (Livingston, 2006). In addition to this, it is believed that ash deposition on the tubes (e.g., in fire tube boilers) causes the corrosion of the heat exchangers. Another difficulty is the deposition that occurs in the flue gas, that is, during the combustion process, large amounts of alkali chlorides accumulate in the flue gas, mainly because elements such as K, Cl, and Na are volatile (Miles et al., 1996).

The ash quantity and quality should be considered mainly from an engineering and technical point of view in order to help in the design of

* Ash can be referred to as bottom ash, cyclone ash, filer ash, and flue dust.

† There is a third type of ash, usually referred to as *boiler slag*, that is, melted ash formed in exhaust stack filter and at the bottom of the boiler.

new hardware with the possibility of preventing or reducing this kind of degradation.

Analysis of the major elements found in ash is important in order to understand the causes of the formation of slag inside the boiler furnace and/or the fouling of the heat transfer surfaces of the boiler. Exposure to gas at a high temperature and to fluctuations of radiant heat, which happens when fused or resolidified molten ash occurs, means that slagging will begin to form. This kind of slagging reduces heat transfer. Therefore, the water side evaporation temperature will consequently be reduced and as a result this will increase the furnace gas exit temperature, as well as that of the superheater (Ireland et al., 2004). Eventual technical problems produced in this manner will raise not only the cost of energy used, but also the cost of hardware maintenance.

The deposition of ash and the possible solution to this can be summarized in the following points:

1. At present, online control of ash deposition can be made through soot blowers.
2. Removal of ash shedding should immediately become part of ash deposition control.
3. The occurrence of low viscosity drips of fused deposits should be minimized either by not reaching the ash melting point and/or reducing the chemical causing it and/or coating the boiler/furnace with nanostructured layers which can resist the formation of this deposit.

The fouling is usually deposited on the tubes within the boiler convective part. As a result, their thickness is gradually reduced (by the falling fly ash particles) (Livingston, 2007). Flying ash is in itself a major problem when it comes to flue gas equipment and emission.

In order to find the ash chemical makeup, the procedure normally starts by gradually heating the sample to an approximate air temperature of 815°C. The ash produced is weighed (as a percentage sample content of ash) then analyzed. The makeup of a standard ash sample, that is, coal, may include: Ca, Na, Mg, P, S, K, Si, Al, and Fe. The major elements of a biomass ash sample may include: Fe, Al, Ca, Mg, K, Si, P, and Na.

3.7 DEGRADATION

Hardware degradation is a major problem when using biomass materials for co-firing with coal. Many researchers and engineers in this field agree that by removing or reducing alkali chloride (Fig. 3.7), corrosion can be minimized (Pronobis, 2005). Alkali chloride is usually found in higher quantities in biomass materials, such as straw (Livingston and Babcock, 2006). By using a particular mixing percentage of biomass materials with coal, this chemical can be converted safely to silicate. Within the flue gases, chlorides can react with

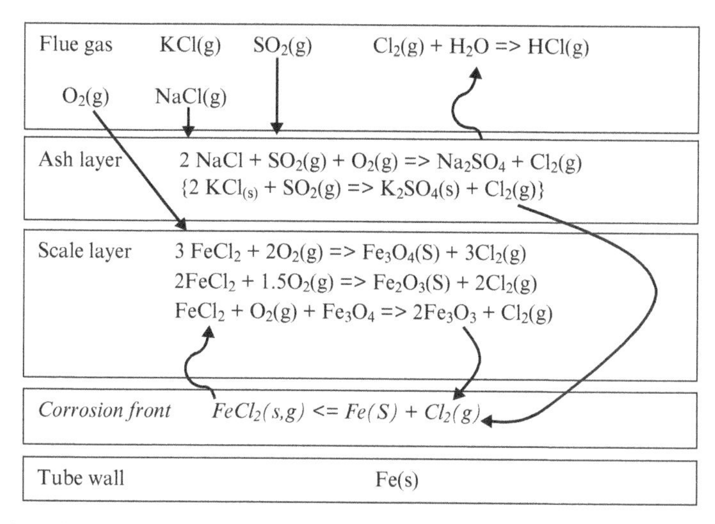

Figure 3.7 Oxidation and corrosion mechanism. *Source*: Adapted from Riedl et al. (1999).

the SO2/SO3 to form sulfates, which follow the release of chlorine (gaseous), as illustrated in Equation (3.1) and Equation (3.2)(Riedl et al., 1999).

$$2NaCl + SO_2 + O_2 \rightarrow Na_2SO_4 + Cl_2 \tag{3.1}$$

$$2KCl + SO_2 \rightarrow K_2SO_4 + Cl_2. \tag{3.2}$$

When using certain materials that can withstand high temperatures as well as chemicals that cause corrosion and degradation, a coating layer (or layers) of nanostructured ceramic and metallic coating can offer improved properties over conventional coating (Joseph et al., 2003). The growing interest worldwide in developing new types of nanostructured coating with high level of hardness and toughness can be applied within the hardware systems in the co-firing mechanism. Nanostructured coatings have a wide range of potential applications, especially those types of coating focused on large commercial boilers used in power generating companies. Continuous improvement in the properties of such coatings has been taking place steadily, but especially so within the last two decades. Many different systems have already been developed. Of particular interest are ceramic/metal nanocomposite systems, such as the MeN/Me coatings proposed by Musil, but with additional changes to make it applicable to a co-firing environment (Joseph et al., 2003) (Table 3.3).

3.8 CONCLUSION

Technical and engineering issues in co-firing can cause a variety of operational problems and be economically costly, especially if not tackled early. By treating

Table 3.3 Hardware systems and fuel mixing: summary

Mills						
Mills category	Grinding mills					Knife/cutting mills
Types	Ball mill (steel balls)	Rod mill (steel rods)	Hammer mill/ swing claw hammer mill	Vertical spindle mill: 1. Bowl mill 2. Babcock and Wilcox mill 3. Rymon ring-roller mill	Foster wheeler	Fabricated—steel rotary knife cutter

Fuel Mixing			
Sources		Electricity and Fuel Sources	
Types	From different sources of energy, for example, electricity sources supplied	For the purpose of co-firing, for example, biomass with coal	From different types of biomass materials

Boilers			
Types		Examples	
Fixed-bed combustion systems	Pile burner (PB)	Stoker grate (SG)	Suspension burner (SB)
Fluidized bed combustion systems	Bubbling fluid bed (BFB)	Circulating fluid bed (CFB)	

Source: Author.

the causes of these technical and engineering challenges using long term solutions, with regular maintenance, the co-firing of coal with biomass can be an easier and cheaper task. A conclusion, therefore, can be highlighted with the following points:

1. Important issues, such as lengthening the storage period for all types of biomass materials used as fuels, will reduce the cost of bio-fuel storage and transportation, as well as making this kind of fuel more acceptable for those who are not yet convinced about the beneficial return in using the co-firing method. Nondegradable or long life storage of biomass materials can benefit the power generating company in many ways both directly and indirectly, particularly by reducing costs.
2. The milling process for biomass should be decided on at least five different factors. These factors include packaging, storage, transportation, fuel feeding, and fuel injection. These milling related factors depend mostly on individual power generating companies' needs, requirements, and operational procedures. The method of injection and whether or not to use one or more boilers are other technical and engineering challenges that should be met to provide long term benefits.
3. Boiler design/selection and their usage is another important factor that should be taken into consideration. This should be done long before operational production takes place.
4. To overcome unwanted high volatility (usually found in biomass materials), one of the successful solutions used in ancient times was simply to add an appropriate percentage of ash, as has been mentioned previously. This will affect the high volatility and consequently, the burning period can last longer.
5. In order to prevent or reduce the corrosive effect of chlorine and alkali contents in the biomass materials, one of the methods that can be used is to apply an internal layer of nanostructured coating to the biomass boiler. Other technical and engineering factors, if they can be applied or modified, should be investigated from long and rich experience connected to various fossil fuels, currently being used at power generating companies (Box 3.2).

Box 3.2

REED CANARY GRASS (RCG)

Reed canary grass (RCG) is part of the plant family named *Poaceae*. RCG is an erect a rhizomatous (coarse) robust perennial type, tall (0.5–2 m), leafy, highly adaptive, high yielding, and with an extensive root system. RCG flowers between June and July and grows mostly during the winter time, which is why referred to it as a "cool-season grass." Apart from being used for pasture, hay, or silage, RCG is used for pulping (paper making) and as a fuel. The grass has good tolerance to drought conditions and tolerance to poor drainage and prolonged flooding. It can be found mostly in the subtropical and temperate regions of the northern hemisphere of North America, Europe, and Asia. It can also be used for pasture, hay, or silage. RCG can be characterized as follows:

1. High yields.
2. More drought resistance than many other types of grasses.
3. Disease resistance.
4. Fast and steady regrowth.
5. Adaptable to natural environment changes (Flood, drought, temperature).
6. Usually slower and more difficult to establish (seed growth) than other type of grasses.
7. It is not suited for short rotation plantation as it takes up to three years to become strong and establish itself.
8. The grass is spread by rhizomes and by seeds.
9. It can be used for land erosion control.

The chemical elements of canary grass are the following:

Carbon	48.27%
Hydrogen	5.22%
Nitrogen	3.36%
Oxygen	43.10%
Sulfur	~0.05%

Source: Author.

REFERENCES

Applachian Hardwood Centre (1998) Overview of wood-fired boiler use in West Virginia. Fact Sheet 16 April, Division of Forestry, West Virginia University. http://www.wvu.edu/~agexten/forestry/fact16.pdf (last accessed December 3, 2013).

Babcock and Wilcox (1992) Atmospheric pressure fluidized-bed boilers. In Steam, 40th ed. Barberton, OH: Babcock and Wilcox, Chap. 16.

Bain RL, Overend RP, Craig KR (1998) Biomass-fired power generation. Fuel Processing Technology 54(1–3):1–16.

Baxter L, Koppejan J (undated) Biomass-coal co-combustion: opportunity for affordable renewable energy. Brigham Young University, Provo, UT and Bioenergy Systems, AH Apeldoorn, The Netherlands. http://www.ieabcc.nl/publications/paper_cofiring.pdf (last accessed December 3, 2013).

Boiler Burner (2013) Boiler introduction. http://www.cleanboiler.org/Eff_Improve/Primer/Boiler_Introduction.asp (last accessed December 3, 2013).

DTI (2003) Our energy future: creating a low carbon economy. Department of Trade and Industry (DTI) Publications Cm 5761, HMSO, London.

DTI (2007) Engineering report: guidance document on biomass co-firing on coal fired power stations. DTI Project 324-2.

Electricity guide.org.uk (2009) Fuel mix information. http://www.electricity-guide.org.uk/fuel-mix.html (last accessed December 3, 2013).

Hansen JL (1992) Fluidized bed combustion of biomass: an overview. Biomass Combustion Conference. Reno, Nevada, US DOE Western Regional Biomass Energy Program, Jan. 28–30, 1992.

Hirayama Y (2006) Retrofitting of existing boilers to biomass boilers. Slides presentation, TechnoSoft Co., Ltd. http://209.85.229.132/search?q=cache:vXEJABEcWloJ:www.apo-tokyo.org/biomassboiler/D1_downloads/presentations/Nepal_Program_DEC2006/Resource_Persons/Hirayama/4_Retrofitting_to_BiomassBoiler.ppt+Retrofitting+of+Existing+Boilers+to+Biomass+Boilers&cd=10&hl=en&ct=clnk&gl=uk (last accessed March 31, 2009).

Ireland SN, Mcgrellis B, Harper N (2004) On the technical and economic issues involved in the co-firing of coal and waste in a conventional PF-fired power station. ScienceDirect 83(7–8):905–915.

Jenkins BM, Baxter LL, Miles TR Jr, Miles TR (1998) Combustion properties of biomass. Fuel Processing Technology 54:17–46. Elsevier. http://www.et.byu.edu/~tom/classes/733/ReadingMaterial/Jenkins-Baxter.pdf (last accessed December 3, 2013).

Joseph MC, Tsotsos C, Leyland A, Mathews A, Baker MA, Kench P, Rebholz C (2003) Characterisation and tribological evaluation of nitrogen: doped molybdenum–copper PVD metallic nanocomposite films. Submitted to Surface and Coatings Technology.

Koppejan J, Baxter L (undated) Global operational status on cofiring biomass and waste with coal. Experience with different cofiring concepts and fuels.

Livingston WR (undated) Advanced biomass co-firing techniques for retrofit and new build projects. Doosan Babcock Energy, slides presentation.

Livingston WR (2006) Biomass ash deposition, erosion and corrosion processes. IEA Task 32/Thermalnet Workshop, Glasgow. http://209.85.229.132/search?q=cache:ALU8KlqDYxUJ:www.thermalnet.co.uk/docs/Biomass%2520ash%2520deposition,%2520erosion%2520and%2520corrosion%2520processes%2520(BL).ppt.pdf+ash+formation+in+biomass+biolers&cd=9&hl=en&ct=clnk&gl=uk (last accessed March 31, 2009).

Livingston WR (2007) Biomass ash characteristics and behaviour in combustion, gasification and pyrolysis systems. Technology and Engineering, Boosan Babcock Energy.

Livingston B, Babcock M (2006) Workshop on ash related issues in biomass combustion. Glasgow, UK.

Mather MP, Freeman MC (2008) Moisture and char reactivity modeling in biomass cofiring boilers. Dinesh Gera, Fluent Inc., National Energy Technology Laboratory, Combustion and Environmental Research Facility. http://www.fluent.com/about/news/newsletters/01v10i1/s3.htm (last accessed September 11, 2008).

Miles TR Jr, Miles TR, Baxter LL (1996) Alkali deposit found in biomass boilers. Vol. II, National Renewable Energy Laboratory, report NREL/TP-433-8142, National Technical Information Service, NTIS (ed.), Springfield, MA.

Mitchell RE, Campbell PA, Ma L, Sørum L (2004) Characterization of coal and biomass conversion behaviors in advanced energy systems. GCEP Technical Report, pp. 202–209.

National Renewable Energy Laboratory (2006) Biomass research, NREL, USA.

Nevalainen H, Jegoroff M, Saastamoinen J, Tourunen A, Jäntti T, Kettunen A, Johnsson F, Niklasson F (2007) Firing of coal and biomass and their mixtures in 50 kW and 12 MW circulating fluidized beds—phenomenon study and comparison of scales. Fuel 86(14):2043–2051.

Perry M, Rosillo-Calle F (2006) Co-firing report: United Kingdom. International Energy Agency (IEA) Bioenergy Task40: Imperial College.

Perry RH, Chilton CH (1973) Chemical Engineers Handbook, 5th ed. New York: McGraw-Hill.

Prairie Lands (2013) Co-firing process. http://www.iowaswitchgrass.com/process.html (last accessed June 14, 2013).

Pronobis M (2005) Evaluation of the influence of biomass co-combustion on boiler furnace slagging by means of fusibility correlations. Institute of Power Engineering and Turbomachinery, Silesian University of Technology.

Quaak P, Kneof H, Stassen HE (1999) Energy from biomass: a review of combustion and gasification technologies. World Bank Technical Paper 422.

Riedl R, Dahl J, Obernberger I, Narodoslawsky M (1999) Corrosion in fire tube boilers of biomass combustion plants. In Proceedings of the China International Corrosion Control Conference '99, paper Nr. 90129, October 1999, Beijing, China, China Chemical Anticorrosion Technology Association (CCATA) (Ed.), Beijing, China.

Soo MI (2006) Technology Options for Reusing Biomass Waste for Energy Recovery. Regional Training Seminar on EE and RE for SMEs in the Greater Mekong Sub-region of ASEAN.

Van Loo S, Koppejan J (2008) The handbook of biomass combustion and co-firing. London: Earthscan.

4

SAMPLES

4.1 SELECTED SAMPLES

4.1.1 Introduction

In this chapter, contributions have been made from the author's own work as well as from various local and international sources. Some of these sources are related to scientific, farming, energy businesses, fuels standard, and energy crop reports. The author's contributions consist of the results of laboratory tests, samples observations (before and after experimental tests), and notes from site visits, such as power stations and energy crop farms.

A number of biomass samples have been chosen for their energy content and for a variety of other factors described in detail in different sections of this chapter, and throughout this book. All the samples have been analyzed according to the needs, aims and objectives of this book. The principle is to select the most suitable biomass samples to be used for co-firing and/or as a fuel in its own right. Looking for the "right" biomass sample can be a difficult task, if all the factors in the methodology are not taken into account.

One of the power generating companies in the United Kingdom provided eight different biomass samples. The author provided seven biomass samples, which brings the total samples used during the laboratory tests to 15 samples (Table 4.1).

The Selection Process of Biomass Materials for the Production of Bio-fuels and Co-firing, First Edition. Najib Altawell.

Table 4.1 List of selected biomass and fossil fuel samples

Sample Number	Selected Sample	Sample Number	Selected Sample	Sample Number	Selected Sample
1	Apple P	7	Natural gas	13	Bituminous coal (Kleinkopje)
2	Barley	8	Niger seed	14	Straw pellets
3	Corn	9	Rapeseed	15	Sunflower BS
4	Crude oil and derivatives	10	Rapeseed meal	16	Sunflower SS
5	Distillers dried corn	11	Reed canary grass	17	Switch grass
6	Miscanthus	12	Rice	18	Wheat

Source: Author.

The biomass samples have been chosen for reasons related to "energy content," "plant growth," "cost," "availability," "ease of production," and environmental issues. They are also related to "transportation" "storage," "processing," "dryness," "by-products," and other factors related to business factors (BF) and scientific and technical factors (S&T). The other samples used are fossil fuel samples (i.e., coal, crude oil "or crude oil derivatives," and natural gas) (Table 4.1). During the test on the these samples, coal was selected as the standard sample, which brings the overall total of the samples analyzed to 18 samples.

4.2 SAMPLES GENERAL DESCRIPTIONS

4.2.1 The Reference Samples

The reference sample used in this book is for the purpose of comparison with the other selected biomass samples. This sample is the South African bituminous coal Kleinkopje. Many power generating companies use this type of coal to generate electricity.

The designs of REA1 methodology allow it to accept any fossil fuel sample (or any sample for that matter) for the purpose of comparison.

4.2.1.1 Keleinkopje South African Bituminous Coal. Presently, coal provides 25% of global primary energy needs, while as a fossil fuel* coal provides

* The prospect of fossil fuel and nuclear energy for the year 2020, according to Winteringham (1992), in relation to global energy reserve—based on the minimum growth of human population, minimum annual energy use per capita, and maximum longevity of all these energy reserves—is estimated as follows: coal, oil, natural gas, and uranium reserve = 10^{23} J, global human population = 10^{10}, annual energy use per capita = 10^{11} J, and combined reserves = 100 years.

approximately 40% of the world's total electricity production (Table 4.2) (IEA, 2012).

Kleinkopje, South African bituminous coal can be described as dense and black (or sometimes dark brown), often with well-defined bands of bright and dull materials. Kleinkopje is a middle-rank coal (between subbituminous and anthracite) formed when additional pressure and heat is applied to lignite (Skhonde et al., 2006). This type of coal is called soft coal.

Bituminous coal has a calorific value ranging from 24 to 33 MJ/kg, that is, around 10,500 to 14,000 Btu/lb. It has a carbon content (on a dry and ash free basis) ranging from 69% to 78% for medium-volatile and from 78% to 86% for low-volatile bituminous coal (Encyclopaedia Britannica, 2008). This is one of the reasons, apart from cost, why this type of coal is widely used as a fuel in steam-electric power generation and for heating. In addition to its use in power applications, it is also a source for the manufacturing of coke.

The percentage value of nitrogen and sulfur in coal is important because when coal is burned, they are released into the atmosphere and become the main cause of acid rain (Fig. 4.1).

Table 4.2 Coal in electricity generation during the year 2011

Poland	87%	Czech Rep.	51%
South Africa	93%	Greece	54%
Australia	78%	United States	45%
China	79%	Germany	41%
Kazakhstan	75%	India	68%

Source: Adapted from IEA (2012).

Figure 4.1 A powder sample of bituminous coal. *Source*: Author.

Table 4.3 The chemical elements of Kleinkopje bituminous coal

Carbon	68%
Hydrogen	4%
Nitrogen	2%
Oxygen	23%
Sulfur	3%

Source: Author.

Some of power stations remove sulfur dioxide via the application of lime slurry. Bituminous coals are graded according to vitrinite reflectance, that is, the measurement of the maturity of organic matter with respect to whether or not it has generated hydrocarbons (Kurkovál et al., 2003). Furthermore, they are graded in relation to plasticity, moisture, volatile matter, and ash content. The laboratory test for Kleinkopje bituminous coal, concerning carbon (and other elements—Table 4.3) has established that around 68% of the total mass of the sample is made from carbon.

Bituminous coal is the fastest-growing power generating fuel on the market, even though it is still a small one in comparison with the sale of its counterparts (Appalachian Blacksmiths Association, 2008). As and when the price of crude oil increases, the demand for coal will increase as well, and consequently, the price of coal in general rises. This is what has happened in recent years; particularly since September 11, 2001,* that is, the price of coal gradually increased from this date onwards (Sohail, 2005). This means that coal is more valuable than ever before, to the extent that old coal mines in the United Kingdom (which were closed years ago) have suddenly become commercially viable. The idea of reviving the commercial life of these old mines has become an attractive business prospect as the demand for additional energy sources is continually rising.

The demand for coal will continue to rise, particularly if harmful emissions from it are significantly reduced, that is, more research and development is taking place for producing what has been termed *clean coal*. The name "King Coal," which is being used in the media, is an appropriate description in a world with an ever-increasing thirst for more sources of energy.

If the present worldwide coal production continues at the same level (Table 4.4), then the global reserve of coal is estimated to last 147 years, mostly recovered from 70 countries around the world (World Coal Institute, 2008).

The value of the Kleinkopje South African bituminous coal as a "standard" for the REA1 methodology is used simply for the purpose of comparison by the energy production companies who require the comparison of biomass energy sources with the value of a fossil fuel.

The values of S&T, as well as BF, have undergone a comparison with the standard reference sample at various stages during the work on this book.

* The United States was attacked on this date.

Table 4.4 Estimated figures in million tons (MT) of the top 10 hard coal producers during 2011

China	3471	Russia	334
USA	1004	Indonesia	376
India	585	Poland	139
Australia	414	Kazakhstan	117
South Africa	253	Germany	189

Source: Adapted from World Coal Association (2013).

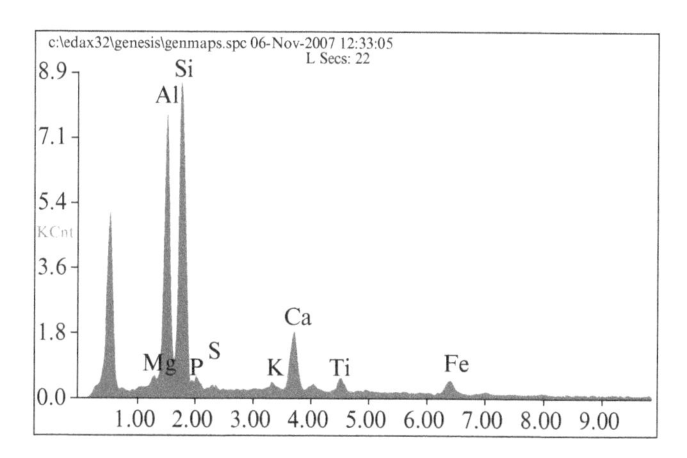

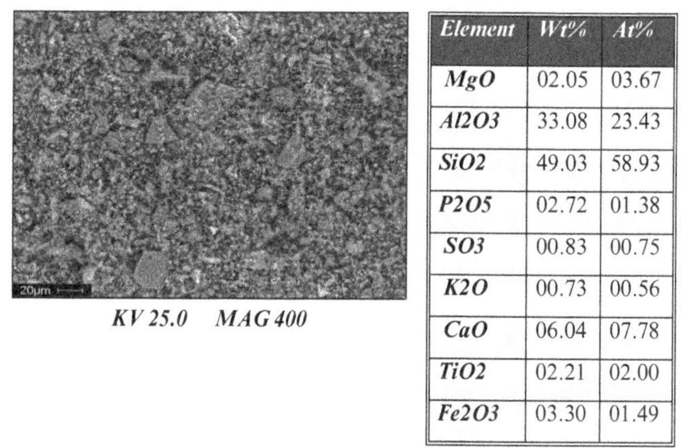

Element	*Wt%*	*At%*
MgO	02.05	03.67
Al2O3	33.08	23.43
SiO2	49.03	58.93
P2O5	02.72	01.38
SO3	00.83	00.75
K2O	00.73	00.56
CaO	06.04	07.78
TiO2	02.21	02.00
Fe2O3	03.30	01.49

Figure 4.2 Coal ash elements plus a magnified image (400×) of the ash. *Source*: Author.

Figure 4.2 illustrates Kleinkopje South African bituminous coal ash composition—with a magnification of 400× of the ash sample images.

The elements of coal ash (the plot obtained via FEI Quanta 600 scanning electron microscope, FEI, Eindhoven, the Netherlands). The x-axis in Figure 4.2 is energy in kev (thousands electron volts) and the y-axis is accumulated counts in kcnt (thousand counts).

4.2.1.2 Crude Oil. The unprocessed fossil fuel oil extracted from underground is referred to as *crude oil*, or sometimes named *petroleum*. The color of crude oil can range from clear to black, and the viscosity can range from close to the viscosity of water up to an almost solid consistency. In all states this kind of fossil fuel is made up mostly from a mixture of hydrocarbons. The hydrocarbon molecule is mainly made up of a chain of hydrogen and carbon atoms. The molecule can be found with various structures and of various lengths, which explains why it can be also found in many different substances, in addition to crude oil.

The origin of crude oil came about when the bodies of small plants and animals were buried and then decayed under layers of earth over a period measured in geological time (Optima Energy Group, 2007).

A variety of plants, animals, and environmental factors in diverse parts of the world can give different shades and colors to the crude oil. Consequently, the color of the crude oil may indicate where it came from, that is, crude oil from around the world can have a different appearance in color, as well as variation in their composition.

The laboratory test for crude oil (in this case North Sea oil) concerning carbon (and other elements—Table 4.5) has established that more than 78% of the total mass of the sample is made up of carbon.

Crude oil main hydrocarbons classes are the following:

Alkenes C_nH_{2n}
Aromatics C_6H_5—Y*
Napthenes (Cycloalkanes) C_nH_{2n}
Paraffins C_nH_{2n+2}.

Some crude oil contains lighter hydrocarbon molecules, while other grades contain much heavier types. When it comes to the market value of crude oil, this value increases correspondingly according to the type of hydrocarbons it contains, that is, the lighter the type of the hydrocarbon molecule, the higher the price per barrel and visa versa (Energy Information Administration, 2009).

Table 4.5 The chemical elements of crude oil

Carbon	78.5%
Hydrogen	11.9%
Nitrogen	0.25%
Oxygen	8.85%
Sulfur	0.5%
Metals	<1000 ppm

Source: Author.

* This molecule connects to the benzene ring.

The value of crude oil as a "standard" for the REA1 methodology is purely for comparison purposes, that is, biomass samples are compared with the value of a crude oil sample (or with one of its derivatives) to see how close they are in chemical composition (S&T) and how much commercial value (BF) they have. These principles are the same as those used with the standard sample "coal," as explained in the previous section.

Crude oil reserves at the present level of production would last approximately 66 years if produced from 41 countries believed to have a sufficient commercial crude oil reserve. In fact, most of these countries, or crude oil reserves, that is, 68% of oil reserves in the world, are in the Middle East (World Coal Institute, 2008).

4.2.1.3 Natural Gas. Natural gas is the cleanest (i.e., emitting the least CO_2) and simplest in comparison with the rest of the fossil fuels (Thompson, 2008). Natural gas is produced when heat and pressure start to break after the molecules of the crude oil at a certain level underneath the earth. Natural gas is composed mainly of methane (CH_4) plus other gases such as ethane (C_2H_6), propane (C_3H_8), butane (C_4H_{10}), pentane (C_5H_{12}), carbon dioxide (CO_2), nitrogen (N_2), and traces of helium (He) and hydrogen sulfide (H_2S) (Table 4.6).

There are two types of natural gas: (a) dry natural gas: natural gas that contains only methane, and (b) wet natural gas: natural gas that contains hydrocarbons other than methane.

Laboratory tests for four types of natural gas have established that the average percentage of methane is 86% of the total mass worldwide (Table 4.6).

Concerning the usage of natural gas at power stations, the positive aspect is that it can be employed directly to generate electricity, that is, directly

Table 4.6 The chemical elements of natural gas

Component	Russian Gas %	Trans Europa Naturgas Pipeline (TENP) Gas %	Ekofisk Gas %	Gronington Gas %	Average Value %
Methane, CH_4	96.6	85.2	84.7	79.3	86.4
Ethane, C_2H_6	1.9	5.3	8.8	2.9	4.7
Propane, C_3H_8	0.5	1.6	3.1	0.5	1.4
Butane, C_4H_{10}	0.1	0.5	0.9	0.2	0.4
Nitrogen, N_2	1.6	5.7	0.4	14.1	5.4
Carbon dioxide, CO_2	0.2	1.5	1.8	2.9	1.6
Pentane, C_5H_{12} and higher hydrocarbons	0.1	0.2	0.3	0.1	0.175

Source: Adapted from Melvin (1988).

turning a turbine using a technique called combined cycle rather than being burned to produce steam to turn the turbine, as is the case with coal and oil (Moore, 1997). For this reason, power generating companies have natural gas at the top of their list for fossil fuels, if cost and availability allow it.

Statistics related to natural gas show that the present reserve is higher than that of oil (total world reserve 6,669,315 trillion cubic feet [TCF] according to the estimation for the year 2010) (Energy Information Administration, 2012).

The value of natural gas as a "standard" for the REA1 methodology comes from the way in which power generating companies compare biomass samples with the general value of the natural gas and how close these samples are both chemically (S&T) and commercially (BF).

Most of the world's natural gas reserve is in the Middle East and Russia: 67% of the world's total reserve is located in these two regions (World Coal Institute, 2008) (Box 4.1).

Box 4.1

ENERGY PROSPECT FOR 2040

The population of the world is expected to reach 9 billion by the year 2040. Three quarters of the world's population will reside in Asia Pacific and Africa. However, by 2040, India will be the country with the highest population in the world. Accordingly, the prospect of fossil fuels and electricity demand for the year 2040 will rise drastically. According to Exxon-Mobil (2013), the energy outlook for the year 2040 has been estimated as follows:

1. Natural gas will be the largest source for generating electricity.
2. The global demand for electricity will grow by 85%.
3. The use of natural gas will grow from the present 1% to 4% of the global transportation fuel mix.
4. A growth of 65% for heavy duty transportation.
5. Non-OECD energy demand will be more than twice that of the OECD.
6. The world's energy demand by domestic and commercial enterprises will be mostly for electricity and natural gas. This will be 60% higher than today's demand.
7. Natural gas will last for 200 years if the present world's demand keeps at the same level.
8. Compared with 2010, the OECD countries will emit 20% less CO_2 by 2040.

4.3 MAIN SAMPLES

4.3.1 Introduction

The use and production of energy crops goes back thousands of years in human history. Depending mostly on basic natural resources, such as sunlight and water, the landscape was made up of wild crop fields, producing varieties of seeds that were consumed by both animals and humans.

The beginning of human dependency on these kind of crops as a source of food eventually led to the first human settlement. As these crops became part of the new settlements' dietary needs, crop cultivation started in earnest and, for many historians, these were the first steps toward the beginning of what has been termed "human civilization" (Table 4.7). In modern times, successful crop production on a commercial scale depends on a variety of factors. Nevertheless, the same main basic principles can be applied to most crops from the first step in the farming process up to the final delivery to the user. Some of these basic factors can be summarized in the following points:

1. Seasonal or annual yields
2. Farm location
3. Tillage practices
4. Irrigation
5. Previous field usage
6. Enterprise size
7. The weather and surrounding environment
8. Natural or man-made disasters
9. Commercial competition and/or profit return
10. Disease and/or insect damage to fields.

Table 4.7 Estimated dates and origin of some of the crops cereals

Cereal	Time	Location
Wheat	7000 BC	Near East
Barley	7000 BC	Near East
Rice	4500 BC	Asia
Maize	4500 BC	Central America
Millet	4000 BC	Africa
Sorghum	4000 BC	Africa
Rye	400 BC	Europe
Oats	AD 100	Europe
Triticale	AD 1930	Russia, Europe

Source: Adapted from McGee (1984).

The balance between energy input and energy output is another important issue regarding energy crops. According to IEA Bioenergy (2013), the energy needed in crop production has been estimated as follows: 50% of the total energy requirement is spent on fertilizer production, 22% on machinery, 15% for transport fuel, and 13% is related to the production/usage of pesticides. Further energy also is needed to process energy crops for the eventual production of bio-fuels for the point of availability to the end user (IEA Bioenergy, 2013).

4.3.2 Crops Basic Composition

Although the chemical composition of plants vary, the basic dry composition can be summarized as follows: the percentage of lignin $[C_{10}H_{12}O_3]_n$ (nonsugar type molecules) is approximately 25%, which holds the cellulose fibers together, while the rest, approximately 75%, is carbohydrates (EUBIA, 2007). These are like chains of a large number of sugar molecules which make up part of the carbohydrate, from which the cellulose $[C_6(H_2O)_5]_n$ and hemicellulose $[C_5(H_2O)_4]_n$ are made. There are other components within the carbohydrates, such as terpenes $[C_{10}H_{16}]$ and triacylglycerols (TAG) $[C_xH_yO_6]$, plus organic proteins. In addition to these, there are inorganic elements, such as K, Se, Si, Mg Ca, and Fe.

When using biomass energy crops in order to generate energy, the thickness of the cell wall in the individual biomass samples should be examined and measured, whenever possible. This thickness is an important factor when it comes to energy output (Himmel et al., 2007), along with the biomass conversion process into ethanol (Mosier et al., 2005).

A selection of outer layer thickness samples have been examined and measured (Table 4.8) as part of the research investigation into the selection process

Table 4.8 Examples of the outer layer thicknesses of various biomass samples

Biomass Type	Outer Layer Thickness of Cross-Section Samples (μm)	Average (μm)
Sunflower SS	240–290	265
Sunflower BS	205–290	248
Date seed	110–124	117
Corn	75–98	86
Rapeseed	79–83	81
Peanut	72–87	80
Niger seed	63–81	72
Peanut shell	45–90	68
Wheat	53–66	60
Barley	50–54	52

Source: Author.

of biomass samples. These samples will form part of the makeup of the new biomass fuel formula (SFS).

Cell wall thickness consists of structural polysaccharides, such as hemicelluloses and cellulose, lignin, and pectic* substances. The thickness of walls differs depending on a number of factors, such as the season, the environment, and the time of harvesting (Fowler et al., 2003).

There are two types of cell walls: primary and secondary. The first wall contains cellulose, which is made up from a large number of glucose molecules (hydrogen-bonded chains) plus other materials, such as hemicellulose.

As the cell growth reaches its end, in certain cells, a secondary wall usually develops in the primary wall (U.S. Department of Energy Office of Science, 2008). The lignin (irregular polymer), which exists in both cell walls, creates a rigidity and protection for the overall cell. As mentioned previously, the cell walls are one of the main ingredients for generating energy when biomass materials are used as a fuel (Himmel et al., 2007).

4.3.3 Crops and Oil Sources

In chemistry literature, vegetable oils are esters (RCOOR′),† that is, a trivalent alcohol (glycerol) bound with three long-chain fatty acids. The sources of vegetable oil can be divided into two sections:

A. Major oils
B. Minor oils.

The following are examples of the two main divisions:

A. Major Oils

1. Soya bean
2. Rapeseed
3. Mustard
4. Sunflower
5. Palm and its fractions
6. Coconut
7. Palm kernel
8. Cottonseed
9. Tallow.

* It is a substance that helps to hold the cells together. Pectin enzyme is used in winemaking.
† Most of the common esters are derived from carboxylic acids (R-COOH).

B. Minor Oils

Tree-borne seed oils (e.g., *Jatropha curcus*, "jatropha or ratanjyot"; *Pongamia pinnata*, "pongamia or karanj"; and *Madhuca latifolia*):

1. Plant seed oil Jatropha
2. Rice bran
3. Watermelon seed
4. Tobacco seed
5. Niger seed
6. Tea seed
7. Jute seed
8. Chili seed.

4.3.4 Oil Quality and Standard

For a high quality vegetable oil to be used as any fuel, though particularly within a combustion engine, the following fuel factors have to be taken into consideration (Sustainable Energy Ireland [SEI], undated):

1. Density
2. Energy content (calorific value)
3. Ignition temperature
4. Viscosity
5. Glycerin content
6. Carbon residue
7. Sulfur content
8. Cetane number (CN)*
9. Iodine number (IN).†

4.3.5 Crops Photosynthesis

Generally speaking, the plant photosynthesis mechanism can be described as the bonding of CO_2 and H_2O molecules together for the purpose of producing sugar molecules (CH_2O). These sugar molecules form the storage of energy which the plant uses for making other compounds (PCC, 2008). In the process, O_2 is produced as a by-product, that is, $CO_2 + 2H_2O + \text{light energy} = (CH_2O) +$

* Cetane number (CN) is a diesel fuel ignition quality measurement, that is, diesel performance rating corresponding to the performance of cetane (colorless, liquid, straight-chain paraffin) in a mixture with methylnaphthalene. This means better combustion is obtained when the cetane number is higher. This kind of test is similar to the octane testing performed on petrol.

† Iodine number (IN) or iodine value is the amount of iodine consumed (or absorbed) by an organic compound, such as unsaturated fat over a given time.

$H_2O + O_2$ or $CO_2 + 2H_2O$ + light energy = biomass + O_2, as described in Section 2.7.

Briefly, there are three different types of photosynthesis:

1. *C3 Plants.* Named C3 as CO_2 is incorporated into a three-carbon compound. Most plants use this type of photosynthesis. It can be more efficient in a moist and cool environment.
2. *C4 Plants.* Named C4 as CO_2 is incorporated into a four-carbon compound. C4 plants that use this type of photosynthesis include several thousand species in a number of plant families. Most are adapted mainly to arid conditions, and they make better use of water and energy.
3. *CAM Plants (*CAM = *Crassulacean Acid Metabolism).* Before photosynthesis, CO_2 is stored in the form of an acid. The plant family name is Crassulaceae (PCC, Plants, 2008). As in type C4, it makes better use of water and energy as the plant adapted itself to harsher environmental conditions.

The selection process with regard to the best type of plant for commercial use as a bio-fuel has concluded that C4 type is the right option as its photosynthesis is more efficient than the other types. The reason that C4 plant type photosynthesis is a better option for the production of biomass fuel is that it reduces the wasteful process of photorespiration and consequently reduces the dependency on other types of resources (i.e., reduces costs). It is also easily adaptable to harsher environments (Heaton et al., 2003). The function of photosynthesis can be described by using two different types of cells, mesophyll and sheath cells in the plant leaf. The C4 plant's photosynthesis occurs with the help of genes that encode photosynthetic enzymes (Nomura et al., 2000). This function is expressed either within the mesophyll or in the bundle sheath cells (Nelson and Langdale, 1992). However, the genes in C3 plants are only expressed in its mesophyll cells. For this reason, C4 photosynthesis is more efficient in its water and nitrogen use (Ehleringer and Monson, 1993). In addition to the above, C4 photosynthesis has the best potential for converting sunlight into biomass energy, which is estimated to be around 40% higher than C3 photosynthesis (Long, 1999). Therefore, energy crops for the purpose of co-firing, as well as for the production of fuel for other purposes, should first consider photosynthesis type. This is important and cannot be ignored from an environmental point of view or from other factors, including: quality, process, commercial aspects, and cost.

4.3.6 Energy Crops Environmental Effect

One of the reported effects in using energy crops as a fuel is the possible stabilization of CO_2 in the atmosphere, which many scientists believe to be one of the main causes of global warming. Since the industrial revolution, the concentration of CO_2 in the atmosphere has increased by 36%, which according to the data obtained from ice cores,* is higher than at any other time during the past 650,000 years (Global Warming: The Rise of CO2 & Warming, 2008). There is a balance needed between the crops used as food for human and animal consumption and their use as a fuel. Short and long term effects on the environment arising from this should be taken seriously. Some concerns have been reported in the media, where energy crops, such as rapeseed and corn (e.g., for the production of biodiesel) have been said to produce up to 70% (from rapeseed) and 50% (from corn) more greenhouse gases than the use of fossil fuels (Smith, 2007). The published claim by Smith in *The Times* newspaper (originated by SRI Consulting) reported that nitrous oxide (N_2O) is 296 times (or 200–300 times as mentioned in the original report) more powerful than carbon dioxide (CO_2) as a greenhouse gas. Corn and rapeseed, used as examples here, together with other types of energy crops, can arguably judged from both sides of the debates as to their benefits or lack thereof. As long as the demand for more and cheaper sources of energy continues to rise, the environmental concern in this respect, for some people, may not always be a top priority (Box 4.2).

4.3.7 Corn (*Zea mays* L.)

4.3.7.1 General Introduction. Corn is a herbaceous annual plant, also known as "maize" and occasionally referred to as "Indian corn." Corn originated from America in two possible locations, Peru and Mexico. Its date of discovery is also uncertain, although some researchers suggest an approximate period of 4600 years ago, others around 8000 years ago (Mangelsdorf, 2008). Corn has certain similarities to wheat and barley; however, more starch can be produced from corn than any other type of crop. This is possibly one of the main reasons why corn is such a vital source of materials in the making of industrial starch (ISI, 2001). Corn can be a tall or short plant

Box 4.2

GLOBAL WARMING: MAIN CONTRIBUTORS (GREENHOUSE)

Water vapour (H_2O)
Carbon dioxide (CO_2)
Methane (CH_4)
Ozone (O_3)
Nitrous oxide (N_2O)

* There is a disagreement among scientists in regard to the accuracy of data obtained from ice cores. This is because change in temperature and the possible growth of bacteria may contaminate the makeup of the ice core.

(from 60 cm up to 6 m) with an erect solid stem, unlike many other types of stem grasses, as they usually tend to be hollow. As a cereal grass, it bears kernels on large ears that usually mature in the later part of the summer. The plant needs warm weather to sustain growth, and is widely cultivated in North and South America in many varieties.

There are two groups of kernel shape:

1. Flint maize (round shape)
2. Dent maize (tooth shaped).

The three most well known types of corn are the following:

A. White maize (flint or dent)
B. Plata maize (flint)
C. Yellow maize (mostly dent).

The most common types of corn include the following:

1. Flint
2. Flour
3. Dent
4. Pop
5. Sweet
6. Waxy
7. Indian
8. Black
9. Salpor
10. Crystalline
11. Starchy
12. Pod (or podcorn).

The maturity of the plant can vary according to the type of corn being cultivated. Certain types can mature within a 2-month period; others may take as long as 11 months (Microsoft Encarta, 2008). The color produced by the corn grains can also differ considerably, ranging from white to yellow to mottled red or red.

As mentioned earlier, corn is an annual crop. There is a Central American perennial type (*Zea diploperennis*) that is believed to be a distant wild relative or teosinte, that is, parent to present day corn (Sullivan, 1982). An attempt was made by an American-Argentine research group to breed the present-day type of corn with a perennial type for the purpose of large increases in the crop production.

Table 4.9 Top 10 corn producers: estimated for February 2014

Country	Metric Tons
USA	353,715,000
China	217,000,000
Brazil	70,000,000
EU-27	64,685,000
Ukraine	30,900,000
Argentina	24,000,000
India	23,000,000
Mexico	21,700,000
Canada	14,200,000
South Africa	13,000,000

Source: USDA (2014).

There are advantages and disadvantages when a comparison is made between perennial and annual crop varieties in that the perennial is much less productive than its annual counterparts (Sullivan, 1982).

4.3.7.2 Characteristics. By considering the usage and importance of corn on the list of crops, corn can be characterized in the following points (FAO, 1990):

a. Corn is the third most important of crops in usage after wheat and rice.
b. 20% of corn production is used for food.
c. 64% of corn production is used for animal feed.
d. 14% is used in processing (value added products).

According to the USDA, world corn production for 2013–14 has been estimated to be around 966.63 million metric tons (USDA, 2014). As can be seen in Table 4.9, the United States is the top producer, representing about 40% of the world's production of corn (Microsoft Encarta, 2008), followed by China and Brazil (USDA, 2014).

The seven types of corn mentioned in the previous section differ from each other when it comes to physical and chemical characteristics and to the type of starch in the endosperm, that is, whether it is horny or floury. These kinds of differences can be attributed to the genotype as well as on environmental influences (Sandhu et al., 2004). The horny or floury type's effect on the endosperm's structure can make a difference in the dry milling industry. This is because a horny endosperm is far more difficult to mill into finer particles, in contrast with the floury, which is usually softer and can be disintegrated much faster (Kikuchi et al., 2006).

The leading type of corn grown in the United States is dent corn, even though sweet corn is very common, which is mostly used for human

Table 4.10 Characteristics of an ideal biomass energy crop: corn

Crop Characteristics	Score
1. Photosynthesis pathway	C4 +
2. Long canopy duration	
3. Perennial or annual (no need for annual tillage or planting)	Annual –
4. No known pests or diseases	
5. Rapid growth in spring to out compete weeds	
6. Sterile; prevent "escape"	
7. Stores carbon in the soil (soil restoration and carbon sequestration tool)	
8. Partitions nutrients back to roots in fall (low fertilizer requirement).	
9. Low nutrient content, that is, <200 mg/MJ nitrogen and sulfur (clean Burning)	
10. High water use efficiency	+
11. Dry down in the field (zero drying costs)	
12. Good winter standing (harvest when needed; "zero" storage costs)	
13. Utilizes existing farm equipment	+
14. Alternative markets (high quality paper, building materials, and fermentation)	+

Source: Data from various sources and adapted in part from Long (1999) and Heaton et al. (2003).

consumption. Unlike other types of corn, sweet corn does not convert the sugar produced during its growth into starch (Microsoft Encarta, 2008).

As a C4 plant type, corn scored 28% (Table 4.10) in accordance to the factors designed describing the characteristics of an ideal biomass energy crop (Box 4.3).

4.3.7.3 Composition. The general composition of corn is that two-thirds of it is starch. Within the kernel, starch is around 73%, while the two macromolecular components of starch are amylose and amylopectin (Watson and Ramstad, 1991). The starch of corn is made up of 75% branched amylopectin and 25% linear amylose (Sandhu et al., 2004). Amylose and amylopectin are two types of glucose polymers commonly found in corn starch. The percentage of glucose, fructose, and sucrose as part of the carbohydrate simple sugars in the kernel is around 1–3% while in common maize, such as a dent or flint, the starch in the endosperm consists of amylose at around 25–30%, while amylopectin is at about 70–75% (FAO, 1992).

The other remaining parts of the corn composition are oil, protein, fiber, and water (Table 4.11). In particular, linoleic acid, $C_{18}H_{32}O_2$ ($C_{19}H_{34}O_2$ methyl linoleate) (acronym: C18:2), and oleic acid, $C_{18}H_{32}O_2$ ($C_{19}H_{36}O_2$ methyl oleate—C18:1), form around 86% of the total makeup of corn oil of which linoleic acid is 58%, that is, around 13% less than that of sunflower seed

Box 4.3

"US BIOFUEL PRODUCTION SHOULD BE SUSPENDED, UN SAYS"

The United Nations (UN) food agency has called on the United States to suspend its production of biofuel ethanol.

Under US law, 40% of the corn harvest must be used to make biofuel, a quota which the UN says could contribute to a food crisis around the world.

A drought and heatwave across the US has destroyed much of the country's corn crop, driving up prices.

The US argues that producing much of its own fuel, rather than importing it, is good for the country.

BBC News (August 10, 2012)

Table 4.11 The composition of corn

Components	%	% Dry Basis
Moisture	9–15	–
Starch	61	72
Protein	8.5	10
Fiber	9.5	11
Oil	4	3.4–5
Ash	–	1.6

Source: Adapted from Satake Corporation U.K. Division (2008).

Table 4.12 The main chemical elements of corn

Carbon	44.83%
Hydrogen	6.39%
Nitrogen	3.76%
Oxygen	45%
Sulfur	~0.02%

Source: Author.

(Section 6.2). The rest of the corn oil is saturated fat, with only 1% alpha linoleic acid.

The laboratory test for corn concerning carbon and nitrogen (and other elements seen in Table 4.12 and Fig. 4.3) has established that around 45% of the total mass of the sample is made up from carbon and 3.76% is nitrogen.

4.3.7.4 Suitability as a Fuel. In the mid-1970s, the production of ethanol in the United States was estimated to be only a few million gallons per year. By

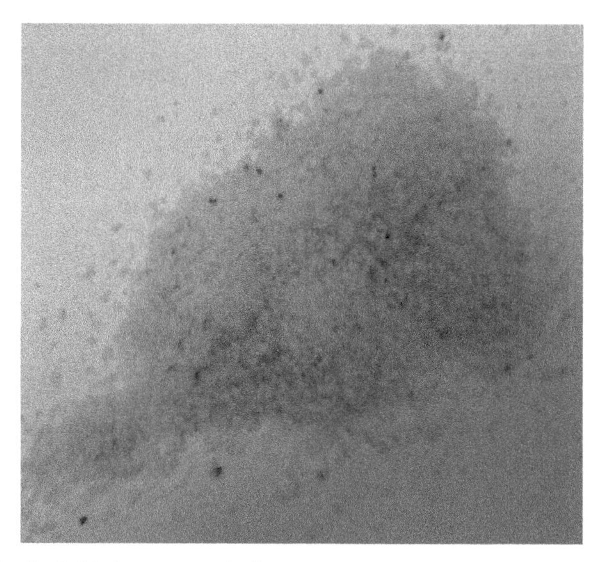

Figure 4.3 Crushed dried corn sample (granules and powder) used during laboratory work. *Source*: Author.

2001, however, the production reached over 1.7 billion gallons (Shapouri et al., 2002). This increase in production meant a higher demand for corn, which in itself changed the balance between farming solely for consumption and the need for corn as a fuel source. This change increased the price of corn drastically. Consequently, the final cost for the production of ethanol had to be increased as well (Leibtag, 2008). This can be seen as a step backwards, that is, a suitable fuel must be not only be technologically and scientifically viable, but also and more importantly, not more expensive to produce than other types of fuel. This means that the total cost involved in producing it and the balance between the energy input and output during production should be at least the same or less than any other type of fuel. This should be the case simply in order to make sense in commercial terms, especially when there is a competition in the same field—while keeping in mind environmental issues and concerns.

There is no doubt that corn can be used as a source of fuel in a co-firing method or for the combustion engine as in the form of ethanol fuel. For ethanol production, there are two common methods used within this type of industry, namely, "wet milling" (WM) and "dry grind" (DG), then the starch obtained can be easily fermented to produce ethanol (C_2H_5OH). Wet milling is meant for the production of a high volume of ethanol and consequently will require massive investment and more time for its establishment. The DG method is far less costly and suitable for a smaller volume of ethanol output (Singh et al., 2001). For a crop such as corn to be used as a fuel, the "green

Table 4.13 The energy of corn tested as a dried matter

Dried matter (12.5% moisture)	Energy (H) J/g
Corn	17,334

Source: Author.

Box 4.4

CORN AND THE UNITED STATES

Strong demand for ethanol production has resulted in higher corn prices and has provided incentives to increase corn acreage. In many cases, farmers have increased corn acreage by adjusting crop rotations between corn and soybeans, which has caused soybean plantings to decrease. Other sources of land for increased corn plantings include cropland used as pasture, reduced fallow, acreage returning to production from expiring Conservation Reserve Program contracts, and shifts from other crops, such as cotton.

USDA (2013a)

light" can only be given if and when the S&T and BF factors can be applied rigorously to that particular type of energy crop.

Corn, as shown in Table 4.11, yields a high percentage of starch (61–72%) with an output energy of 17,334 J/g (Table 4.13). This makes it a good candidate for use in the production of ethanol and/or in a co-firing method used directly or indirectly in the form of powder or pellets. In the United States, there is already a well-established industry converting corn to ethanol. Investments and incentives to do so by the federal government (as well the private sector) mean billions of dollars are presently invested in this field.

According to U.S. Department of Agriculture (USDA), corn presently is the most prevalent energy crops being cultivated in the United States.

In the REA1 methodology listing, corn was at number 11, while distillers dried corn was at number thirteen (see Chapter 9, Section 9.6). This means that corn can be a good source for the production of ethanol, and possibly this is how it should be used to produce a biomass fuel, as long as a balance is found between producing corn as food and farming it solely as a source of fuel. Having said that, using corn directly as a fuel in co-firing and/or on its own (if compared with the other fourteen biomass tested samples), might not be the right option, simply because the test has already shown that there are other energy crops more suitable both economically and environmentally in the case of co-firing than corn (see methodology and results sections) (Box 4.4).

4.3.7.5 Corn Ash Composition. Figure 4.4 illustrates a corn ash composition with a 400× magnification of the ash sample image.

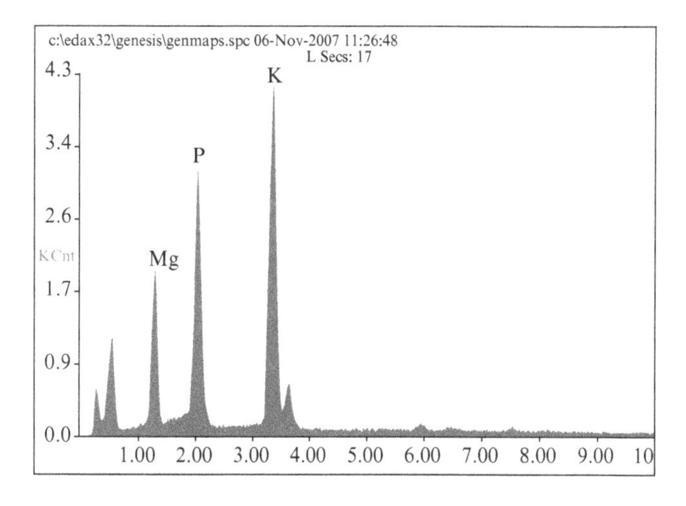

Element	Wt%	At%
MgO	23.59	47.09
P2O5	42.95	24.34
K2O	33.46	28.57

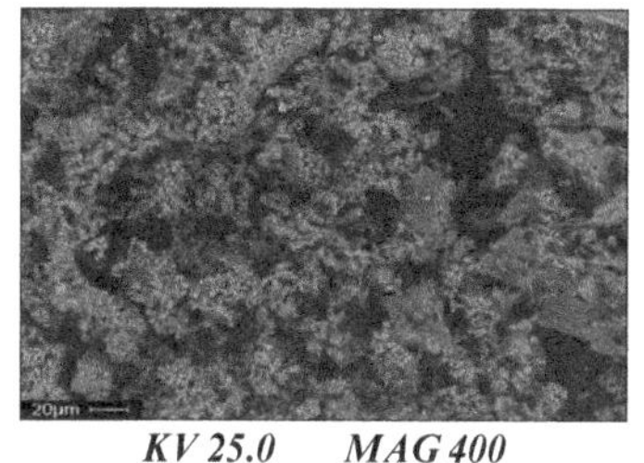

Figure 4.4 Corn ash elements plus a magnified image (400×) of the ash. *Source*: Author.

The elements of corn ash (the plot obtained via FEI Quanta 600 scanning electron microscope). The x-axis in the above figure is energy in kev (thousands electron volts) and the y-axis is accumulated counts in kcnt (thousand counts).

4.3.8 Wheat (*Triticum aestivum* L.)

4.3.8.1 General Introduction. Wheat was first cultivated in an area currently known as the Middle East, probably within the boundaries of the Fertile Crescent (Fig. 4.5 and Fig. 4.6). Wild wheat species (einkorn) was compared with the domesticated wheat (emmer) and provided the possible location of the first human domestication of wheat. This area is now known as Diyar Bakir (Diyarbakir) in southeastern Turkey, around 10,000 years ago (Dubcovsky and Dvorak, 2007).

Wheat is an annual herbaceous grass with erect flower spikes and light brown grains with an average height of 1.2 m.

Wheat is cultivated around the world mainly as a food product (a low grade of wheat/flour or bran is usually used as feed for livestock). Other uses are in

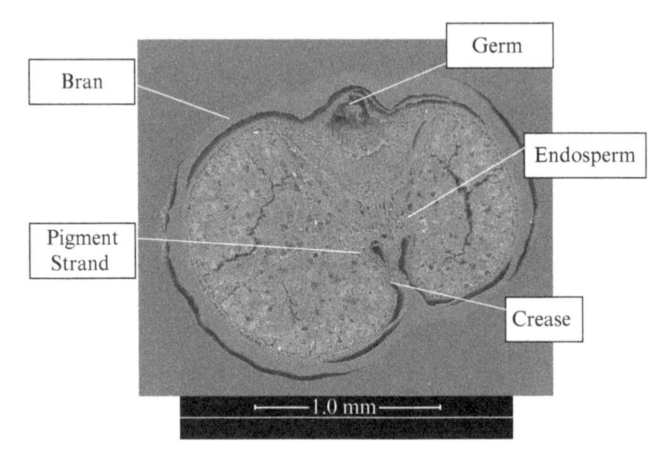

Figure 4.5 Electronic microscope image showing a cross section of a dried wheat grain sample. *Source*: Author.

the making of industrial alcohol as well as alcoholic drinks, such as beer and whiskey (ISI, 2001). Globally, in terms of production, it is the second-largest cereal crop after maize, the third being rice (FAO, 2013a). Wheat contains moderate protein and mineral content. The most widely cultivated types of wheat are common wheat (*Triticum aestivum*), durum, and club wheat. In scientific terms, wheat is classified by the number of chromosomes within the cell. These kinds of classifications come under three different types (ISI, 2001):

1. Diploid (or einkorn): 14 chromosomes
2. Tetraploid (or emmer): 28 chromosomes
3. Hexaploid (or bread wheat): 42 chromosomes.

Generally speaking, wheat is divided into red and white wheat in relation to the outer skin color of the grain. At the same time, wheat can be classified into two main categories: (1) spring wheat; (2) winter wheat. From these, there are other classifications (mt.gov, 2006).

4.3.8.2 Characteristics. North European and North American wheat is mostly the spring and winter varieties. These can be distinguished by the color of the grains, as white wheat is more likely to be planted during the winter and red wheat usually planted during the spring. Another type of wheat closely related to the common wheat is club wheat, described as having compact spelta and spikes as the glumes enclose the grain. Club wheat is characterized as "hard" grain, mostly as a result of the hardness of the content related to high gluten (ISI, 2001). For quality and grade classifications, there are three different types of wheat available on the market (Farm Direct, 2001):

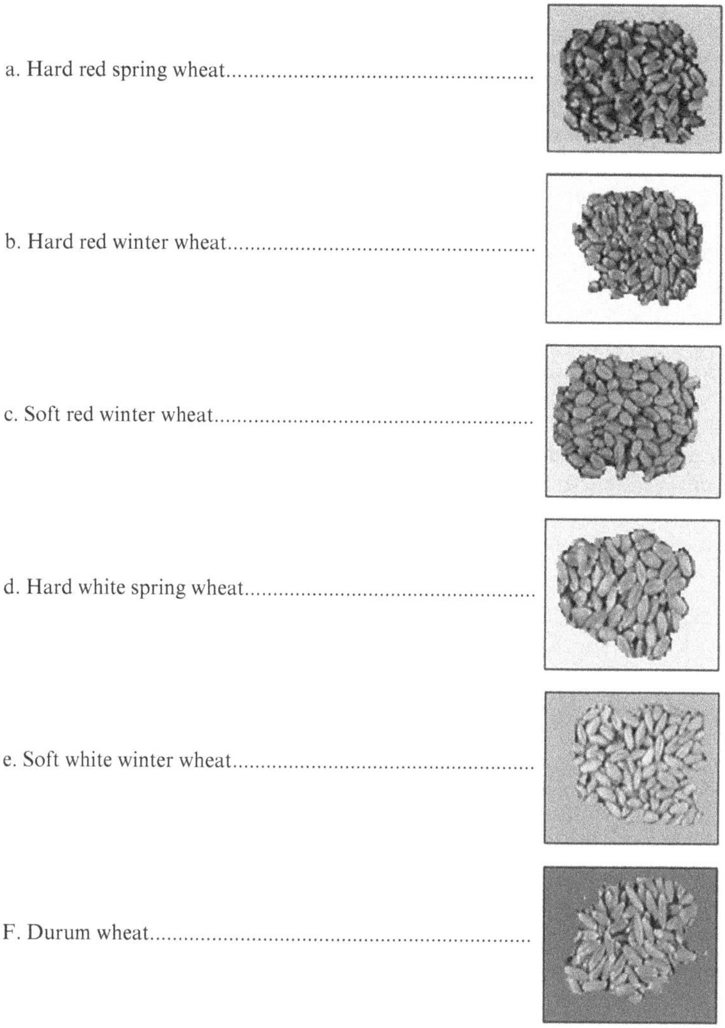

a. Hard red spring wheat...

b. Hard red winter wheat...

c. Soft red winter wheat...

d. Hard white spring wheat...

e. Soft white winter wheat...

F. Durum wheat...

Figure 4.6 Images of various types of wheat. Adapted from various sources.

A. *Class 1.* Hard extensible gluten with low harvesting yields. Used mostly for bread making. Less capable of withstanding changing in weather and/or in environmental conditions (high risk).
B. *Class 2.* The quality is lower than "class 1," but yields can be higher (high risk);
C. *Class 3.* Quality is lower than the above two classes but with high yields. Used mostly in industry and as animal feed (low risk).

Table 4.14 Characteristics of an ideal biomass energy crop: wheat

Crop Characteristics	Score
1. Photosynthesis pathway	C3 –
2. Long canopy duration	
3. Perennial or annual (no need for annual tillage or planting)	Annual –
4. No known pests or diseases	
5. Rapid growth in spring to out compete weeds	
6. Sterile; prevent "escape"	
7. Stores carbon in the soil (soil restoration and carbon sequestration tool)	
8. Partitions nutrients back to roots in fall (low fertilizer requirement).	
9. Low nutrient content, that is, <200 mg/MJ nitrogen and sulfur (clean burning)	
10. High water use efficiency	
11. Dry down in the field (zero drying costs)	+
12. Good winter standing (harvest when needed; "zero" storage costs)	
13. Utilizes existing farm equipment	+
14. Alternative markets (high quality paper, building materials, and fermentation)	+

Source: Data from various sources and adapted in part from (Long, 1999) and (Heaton et al., 2003).

Clearly, if wheat is selected to be used as a source of fuel then "class 3" would be the choice, as it is the cheapest to purchase and can withstand harsher environmental conditions while providing high yields. EU-27 is the world's top producer of wheat. In ten years (i.e., the average of the crop during the years 1997–1998 to 2006–2007) total production of wheat was around 123,000 million tons. For wheat production at the time of compiling this book, the following estimated/projected figures in MT during February 2014 illustrate: USA, 57,961,000; China, 122,000,000; Brazil, 70,000,000; EU-27, 142,866,000; Ukraine, 22,278,000; Argentina, 24,000,000; India, 92,460,000; Canada, 37,500,000; Australia, 26,500,000; Pakistan, 24,000,000 (USDA, 2014). As a C3 plant type, wheat scored in point 11, 13, and 14 (Table 4.14) according to the profile of an ideal biomass energy crop.

4.3.8.3 Composition. The main makeup of wheat's chemical components is a high percentage of starchy endosperm (Fig. 4.5), plus proteins (Michniewicz et al., 1990). The wheat bran itself contains a high percentage of protein, cellulose, ash, and hemicelluloses, while the makeup of a wheat kernel is 13–17% bran, 2–3% germ, and 81–84% endosperm (Wang, 2003). As dry matter wheat flour is made up mainly of the kernel starchy endosperm. This is made up of 8–18% protein, approximately 70–80% carbohydrates (Table 4.15), 1.5–2.5% lipids, and 2–3% polysaccharides (nonstarch) (Wang, 2003). Depending on weather conditions during harvesting, the moisture content of

Table 4.15 The chemical makeup of a whole wheat grain

Composition	Whole Grain
Proteins	16
Fats	2
Carbohydrates	68
Fibers	11
Minerals (ash)	1.8
Other components	1.2
Total	100

Source: Paredes-Lopez (2000).

Table 4.16 The chemical elements of wheat

Carbon	39.67%
Hydrogen	6.20%
Nitrogen	1.46%
Oxygen	52.60%
Sulfur	~0.07%

Source: Author.

wheat can range from 12–18%, when farmed on a large commercial basis. The makeup of wheat's linoleic acid ($C_{18}H_{32}O_2$) is between 55% and 60%, which forms the highest percentage of the fatty acid. The other elements, according to the same author, are palmitic acid ($C_{16}H_{32}O_2$) 13–20%, stearic acid ($C_{18}H_{36}O_2$) maximum 2%, oleic acid ($C_{18}H_{34}O_2$) 13–21%, and 3–4% of the chemical component unsaponifiables. Carbon and nitrogen in a dried milled sample form around 40% and 1.46%, respectively, of the total mass of wheat grain (Table 4.16).

4.3.8.4 Suitability as a Fuel. Grain and straw wheat is the subject of commercial and scientific energy research into biomass across the globe, similar to the corn energy field of work presently taking place worldwide (Babcock, 2008). For example, during the year 2000, there were around 44 million tons of wheat straw in Europe with 1.7 million tons in Sweden alone (Olsson, 2006). Worldwide, there are more than 530 million tons of wheat straw produced annually. This is because around 1 kg of wheat straw is produced from ~1.3 kg of wheat (Ruiz et al., 2012) meaning that there is no shortage of biomass materials from wheat (usually produced in the form of pellets) in Europe, Sweden and in many parts of the world. By using straw rather than wheat grains, cost will be much lower during the early fuel production process; however, the output energy obtained will also be far less than when using actual wheat grains (Table 4.17 and Table 4.18).

Table 4.17 Wheat straw calorific value and quality as a fuel for domestic use

Sample	Gross Calorific Value (MJ/kg)	As Fuel (Domestic Use)
Wheat straw	17.20	Poor

Source: Hollander (1992).

Table 4.18 Wheat grains calorific value, dry matter with 14% moisture

Sample	Energy (MJ/g %)	Moisture (%)
Wheat grain	15.128	14

Source: Author.

It is possible to combust wheat straw pellets in adapted residential boilers with low emissions of carbon monoxide (CO), nitrogen oxides (NO_x), and with relatively low emissions of sulfur dioxide (SO_2) (Fløjgaard Kristensen et al., 1999).

Wheat straw is a comparatively good fuel as it has relatively low emissions during combustion. However, the emissions of polycyclic aromatic hydrocarbons naphthalene and phenanthrene are higher from straw wheat than other types of crops (Olsson, 2006).

To obtain a better quality fuel from wheat grain (e.g., during the ethanol process), the grain should contain a higher percentage of starch and a lower percentage of protein, such those found in soft white wheat. White wheat is close to corn in its protein content, that is, at around 8–9%. On the other hand, red spring wheat has 13–15% protein. In consequence, corn has a higher percentage of starch than any other type of wheat with lower protein, as reported in the previous section, simply because corn has a much larger kernel. It is also reported that winter wheat may have a higher yield of 30–40% in comparison to spring wheat.

4.3.8.5 Wheat Ash Composition. Figure 4.7 is an illustration of wheat ash composition. The elements of wheat ash (the plot obtained via FEI Quanta 600 scanning electron microscope). The *x*-axis in Figure 4.7 is energy in kev (thousands electron volts) and the *y*-axis is accumulated counts in kcnt (thousand counts).

4.3.9 Miscanthus (*Miscanthus sinensis*)

4.3.9.1 General Introduction. Miscanthus has a variety of names, such as Asian elephant grass and Chinese silvergrass. The name is mostly dependent on the geographical origin of the plant itself. The scientific name for

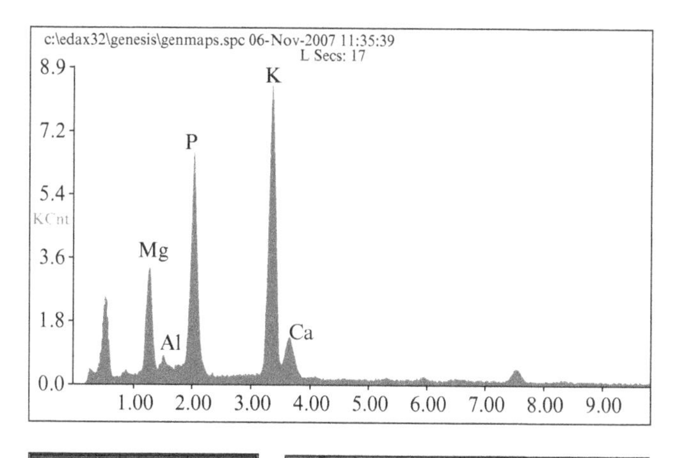

Element	Wt%	At%
MgO	19.54	40.25
Al2O3	02.59	02.11
P2O5	42.69	24.98
K2O	32.43	28.59
CaO	02.75	04.07

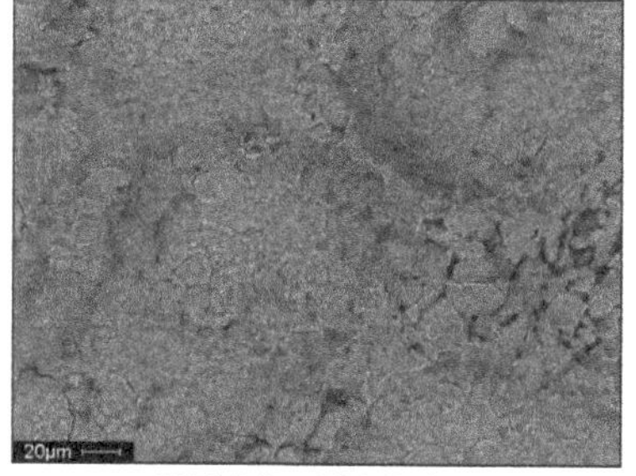

KV 25.0 MAG 400

Figure 4.7 Wheat ash elements plus a magnified image (400×) of the ash. *Source*: Author.

miscanthus is *Miscanthus sinensis Andes*. The commonly accepted historical origin of miscanthus is Asia (Southeast Exotic Pest Plant Council, 2008).

As an ornamental pot plantation, Europe's first miscanthus cultivation took place during the 1930s, when it was imported from Japan. From the early 1980s, field trials in Northern Europe have been carried out for the purpose of using miscanthus for the production of bio-fuels (Scurlock, 1999).

The plant can be described as woody, perennial, and is divided into 15 species of perennial grasses (Defra, 2007). Miscanthus can grow at a high rate and the plant height range from 2.5to 4 m. This in itself can make it a good example of a fast-growing rotational plant suitable for use as an energy crop.

After planting during spring time miscanthus reaches maturity within 3–4 years and can remain in the ground for at least 15–20 years. Apart from its first year, miscanthus grows fastest during the summertime and dies in the autumn/winter. The roots of the plant start to grow during the spring producing a new harvest for the following year. The mature miscanthus plant provides regular high growth yearly. This means it can provide a regular income to the farmers while providing harvest for the power generating companies. Miscanthus can grow successfully in the United Kingdom and can provide part of the biomass contribution in generating electricity, particularly as it has been reported that 22,000 tons/year from this plant can provide 2000 homes with electricity (Defra, 2007).

4.3.9.2 Characteristics. Miscanthus is characterized by a high rate of photosynthesis;for this reason, the plant is listed under the C4 plant type, that is, uses fewer elements than type C3 plant. This means miscanthus is highly efficient at processing light, water, and nitrogen (Scurlock, 1999). However, for a maximum growth and high productivity, the plant may need large amounts of water to achieve its optimum potential. Through the propagation of a rhizome cutting, miscanthus can establish itself and grow easily. This root development means there is an additional sequestration of carbon underneath the soil.

As a C4 plant type, miscanthus scored 100% (Table 4.19) according to the factors designed to describe the characteristics of an ideal biomass energy crop. Giant miscanthus seeds are reported to be sterile, meaning that the plant should be transplanted from rhizomes. In addition to this, miscanthus is slow to establish from clones (Baldwin and Holmberg, 2006).

Europe's miscanthus spring harvest yields can range from 2 to 44 dry tons/ha (Lewandowski, 2003), while the yields in the southern part of Europe can

Table 4.19 Characteristics of an ideal biomass energy crop Miscanthus

Crop Characteristics	Score
1. Photosynthesis pathway	C4+
2. Long canopy duration	+
3. Perennial or annual (no need for annual tillage or planting)	Perennial +
4. No known pests or diseases	+
5. Rapid growth in spring to out compete weeds	+
6. Sterile; prevent "escape"	+
7. Stores carbon in the soil (soil restoration and carbon sequestration tool)	+
8. Partitions nutrients back to roots in fall (low fertilizer requirement).	+
9. Low nutrient content, that is, <200 mg/MJ nitrogen and sulfur clean burning	+
10. High water use efficiency	+
11. Dry down in the field ("zero" drying costs)	+
12. Good winter standing (harvest when needed; "zero" storage costs)	+
13. Utilizes existing farm equipment	+
14. Alternative markets (high quality paper, building materials, and fermentation)	+

Sources: Data from various sources and adapted in part from Long (1999) and Heaton et al. (2003).

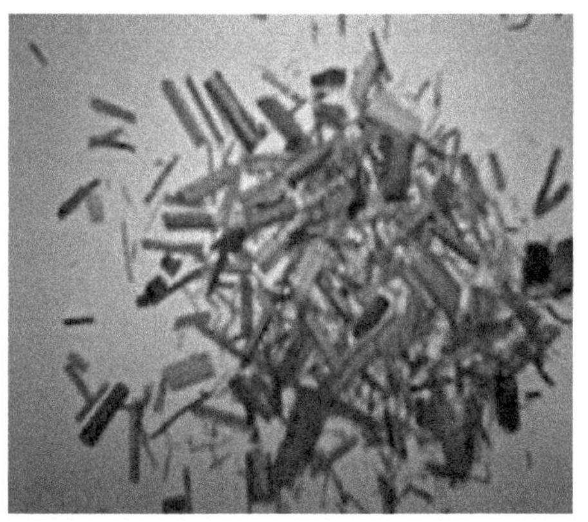

Figure 4.8 Two forms of dry crushed miscanthus samples used during the laboratory tests. *Source*: Author.

Table 4.20 Composition of oven-dried weight of *Miscanthus sinesis* stalk

Component	Percent
Ash	0.7
Water extractives	3.1
Organic extractives	9.1
Klason lignin	19.9
Hemicelluloses (xilanes)	21.1
Cellulose	42.6

Sources: Barba (2002) and Velasquez et al. (2003).

be highest, mainly because of the warmer climate and longer days (Clifton-Brown et al., 2001).

4.3.9.3 Composition. Cellulose $[C_6(H_2O)_5]_n$ and hemicelluloses $[C_5(H_2O)_4]_n$ form the largest percentage of a dry miscanthus sample (Fig. 4.8). Other compositions, such as ash, water extractives, and organic extractives can range from 0.7% to 9.1%, respectively (Table 4.20). Klason lignin forms about 20% of the total sample. The concentration of minerals in an early spring miscanthus harvest is low, ranging from 0.09% (the lowest) up to 1.12% (the highest) (Lewandowski and Kahnt, 1993). Miscanthus ash reportedly includes 30–40% SiO_2, 20–25% K_2O, and approximately 5% of P_2O_5, CaO, and MgO (Hallgren and Oskarsson, 1998; Moilanen et al., 1996).

The chemical composition of miscanthus: carbon, nitrogen, sulfur, oxygen, and hydrogen (Table 4.21) is relatively close to that of other types of energy crops. Changes in chemical element percentage values is dependent on the season it is being harvested, the soil quality, compositions, the amount of water,

Table 4.21 The chemical elements of *Miscanthus*

Element	Value (%)
Carbon	46.24
Hydrogen	5.45
Nitrogen	2.86
Oxygen	45.43
Sulfur	~0.02

Source: Author.

Table 4.22 Gross heat and net energy obtained from *Miscanthus floridulus* sample: with ~12% moisture

Factors	*Miscanthus floridulus*
Gross heat	18.5 GJ/Mg
Net energy	4150 cal/g
Moisture	12%
Ash	1.5–4.5%
Sulfur	0.1

Source: Baldwin and Holmberg (2006).

and the light available. Carbon and nitrogen in a dried crushed sample form around 46% and 3%, respectively, of the total mass (Table 4.21).

4.3.9.4 Suitability as a Fuel. Many of the scientific reports concerning miscanthus emphasize its high potential for use as a fuel on its own and/or during the co-firing process. Reports have indicated that miscanthus can be successfully burned on a commercial scale in a co-firing process with coal, for example, in Denmark, by employing a circulating 78-MW fluidized bed combustor (Scurlock, 1999).

The net calorific value of dry miscanthus (calculated in Table 4.22) is estimated to be close to 17 MJ/kg, that is, using 20 tons of dry miscanthus can be equal to using 8 tons of coal (Defra, 2007).

Compared with oil, miscanthus has a crop yield of 12t DM ha-1 and energy equivalent of 30–35 barrels of oil per hectare (Clifton-Brown and Valentine, 2007). Despite this, the energy balance, that is, the overall energy output/input ratio, can be as low as 1.1 when it comes to co-firing miscanthus with coal. This is believed to be the case because of the very high energy requirement related to fuel pulverization (Scurlock, 1999).

In regards to by-products such as ash (Fig. 4.10), it appears that in most cases related to the co-firing of biomass with coal (including the use of miscanthus), around 95% of the produced ash is in the form of fly ash. The rest is bottom ash, mostly left over after combustion (Jones and Walsh, 2001). Depending on the type of miscanthus being used and the percentage of dryness

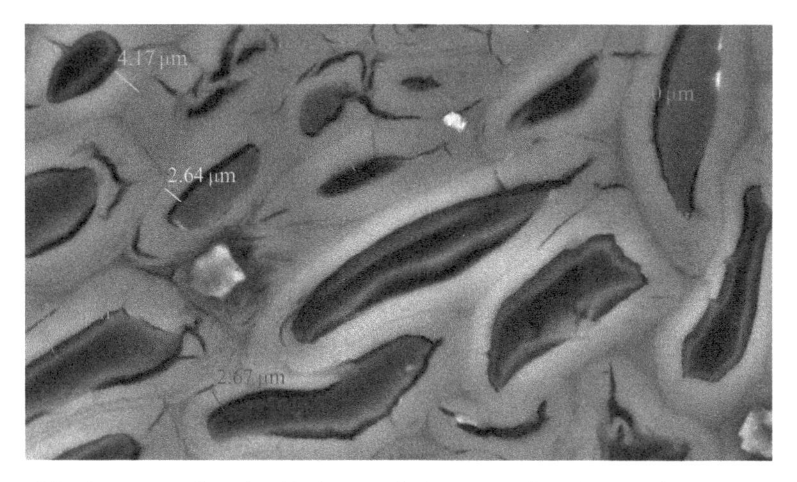

Figure 4.9 A cross section of a dried pressed miscanthus. Sample magnified 2200× with an average cell wall thickness of 3.63 μm. *Source*: Author.

Box 4.5

SEVEN TO THIRTEEN TONS/HECTARE

Depending on a number of factors, such as soil type, pH, topography, field slop and the level of nutrient in the soil, it has been estimated that around seven tonnes per hectare of miscanthus (with 20% moisture) can be obtained during the first harvest, i.e. two years after plantation. Afterward, miscanthus yields can rise up to 9–13 tonnes/hectare under similar conditions.

Teagasc, Agri-Food and Bioscience Institute (2011)

and condition of the sample, the average thickness of an individual cell wall can range from 0.71 to 4.50 μm (Fig. 4.9). The breakdown of carbohydrates (celluloses) locked in the cell walls of plants provides the main source of energy, that is, a cell wall contains long chains of sugars or polysaccharides, which make up ethanol, one of the main bio-fuels (U.S. Department of Energy Office of Science, 2008) (Box 4.5 and Box 4.6).

4.3.9.5 Miscanthus Ash Composition. Figure 4.10 is an illustration of miscanthus ash composition.

The elements of miscanthus ash (the plot obtained via FEI Quanta 600 scanning electron microscope). The x-axis in the above figure is energy in kev (thousands electron volts), and the y-axis is accumulated counts in kcnt (thousand counts).

Box 4.6

FOREST AREA

forest area in industrialized regions will increase between 2000 and 2050 by about 60 to 230 million ha. At the same time, the forest area in the developing regions will decrease by about 200 to 490 million ha. In addition to the decreasing forest area globally, forests are severely affected by disturbances such as forest fires, pests (insects and diseases) and climatic events including drought, wind, snow, ice, and floods. All of these factors have also carbon balance implications.

IPCC (2007)

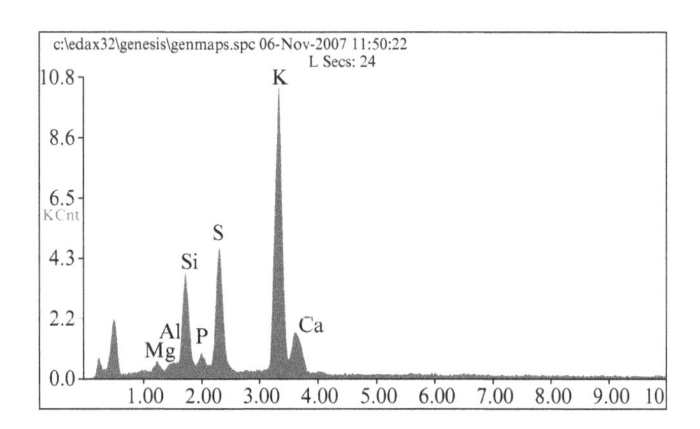

Element	*Wt%*	*At%*
MgO	02.67	05.15
Al2O3	02.18	01.66
SiO2	19.74	25.54
P2O5	04.37	02.39
SO3	29.30	28.45
K2O	37.49	30.93
CaO	04.24	05.88

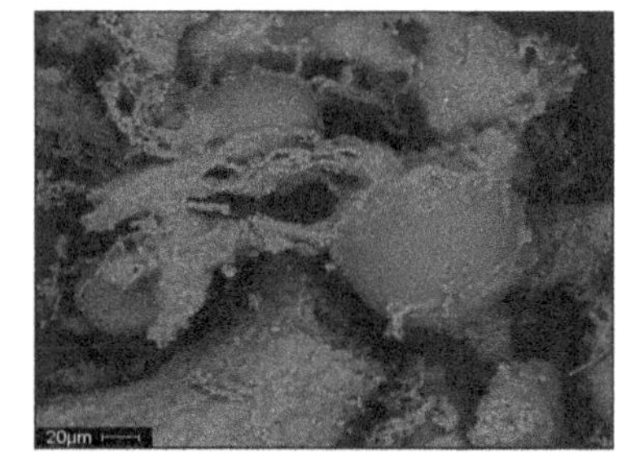

KV 25.0 MAG 400

Figure 4.10 Miscanthus ash elements plus a magnified image (400×) of the ash. *Source*: Author.

4.3.10 Rice (*Oryza sativa*)

4.3.10.1 General Introduction. Rice is one of the oldest sources of food used by humans and dates back to a few thousand years BC, as archaeologists in India have discovered.

The common rice used around the world comprises of two species, *Oryza sativa* and *Oryza galberrima*. These species originate from the southern part of Asia and the southeastern part of Africa.

As herbaceous annual or perennial rhizomatous marsh grass, rice thrives in warmer climates. There are over 40,000 varieties of cultivated rice from the grass species *Oryza sativa* (The Rice Association, 2008).

Depending on the geographical location and the fertility of the soil, the rice plant can grow from 1 to 1.8 m in height, while floating rice may range from 0.6 to 6 m (UNCTAD, Rice, 2008). Rice develops a main stem and a number of tillers, which have a panicle* measuring around 20–30 cm in width. The grains forms from 50 to 300 flowers produced on each panicle. Around 85% of the rice produced worldwide is directly used for human consumption (International Rice Research Institute, 2002). Rice can also be found in cereals, snack foods, brewed beverages, flour, oil, syrup, and religious ceremonies, to name but a few other uses.

General classification of the main rice types can be made as follows:

1. Brown rice (husked rice)
2. White rice
3. Red rice
4. Black rice
5. Arborio rice
6. Aromatic rice.

Rising production of rice worldwide is reflected in parallel to the increase in human consumption of it† (Fig. 4.11). There is also rising demand for this kind of crop to be used as an alternative source of fuel (Oryza.com, 2008). The media has reported that top rice producers, such as China and India (Table 4.23), are finding it difficult to keep their rice export at the same level. In fact, during the first quarter of 2008, India banned the export of nonbasmati rice, simply in order to manage the supply and reduce rising prices of rice for the local population (BBC News—http://news.bbc.co.uk/news, 2008).

* A panicle occurs when the branches of a raceme branch are also racemes. Example: yucca.

† "The first Regular Session of the International Rice Commission was held in Bangkok, Thailand in 1949. A regular Session of the Commission has been held every 4 years since then to review progress and advise on adjustments in national rice programmes" (FAO, 2013a).

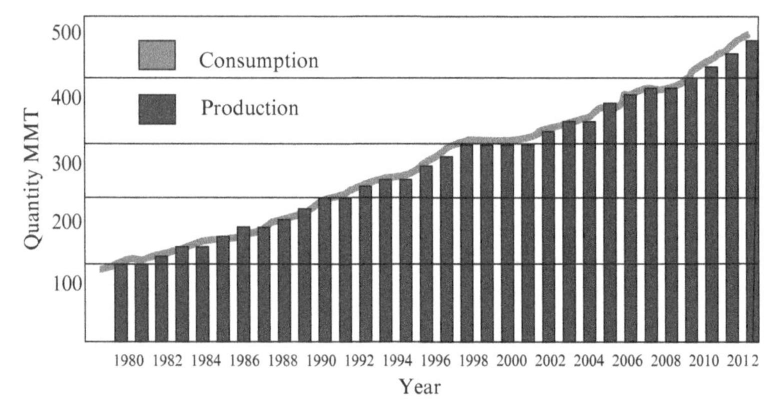

Figure 4.11 World milled rice production and consumption (MMT) from 1980 to 2012. *Source*: Data from USDA (2013b).

4.3.10.2 Characteristics. As a semiaquatic plant, rice can be grown much more easily in the tropics than in any other type of environment. This type of plant is usually referred to as an "autogame" grass (UNCTAD, Rice, 2008). As mentioned in the previous section, the most common type of rice used around the world is the Oryza. There are, however, no less than 20 different varieties of species belonging to the Oryza type. Paddy, brown, and milled rice come under three subsequent grain classifications, based on length to width ratio of the grain, that is, short, medium, and long kernel (Cooperative Extension Service, 2000). As a C3 plant type, rice scored in point 3, 13, and 14 points (Table 4.24) according to factors related to the characteristics of an ideal biomass energy crop.

4.3.10.3 Composition. As a dry matter, the percentage of starch in milled rice can be close to 90% (Shon et al., 2005). The starch itself is a polymer of glucose plus amylose in one linear of glucose polymer. In this polymer, the amylose has a content percentage value of 15–35% (Cereal Knowledge Bank, 2008).

Comparing the composition of husked rice with polished rice, the percentage of carbohydrates is higher in polished rice than husked rice (Table 4.25). The variations in the percentage value between the two types can vary from one element to another, as indicated in the same table.

Carbon and nitrogen in a polished dried rice sample form around 34% and 3%, respectively, of the total mass (Table 4.26).

4.3.10.4 Suitability as a Fuel. Rice husk ($C_{4.1}H_{5.8}O_{2.8}$) comes only second to bagasse ($C_{3.7}H_{6.4}O_3$) as a source of biomass energy (Aggarwal, 2003). Rice mills produce large quantities of rice husks, which make it an attractive source of energy, mainly because the initial cost involved is very low. From

Table 4.23 Major world rice producers (2010–2013)

	Production (MT)
India (Oct/Sept)	
2010/11	96.0
2011/12 est.	105.3
2012/13 f'cast.	100.0
Pakistan (Nov/Oct)	
2010/11	4.9
2011/12 est.	6.4
2012 f'cast.	6.7
Thailand (Jan/Dec)	
2010/11	20.3
2011/12 est.	20.5
2012 f'cast.	20.7
United States (Aug/July)	
2010/11	7.6
2011/12 est.	5.9
2012 f'cast.	6.4
Vietnam (Jan/Dec)	
2010/11	26.3
2011/12 est.	26.9
2012 f'cast.	27.5
Bangladesh (July/June)	
2010/11	31.7
2011/12 est.	33.7
2012/13 f'cast	34.0
	(33.8)
China (Jan/Dec)	
2010/11	137.0
2011/12 est.	140.5
2012/13 f'cast	142.5
Indonesia (Jan/Dec)	
2010/11	35.5
2011/12 est.	36.4
2012/13 f'cast	36.8
	(37.0)
Philippines (July/June)	
2010/11	10.5
2011/12 est.	11.3
2012/13 f'cast	11.8
	(11.0)

Source: Data from International Grain Council (2013).

Table 4.24 Characteristics of an ideal biomass energy crop: rice

Crop Characteristics	Score
1. Photosynthesis pathway	C3 –
2. Long canopy duration	
3. Perennial or annual (rhizomatous)	Both +
4. No known pests or diseases	
5. Rapid growth in spring to out compete weeds	
6. Sterile; prevent "escape"	
7. Stores carbon in the soil (soil restoration and carbon sequestration tool)	
8. Partitions nutrients back to roots in fall (low fertilizer requirement).	
9. Low nutrient content, that is, <200 mg/MJ nitrogen and sulfur (clean burning)	
10. High water use efficiency	
11. Dry down in the field (zero drying costs)	
12. Good winter standing (harvest when needed; "zero" storage costs)	
13. Utilizes existing farm equipment	+
14. Alternative markets (high quality paper, building materials, and fermentation)	+

Source: Data from various sources and adapted in part from Long (1999) and Heaton et al. (2003).

Table 4.25 The chemical composition of rice measured on a dry weight basis, redesigned with additional columns

Index	Husked Rice %	Polished Rice %	Percentage Value Difference
Protein	8.75	7.92	0.83
Lipids	1.80	0.60	1.2
Carbohydrates	84.80	87.60 (~85% starch)	2.8
Ash	1.3	0.60	0.7

Source: Muzafaroy and Mazhidov (1997) and Juliano (1993).

Table 4.26 The chemical elements of rice

Carbon	34.45%
Hydrogen	5.78%
Nitrogen	2.97%
Oxygen	56.76%
Sulfur	~0.04%

Source: Author.

Table 4.27 Rice straw and husks calorific value and quality as a fuel for domestic use

Sample	Gross Calorific Value (MJ/kg)	As Fuel (Domestic Use)
Rice straw	15.00	Poor
Rice husks	15.50	Poor

Source: Hollander (1992).

measurements of the final output for domestic use from rice husks (as well as rice straw $C_4H_{6.3}O_{2.8}$), their use as a fuel is rated as "poor" (Table 4.27) (Hollander, 1992). On the other hand, rice grains for energy production offer a higher percentage of energy output. This is because of the higher percentage of carbohydrates in general (or starch—converted into sugar) (Table 4.25). The percentage of carbohydrates is one of the important factors in obtaining a quality bio-fuel needed on the market.

The principles of producing bio-fuel from corn compared with rice, or any other type of energy crop, can be partly determined by the amount of carbohydrates (or starch content) within the individual grain. A question can be raised, therefore, as to whether or not the starch content of rice (which can be up to 85% in polished rice) (Table 4.25) would make rice grains a better source of fuel when compared, for example, with corn grains (starch in corn is approximately 61–72%). The answer depends entirely on a number of factors/subfactors within BF and S&T, as well as the overall final cost to the end user.

Regarding rice bran oil (RBO), the possibility of using it for the production of bio-diesel is already part of same research and applications (Lin et al., 2008). RBO is defined as the by-product collected after the grinding process of paddy rice and can be used commercially as a bio-fuel (Barnwal and Sharma, 2005). However, a good deal of work is needed in relation to the development of RBO for SI (spark ignition engines) and CI (compression ignition engines) if this type of fuel is to compete with or replace diesel fuel (Gattamaneni et al., 2008).

To summarize, rice can be a good source of energy as fuel but more work is needed, in particular in areas related to cost and emissions. Ultimately, rice can be used as a fuel only if and when the balance between rice demands for human consumption and the demand for its use as fuel are socially and economically acceptable (Box 4.7).

4.3.10.5 Rice Ash Composition. Figure 4.12 is an illustration of rice ash composition. The elements of rice ash (the plot obtained via FEI Quanta 600 scanning electron microscope). The *x*-axis in the above figure is energy in kev (thousands electron volts), and the *y*-axis is accumulated counts in kcnt (thousand counts).

Box 4.7

RICE HUSK

- Widely available.
- Globally, there are approximately 110 million tons of rice husks per year.
- Rice husk is difficult to ignite.
- The ash from rice husks has a low melting point, and as a result, slagging and fouling is an issue when used as a fuel.
- Energy—approximately 14 GJ/ton.
- Decomposition is slow due to the high content of silica (SiO_2), which can be a positive aspect with regard to storage.
- The ash content is high—it can be up to 26% (the ash is made up of around 85–90% amorphous silica).
- Low bulk density, that is, there are storage and transportation issues.

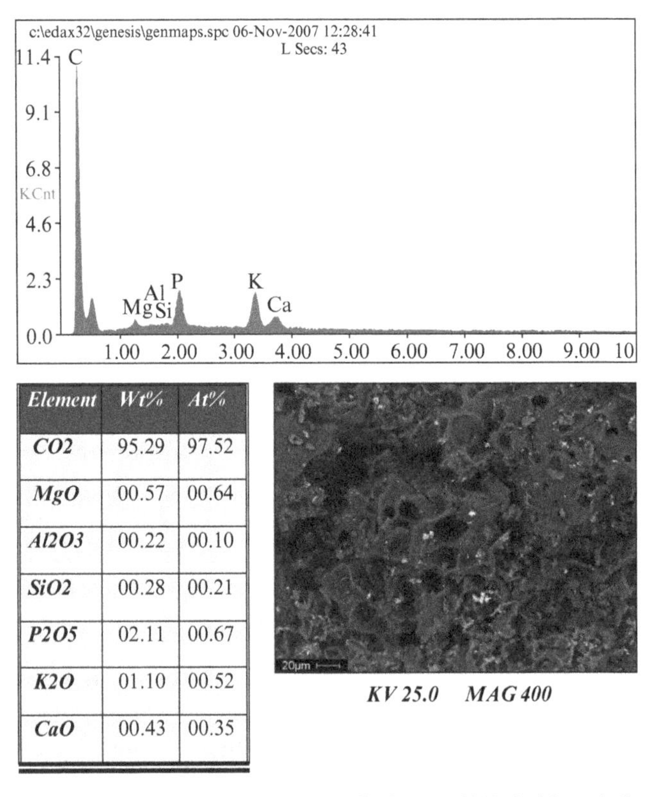

Element	Wt%	At%
CO2	95.29	97.52
MgO	00.57	00.64
Al2O3	00.22	00.10
SiO2	00.28	00.21
P2O5	02.11	00.67
K2O	01.10	00.52
CaO	00.43	00.35

Figure 4.12 Miscanthus ash elements plus a magnified image (400×) of the ash. *Source*: Author.

4.3.11 Barley (*Hordeum vulgare* subsp.)

4.3.11.1 General Introduction. Barley has been found in archeological sites dating back 9000 to 10,000 years within the region known as the Fertile Crescent* (Lorenz and Kulp, 1990). It is possible, therefore, that the present four-row† barley type is descended from wild two-row barely. Different varieties of barely are classified by the number of grain rows on the head, that is, two, four, or six (Gramene.org, 2009).

A native of Asia, barley grows easily in colder climates, producing small spherical grains, and can be cultivated in various locations across the world (e.g., Russia, Australia, the United States, Canada, Argentina, EU-27, Ukraine, and the Middle East). Barley is considered a source of food in some parts of the globe (e.g., Ethiopia and part of the Middle East). In the United Kingdom, it is mostly used for making beer, malt drinks, and whiskey. However, most barley produced in the United Kingdom and in other parts of the world is used as animal feed.

According to FAO, during the year 2005, the amount of barley produced and the size of the cultivated crop ranks fourth worldwide after wheat, corn, and rice. This is still the case during 2012/2013 despite the fact that recent worldwide grain production is low. Present global consumption exceeds grain production during 2012 (Earth Policy Institute, 2013).

When they are clean, barley seeds are usually bright yellow-white (or greyish), depending on the type of barley and the conditions/location they originated from.

In comparison with wheat, barley has a shorter growing season and can achieve better results in poorer environments. This is why barley can be found in drought-prone thin acid soils and at higher altitudes (Farm Direct, Barley, 2008). The germination time is anywhere from 1 to 3 days.

Regardless of the names attached to barley during the season of plantation, such as "winter barley" and "spring barley," there are basically three main types of barley worldwide:

a. Hulled (six-row)
b. Hulled (two-row)
c. Hull-less/hulless.

These are regarded as subgroups of one species, *Hordeum sativum* or *Hordeum vulgare* (Kent and Evers, 1992).

Around 1.1 million hectares is used to cultivate barley in the United Kingdom, which make it the second most farmed arable crop in the United Kingdom after wheat. The United Kingdom produces about 6.5 million tons

* A region in the Middle East extending from the Persian Gulf (Arabian Gulf) to the northern part of Egypt, that is, Iraq, Syria, Lebanon, Jordan, Palestine/Israel, and part of Egypt.
† The four-row has been identified as a loose six-row barley.

yearly. Around 1.5 million tons of this is for export, 2 million tons for the brewing and distilling industries, and 3 million tons mostly for animal feed (UK Agriculture, 2008).

Present world barley production is declining in comparison with previous decades (Karvy Comtrade Ltd., 2008). Previous statistical reports from the United Nation Food and the Agriculture Organization (FAO) reported that there was a decline in the world barley production by a number of top world barley producers. Most of the countries that produced barley on a large commercial scale, such as the United States, have switched to corn production instead of barley, mainly for recent higher prices on the international market. Corn also has a shorter season when compared with barley. Total world production during 2011/2012 = 134,227 thousand metric tons (TMT), while the projected world total production for March 2012/2013 = 130,209 TMT (Table 4.28) (USDA, 2013d).

4.3.11.2 Characteristics. Like wheat, barley is a grass that produces swollen grains suitable for the production of bread. One of the characteristics of barley as a crop is that it has less stiffness than its counterpart wheat, which means it can yield less. If barley's fertilization is increased, then the yield may increase, but because of the lack of stiffness, it tends to fall flat (Farm Direct, Barley,

Table 4.28 World barley producers and world total production: for the year 2011/2013, estimated in thousand metric tons

Countries	2011/2012	2012/2013 February	2012/2013 March
Algeria	1350	1700	1700
Argentina	4500	5500	5500
Australia	8349	7000	7000
Belarus	2013	2100	2100
Canada	7892	8010	8010
China	2500	2600	2600
Ethiopia	1592	1580	1580
EU-27	51,449	54,407	54,407
India	1660	1610	1620
Iran	2900	3400	3400
Kazakhstan	2593	1500	1500
Morocco	2340	1100	1100
Russia	16,938	13,950	13,950
Turkey	7000	5500	5500
Ukraine	9098	6935	6935
Others	8661	8511	8511
Subtotal	130,835	125,403	125,413
United States	3392	4796	4796
World total	134,227	130,199	130,209

Source: Data from USDA (2013c).

2008). The same source also reported that barley is susceptible to mildew.* However, barley is considered as one of the most adaptable crops in soils containing saline and alkaline (WCC, 1999). This means planting barley in saline and alkaline soils can conserve soil and water and consequently improve conditions for other types of crop plantations, especially when barley is used as part of other crops, that is, using the rotational plantation method.†

Processing barley can be an issue in relation to animal feed, food preparations, or biomass fuel requirements that require the separation of the bran from barley flour. Therefore, a certain process is needed. This is because barley bran (unlike wheat) shatters quickly during the milling process (Jadhav et al., 1998). For this reason, barley flour usually contains a higher ash percentage and is consequently, darker in color.

The hardness of the barley grain may determine the possibilities of its use and the type of processing required within the barley industry. Some of the initial breeding and selection of various types of barley was conducted for the malting process. This selection process chose softer grains rather than harder endosperm textures. The long tradition of breeding softer barely grains for the purpose of malting resulted in producing grains that are much softer than wheat grains. To use barley as a source of biomass energy, such as within a co-firing environment, harder grains are the obvious option. This is because harder grains produce better output during the milling process, and therefore form an important part of the co-firing system.

In relation to the characteristics for an ideal biomass energy crop, barley scored 10, 11, 13, and 14 points (Table 4.29).

4.3.11.3 Composition. The main components of the barley grain are starch, protein, and fiber. Barley contains about 55–60% starch, which is relatively close to certain types of corn. Analysis of barley (barley ripe seeds and barley meal) is described in Gmelin (1817), where a variety of barley compositions are detailed (Table 4.30).

Other elements, apart from protein and β-glucan, such as minerals and lipids, are reported to be around 2% (Table 4.31). Analysis of crushed dried barley straw at a temperature of 55°C reported that of the total weight, 37.5% is made up of cellulose, 36.1% hemicelluloses, 16.6% lignin, 4.8% ash (silica 1.9%), and 2.5% wax (Sun et al., 2002). Compared with corn, barley has less fat and, as with any energy crop plant (or any plant for that matter), the mineral content may depend on the growing season, soil zone, soil type, and soil fertility (Kulp and Ponte, 2000).

The "hull-less" type of barley has more starch than the "hulled" variety but with a lower fiber content. However, compared with corn, the total output of

* Powdery mildew is a type of fungus that attack leaves and other parts of the plant. It is a common disease among annual and perennial grasses, particularly barley and wheat.

† The rotational plantation method is the plantation of different crops on the same land from season to season.

Table 4.29 Characteristics of an ideal biomass energy crop: barley

Crop Characteristics	Score
1. Photosynthesis pathway	C3 –
2. Long canopy duration	
3. Perennial or annual (no need for annual tillage or planting)	Annual –
4. No known pests or diseases	
5. Rapid growth in spring to out compete weeds	
6. Sterile; prevent "escape"	
7. Stores carbon in the soil (soil restoration and carbon sequestration tool)	
8. Partitions nutrients back to roots in fall (low fertiliser requirement).	
9. Low nutrient content, that is, <200 mg/MJ nitrogen and sulfur (clean burning)	
10. High water use efficiency	+
11. Dry down in the field ("zero" drying costs)	+
12. Good winter standing (harvest when needed; "zero" storage costs)	
13. Utilizes existing farm equipment	+
14. Alternative markets (high quality paper, building materials, and fermentation)	+

Source: Data from various sources and adapted in part from Long (1999) and Heaton et al. (2003).

Table 4.30 Composition of barley according to Einhof

Ripe Seeds		Barley-meal	
Meal	70.05	Starch	67.18
Husk	18.75	Fibrous matter (gluten, starch, and lignin)	7.29
Moisture	11.20	Gum	4.62
	100.00	Sugar	5.21
		Gluten	3.52
		Albumen	1.15
		Phosphate of lime with albumen	0.24
		Moisture	9.37
		Loss	1.42
		Total	100.00

Source: Adapted from Gmelin (1817).

Table 4.31 Barley grain chemical composition

Element	Starch %	Protein %	β-glucan %	Lipid %	Minerals %
Average value	55–60	13	6	2	2–3

Source: Adapted from Baik and Ullrich (2008).

Table 4.32 The chemical elements of barley

Carbon	41.81%
Hydrogen	6.37%
Nitrogen	2.59%
Oxygen	49.20%
Sulfur	~0.03%

Source: Author.

energy and the percentage of fiber is around 95% of the total energy of corn but higher in fiber (Ingredients101.com 2008).

Finally, carbon and nitrogen, in a dried barley sample, form around 42% and 2.59%, respectively, of the total mass (Table 4.32).

4.3.11.4 Suitability as a Fuel. Interest in obtaining fuel from barley is gathering momentum in recent years, despite the present decline in barley production worldwide. In fact, one of the main sources for the production of ethanol in Europe is largely dependent on barley, while the usage of barley in the United States for the same purpose is close to zero among the 87 or so operating ethanol plants (Hicks et al., 2005). The reason for this is attributed to the present popularity and demand for corn. Furthermore, hulled barley is more difficult to process in an ethanol plant compared with corn. Some of the attributed difficulties can be summarized in the following points:

1. Higher viscosity during fermentation.
2. Higher fiber content
3. Lower starch content
4. Lower ethanol yield
5. The initial processing cost of abrasive hulled barley
6. Distillers dried grains may not be suitable to feed monogastric animals.

The net calorific value of barley straw calculated on a dry basis with 15% moisture is 17.5 MJ/kg (FAO Forestry Department, UBET, 2004).

The figure is slightly more than the estimated calorific value of miscanthus (see miscanthus, Section 4.3.9). However, the same report did not mention how much energy (or the level of cost) was used to convert it into dry matter (around 15% of moisture is left as a standard).

According to laboratory data obtained during the work on the present sample, energy from a dried pot barley sample is 15,740 J/g at a moisture of 11% (Table 4.33), which is close to those of other energy crops, such as wheat and rice grains (see Chapter 9, Sections 9.3.2.19, 9.3.2.20, 9.3.2.29, and 9.3.2.30).

In another report concerning barley grains as a fuel, a comparison is made with wood pellets and wood chips. NIFES Consulting Group stated: "Barley at 15% moisture content has a net calorific value of 4.2MWh/te (15.1 GJ/te)

Table 4.33 Energy content of a dried pot barley sample

Sample (Dry Matter)	Energy (H) J/g	Moisture %
Pot Barley	15,740	11

Source: Author.

Box 4.8

BIO-ENERGY CROPS

Among bioenergy crops, rapeseed is the most widespread accounting for around 80–85% of the total energy crop area. Low portions are covered by sunflower, maize, rye, wheat and sugar beet. A high concentration of these areas can be found in Germany (almost 60% of the total EU-27 energy crops area), France (more than 25%), the UK (8%). Large areas can also be found in Poland, Czech Republic, Sweden, Spain and Italy.

The main energy crops cultivated for solid biofuels in EU27 are: miscanthus in UK, Germany, Spain and Portugal; willow in UK, Sweden and Germany; reed canary grass in Finland and Sweden; poplar in Italy and Spain.

Panoutsou et al. (2011)

compared to wood-pellets at 12% moisture content and 4.6MWh/te (16.6GJ/te) and wood-chips at 3.33MWh/te (12GJ/te) at 35% moisture content" (NIFES Consulting Group, 2008) (Box 4.8).

4.3.11.5 Pot Barley Ash Composition. Figure 4.13 is an illustration of pot barley ash composition.

The elements of pot barley ash (the plot obtained via *FEI Quanta 600 Scanning Electron Microscope*). The *x*-axis in Figure 4.13 is energy in kev (thousands electron volts) and the *y*-axis is accumulated counts in kcnt (thousand counts).

4.3.12 Sunflower (*Helianthus annuus*)

4.3.12.1 General Introduction. The sunflower name originates from the Greek language using the Greek words *helios*, meaning "sun," and *anthos*, meaning "flower." The origin of the sunflower is reportedly in South America, where the wild native sunflower was first discovered by Francisco Pizarro in Tahuantinsuyo, Peru. However, the National Sunflower Association, and a number of other sources, attribute the origin to North America (National Sunflower Association, 2008). Sunflowers were introduced to Europe later for ornamental purposes by the Spanish during the sixteenth century. Reportedly, the commercialization of the sunflower started in Russia (NDSU, 2007).

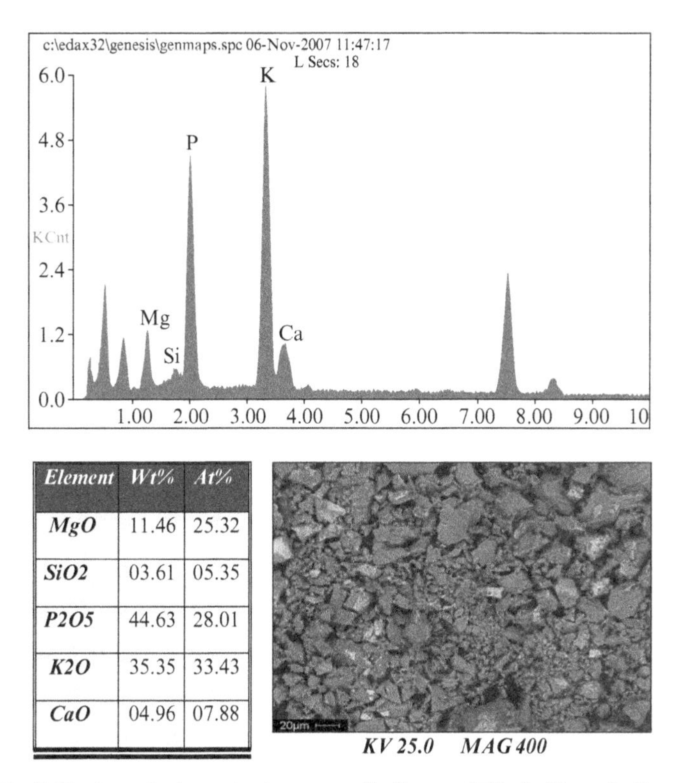

Element	*Wt%*	*At%*
MgO	11.46	25.32
SiO2	03.61	05.35
P2O5	44.63	28.01
K2O	35.35	33.43
CaO	04.96	07.88

Figure 4.13 Pot barley ash elements plus a magnified image (400×) of the ash. *Source*: Author.

As members of the genus *Helianthus*, many of them are perennials but on occasion they are referred to as *Helianthus annuus*, which is the most common scientific name for the sunflower annual plant. The sunflower family (*Asteraceae*) is one of the largest known plant families (Wayne's Word, 2000). It has also been reported by the same source that Professor James L. Reveal of the University of Maryland said that the family contains around 1550 genera and 24,000 species. Accordingly, the sunflower family comes third in the ranking of species numbers after the orchid family (*Orchidaceae*, approximately 20,000 species) and the legume family (*Fabaceae*, approximately 18,000 species). The sunflower family is made up from a vast number of different species, consisting of annuals, perennials, stem succulents, vines, shrubs, and trees (Wayne's Word, 2000).

There are a large number of sunflower varieties, including Soraya, Strawberry Blonde, Sunny Hybrid, Taiyo, Tarahumara, Teddy Bear, Titan, Valentine, Velvet Queen, American Giant Hybrid, Arikara, Autumn Beauty, Aztec Sun, Black Oil, Dwarf Sunspot, Evening Sun, Giant Primrose, Indian Blanket Hybrid, Irish Eyes, Italian White, Kong Hybrid, Large Gray Stripe, Lemon

Figure 4.14 A black or striped sunflower seed is sometimes referred to as an achene. The edible part of the dehulled seed is called the sunflower kernel. *Source*: Author.

Queen, Mammoth Sunflower, Mongolian Giant, Orange Sun, Red Sun, Ring of Fire, and Rostov.

Commercially, there are two main types of sunflower seeds; these are black seeds and striped seeds (Fig. 4.14). In commercial terms, the classification of seeds may depend largely on the pattern of the husks that is, the black seeds are referred to as "black oil sunflower seeds," or "oilseed." If the seeds are striped, then they are called "striped sunflower seeds," or "nonoilseed" or "stripers." The black seed, or oilseed, as the name suggests, is very high in oil and used mostly for oil extraction and meal, as well as being a popular option for bird feeders. The striped seeds, or nonoilseeds, are mostly used in food production. In addition to these two types, there is a third sunflower seed, which that is whitish in hue (white). No actual report of the commercialization of this type has been found.

It is estimated that there are approximately 25 million tons of sunflower seed globally. In the same market, oil from sunflower seeds is approximately 10 million tons (Commodity Online, 2008).

The present top world producers of sunflower seeds, according to the FAO statistics, are Russia followed by Ukraine (Table 4.34). Despite the small quantity of sunflower seeds exported by the United Kingdom, the United Kingdom is listed as an actual exporter during 2006. However, the United Kingdom imported around 51,257,565 kg of sunflower seeds in comparison with 404,331 kg exported during the same year (Table 4.35).

During the year 2003/4, the world's largest importers of sunflower seeds were the Netherlands (approximately 17.4% of world imports), Turkey (17.4%), and Spain (10.2%) (Commodity Online, 2008).

4.3.12.2 Characteristics. One of the main reasons for the sunflower's popularity in many parts of the world is its ability to easily manage drought

Table 4.34 Sunflower seeds production 2011/2012 in thousand metric tons (TMT)

Country	TMT/Year
Russia	9,500
Ukraine	9,359
Argentina	3,720
China	1,720
India	620
USA	925
Turkey	940
European Union	8,317
Other Europe	400
Other	3,981
Total	39,473

Source: National Sunflower Association (2013).

Table 4.35 Example of the U.K. export and import of sunflower seeds during 2006

Flow	Quantity in kg
Import	51,257,565
Export	404,331

Source: Adapted and redesigned from UNCTAD (2008).

conditions in comparison with other types of crops (Agri Publication, 2008). Sunflower seeds are also considered to be one of the top sources of vegetable oil. Measuring the content of oil from a single sunflower seed (with hull) shows that it contains approximately between 40% and 50%. Using kernel only, the content is approximately 50–60% (Ienica.net, 2003).

Classifications of sunflower oil are made according to the content of oleic acid ($C_{18}H_{34}O_2$). For this reason, the trading in sunflower oil can be described as:

1. NuSun
2. High oleic
3. Linoleic (regular olieic).

Type 1 (NuSun) contains approximately 65% oleic acid, and type number 2 (high oleic) may contain around 80% oleic acid, while number 3 (linoleic) usually contains around 69% oleic acid (Soyatech, 2008). There is a similarity with regard to the content of linoleic acid between oil made from sunflower and corn or soybean as all of them contain high levels of this element. Oil with a high level of oleic acid is more suitable for use in high temperature environments, as this element reduces the possibility of oxidation at a lower temperature. Such characteristics can give sunflower oil a longer shelf life and make it easy to be developed for use in combustible engines as in the case of diesel oil.

Reports concerning developments in Olefin metathesis create new possibilities for the conversion of vegetable oils, such as sunflower oil, into varieties of useful different applications (Marvey, 2008). These applications can range from industrial use within the field of combustion engines to all aspects of power generating.

In relation to the characteristics of an ideal biomass energy crop, sunflower seed scored 5, 10, 11, 13, and 14 points (Table 4.36) (Box 4.9).

Table 4.36 Characteristics of an ideal biomass energy crop: sunflower seed

Crop Characteristics	Score
1. Photosynthesis pathway	C3 –
2. Long canopy duration	
3. Perennial/biennial/annual (no need for annual tillage or planting)	Annual –
4. No known pests or diseases	
5. Rapid growth in spring to out compete weeds	+
6. Sterile; prevent "escape"	
7. Stores carbon in the soil (soil restoration and carbon sequestration tool)	
8. Partitions nutrients back to roots in fall (low fertilizer requirement).	
9. Low nutrient content, that is, <200 mg/MJ nitrogen and sulfur (clean burning)	
10. High water use efficiency	+
11. Dry down in the field ("zero" drying costs)	+
12. Good winter standing (harvest when needed; "zero" storage costs)	
13. Utilizes existing farm equipment	+
14. Alternative markets (high quality paper, building materials, and fermentation).	+

Source: Data from various sources and adapted in part from Long (1999) and Heaton et al. (2003).

Box 4.9

BIOMASS AND ELECTRICITY SUPPLY IN INDIA

There are around 400 million people in India without regular access to an electricity supply.

There are around 855 million people still reliant on the traditional use of biomass materials, as a way for cooking their food.

Table 4.37 The composition of sunflower seed

Constituent	Percentage %
Protein	20.8
Lipid	54.8
Carbohydrates	18.4
Ash	3.9

Source: Gopalan et al. (1982) Nutritive value of Indian foods, cited in Department of Agriculture, Sri Lanka (2008).

4.3.12.3 Composition. The chemical composition of sunflower seed can be compared with that of the groundnut composition. This clearly depends on its genetic makeup along with environmental conditions (Department of Agriculture, Sri Lanka, 2008). The main elements of the chemical composition are made up of acids under the following names:

1. Palmitic ($C_{16}H_{32}O_2$)
2. Stearic ($C_{18}H_{36}O_2$)
3. Oleic ($C_{18}H_{34}O_2$)
4. Linoleic ($C_{18}H_{32}O_2$).

Sunflower oil, like most other vegetable oils, contains a high percentage of unsaturated fatty acids. It contains more than other types of seeds, such as peanut and soybean (Table 4.37) (Department of Agriculture, Sri Lanka, 2008). For example, the concentration of lipids within different parts of the sunflower seed can vary considerably. The hull contains fewer lipids than the rest of the seed, while the kernel usually contains the highest percentage, around 87%. The embryo itself may have around 74% lipids (Department of Agriculture, Sri Lanka, 2008).

Sunflower seed oil as a nonvolatile oil is made up of over 90% oleic and linoleic acids, that is, approximately 25.1% oleic, approximately 66.2% linoleic, plus other contents, such as protein, making up about 20–30% (Ienica.net, 2003).

Different sampling methods for sunflower seeds may give a slightly different data readings. This is illustrated from the measurement of the same chemical composition of sunflower seeds obtained by the Laboratory of Oils (TEL of Thessaloninki) using mass spectroscopy (Table 4.38) (Triandafyllis et al., 2003).

Laboratory tests carried out on the black and the striped sunflower seeds show that the amount of carbon and nitrogen in the black seed is approximately 65% and 6%, respectively, while the striped seed contains approximately 67% carbon and 7% nitrogen (Table 4.39 and Table 4.40).

Table 4.38 The four main elements with their percentage values of the sunflower oil chemical makeup

Elements	%
Palmitic acid (C_{16} H_{32} O_2)	8.5
Stearic acid (C_{18} H_{36} O_2)	5.3
Oleic acid (C_{18} H_{34} O_2)	21.5
Linoleic acid (C_{18} H_{32} O_2)	65

Source: Triandafyllis et al. (2003).

Table 4.39 The chemical elements of black sunflower seed

Carbon	64.68%
Hydrogen	9.33%
Nitrogen	5.60%
Oxygen	20.33%
Sulfur	~0.06%

Source: Author.

Table 4.40 The chemical elements of striped sunflower seed

Carbon	67.31%
Hydrogen	9.48%
Nitrogen	6.80%
Oxygen	16.38%
Sulfur	~0.03%

Source: Author.

4.3.12.4 Suitability as a Fuel. Sunflower oil is still not yet fully commercially developed in order to be used as a fuel on its own. This can be attributed mainly to the cost and possibly to international and/or local supplies (Ienica.net, 2003). The delay can be related to the chemical composition of the oil itself. The range of carbon chain is C16–C18, which may not make them reliable on industrial basis, in comparison with other types of oils, for example, epoxy and hydroxyl (Marvey et al., 2008).

Additional research and development, therefore, can be useful in furthering the use of sunflower oil within the industrial energy fields. Sunflower oil that contains high oleic hybrids is, however, suitable for biodiesel production according to the standards in Argentina and the United States.

Table 4.41 Physical and chemical properties of sunflower oil and diesel

Physical Factors and Elements	Sunflower Seed Oil	Diesel
Density at 5°C (kg/m^3)	923	848
Density at 35°C (kg/m^3)	912	635
Viscosity 40°C (cSt)	32.6	2.7
Viscosity 100°C (cSt)	7.7	1.2
%S	0.003	0.07
%C	79.3	86.6
%H	11.1	12.3
%O	9.57	1.03
HHV (kcal/kg)	9,414	10,735
LHV (kcal/kg)	8,851	10,112

Source: Adapted from Reinhold (2002).

Table 4.42 Chemical property factors and higher heating values (HHVs) of biodiesel and petrodiesel fuels

Factors listing	Factors	Biodiesel (Methyl Ester)	Diesel
1	Ash (wt%)	0.002–0.036	0.006–0.010
2	Sulfur (wt%)	0.006–0.020	0.020–0.050
3	Nitrogen (wt%)	0.002–0.007	0.0001–0.003
4	Aromatics (vol%)	0	28–38
5	Iodine number	65–156	0
6	HHV (MJ/kg)	39.2–40.6	45.1–45.6

Source: Adapted from Demirbas (2007).

When a comparison is made between sunflower oil and diesel oil, the density and viscosity of both (which form important factors related to the use of oil as a fuel) can be problematic (Table 4.41). Factors listing 1–3 in Table 4.42 are close in values to biodiesel and diesel fuels, while factors listing from 4 to 6 clearly mark the difference between these two types of bio and fossil fuels (Demirbas, 2007).

To improve sunflower oil so that it can be used more in industrial applications, particularly as a fuel, a number of factors have to be considered, mostly related to farming aspects and the environment in general. Regardless of genetic makeup, the environment have a significant effect on growth and seed production. This kind of effect may include the chemical composition of the fatty acids (oil quality, e.g., higher or lower oleic acid), as well as plant yields in general. Factors, therefore, which may influence sunflower oil quality (Flagella et al., 2006), especially in the case of the production of sunflower oil for industrial use, that is, a higher percentage of oleic acid is needed for more efficient oil oxidization (Bondioli and Folegatti, 1996), as is the case in the production of biodiesel. These factors may include the following:

Box 4.10

STRAIGHT VEGETABLE OIL (SVO)

The long-term effect of using SVO in modern diesel engines that are equipped with catalytic converters or filter traps is also a matter of concern. Buildup of fuel in the lubricant is more significant in these engines—even for petroleum diesel—and would likely be severe with SVO. In general, these systems were not originally designed with SVO in mind and can be seriously damaged or poisoned by out-of-spec or contaminated fuel.

USDE (2013)

1. Genotype
2. Temperature
3. Seasonal plantation
4. Irrigation.

Finally, in a co-firing system, sunflower seeds can be used directly or indirectly. This can be done by using sunflower seed hulls (or the whole seed, if this practice is economically viable) in the form of pellets or a dried crushed material. In a research experiment reported by the National Sunflower Association, 50 tons of sunflower hulls were co-fired with coal (a study completed at the Energy and Environmental Research Centre, University of North Dakota) in a ratio of 75% coal and 25% sunflower hulls. In this study, burning sunflower hulls along with coal helped to reduce airborne emissions of sulfates and nitrates. In addition, the same report concluded that co-firing hulls with coal requires no additional processing (Box 4.10).

4.3.12.5 Sunflower Seeds Ash Composition. Figure 4.15 and Figure 4.16 are illustrations of sunflower seeds composition.

The elements of sunflower seeds ash (the plot obtained via FEI Quanta 600 scanning electron microscope). The *x*-axis in the above figure is energy in kev (thousands electron volts) and the *y*-axis is accumulated counts in kcnt (thousand counts).

The elements of sunflower black seeds ash (the plot obtained via FEI Quanta 600 scanning electron microscope). The *x*-axis in the above figure is energy in kev (thousands electron volts), and the *y*-axis is accumulated counts in kcnt (thousand counts).

4.3.13 Niger Seed (*Guizotia abyssinica*)

4.3.13.1 General Introduction. Niger seed is an annual/biennial herbaceous herb cultivated mostly for its oil and for the production of birds' feeding stock (Fig. 4.17). The plant has various names, such as Blackseed, Ramtilla,

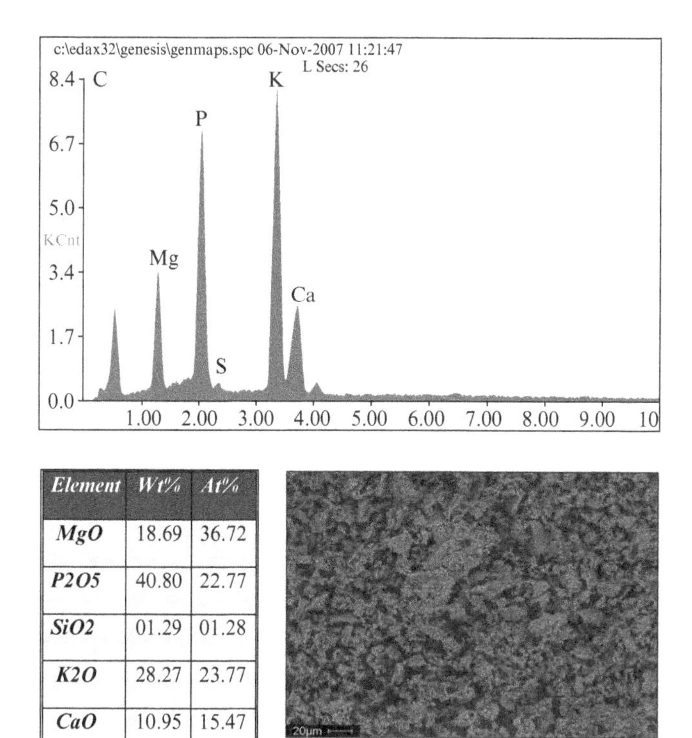

Element	Wt%	At%
MgO	18.69	36.72
P2O5	40.80	22.77
SiO2	01.29	01.28
K2O	28.27	23.77
CaO	10.95	15.47

Figure 4.15 Striped sunflower seed ash elements plus a magnified image (400×) of the ash. *Source*: Author.

Noog, Inga seed, and Niger (Nyjer) seed. It is believed that the origin of niger seed is Africa, where first domestication began somewhere in an area that stretched from Ethiopia to Malawi. However, Ethiopia is the country where many historians and researchers believe that the first cultivation of niger seed originated (Quinn and Myers, 2002). The species of *Guizotia abyssinca* are closely related to the *Guizotia schimperi* species. In fact, the *Guizotia schimperi* is the progenitor of *Guizotia abyssinicia*. However, other reports consider *Guizotia scabra* to be a subspecies (van der Vossen and Mkamilo, 2006).

Niger seed is very adaptable to a variety of soils, but the growth can be poor when planted in a gravely and/or light sandy soil, even though it can be suitable to grow on acid soils with very low fertility. The plant can grow at a high altitude (e.g., more than 2500 m in East Africa) or on hilly slopes with low fertility soils (usually caused by erosion). Field trials have shown that niger seed can produce a better yield when cultivated on a lower ground rather than those planted at higher altitude (Duke, 1983).

Germination usually begins in about 2days after sowing. Flowering happens between the months of June and August, while the seeds ripen between August and September/October.

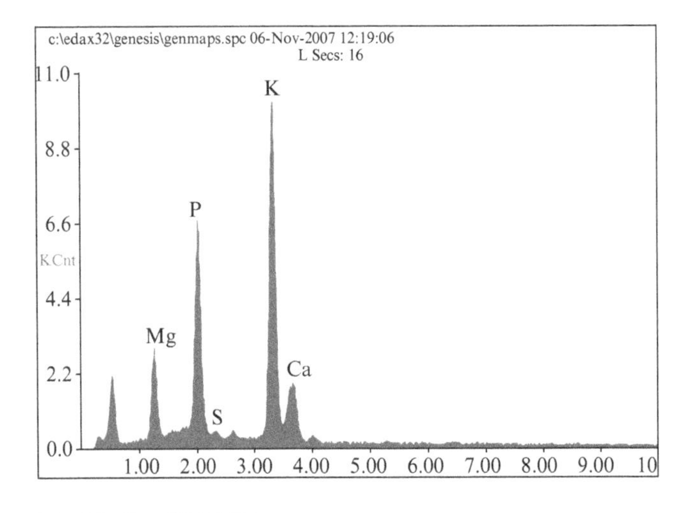

Element	*Wt%*	*At%*
MgO	16.35	33.55
P2O5	38.93	22.69
SiO2	01.17	01.21
K2O	36.31	31.88
CaO	07.24	10.68

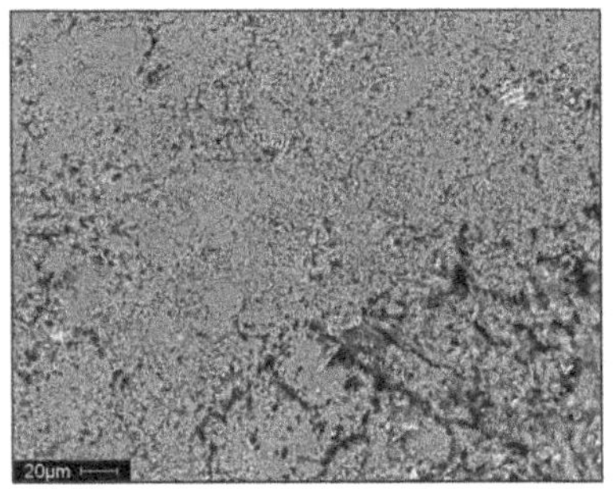

Figure 4.16 Black sunflower seed ash elements plus a magnified image (400×) of the ash. *Source*: Author.

One of the advantages niger seed has over similar types of crops is that it has the ability to tolerate maritime exposure.

The top producers of niger seed are India (southern part) and Ethiopia. India's production ranges between 160,000 to 250,000 MT/year, and from this figure around 23,600 MT is exported. The Ethiopian type of niger seed contains more linoleic acid and therefore is considered to be superior in quality to the Indian type (Roecklein and Leung, 1987).

Imported niger seed must be heat sterilized first as a protective measure from pest, weeds, and other types of contamination, such as the presence of dodder seeds. This preventative procedure increases the cost of purchasing on the international market, in comparison to some other types of energy crops. This extra cost can make the overall cost less attractive for the bio-fuel industries and power generating companies.

4.3.13.2 Characteristics. Niger seed is a stout, erect branched plant with an annual growth from 0.5 to up to 2 m. Reportedly, it can tolerate disease,

Figure 4.17 Sample of dry niger seeds. *Source*: Author.

grazing, insects, laterites, poor soil, and slopes. It is a good crop for rotation with corn or wheat. To ensure a better harvest of seed, it should be rogued off of type plants 3 months after sowing and before the plant starts flowering (Duke, 1983).

Niger seed flowers need to be pollinated by insects, as the plant is self-sterile. The plant flowers themselves are hermaphrodite, that is with male and female organs (Plants for a Future: Database Search Results, 2008).

Niger seed from Ethiopia has more varieties than the counterpart from India, thus three different Ethiopian types exist, according to the type of land grown on and/or length of the maturity period (van der Vossen and Mkamilo, 2006). These types are the following:

1. Abat noug (late maturing; highland type)
2. Mesno noug (short season type)
3. Bunegne noug (lowland type).

After emergence, the Ethiopian type matures within a period, which can range from 120 to 180 days. The Indian type mature in a much shorter period, ranging from around 75 to 120 days.

As a C3 plant type, niger seed scored in 4, 5, 10, 11,13, and 14 points (Table 4.43) according to the factors related to the characteristics of an ideal biomass energy crop.

Table 4.43 Characteristics of an ideal biomass energy crop: niger seed

Crop Characteristics	Score
1. Photosynthesis pathway	C3 –
2. Long canopy duration	
3. Perennial/biennial/annual/biannual (no need for annual tillage or planting)	Annual –
4. No known pests or diseases/tolerate disease	+
5. Rapid growth in spring to out compete weeds	+
6. Sterile; prevent "escape"	
7. Stores carbon in the soil (soil restoration and carbon sequestration tool)	
8. Partitions nutrients back to roots in fall (low fertilizer requirement).	
9. Low nutrient content that is <200 mg/MJ nitrogen and sulfur (clean burning)	
10. High water use efficiency	+
11. Dry down in the field ("zero" drying costs)	+
12. Good winter standing (harvest when needed; "zero" storage costs)	
13. Utilizes existing farm equipment	+
14. Alternative markets (high quality paper, building materials and fermentation)	+

Source: Data from various sources and adapted in part from Long (1999) and Heaton et al. (2003).

Table 4.44 The chemical makeup of niger seed/100 g

Fiber	Calories	H_2O	Protein	Fat	Carbohydrate	Ash
13.5	483	6.2–7.8	17.3–19.4	31.3–33.9	34.2–39.7	1.8–8.4

Source: Adapted in part from Duke (1983).

4.3.13.3 Composition. According to Hager's Handbook (cited at Duke, 1983), the oil content in niger seed is made up from glycerides of oleic, linoleic, palmitic, myristic, and physetolic acids to the value of 35–40%. Table 4.44 contains data related to 100 g of niger seed, where fat is counted for approximately 33% of the content. Analyzing the fatty acid, the linoleic acid is approximately 53% (Table 4.45) which is much less than the linoleic acid of the sunflower seed oil, estimated at around 65% (Table 4.38).

In a different test, the saturated acid (palmitic acid) of an individual seed was on average approximately 13%, oleic acid averaged 35%, and linoleic acid averaged 53%. The overall oil content is approximately 40%, with protein forming around 20% (Roecklein and Leung, 1987), which is close to the figures in Table 4.46. The percentage of carbon and nitrogen in an individual seed is 59% and 6%, respectively (Table 4.46).

4.3.13.4 Suitability as a Fuel. The extraction of oil from niger seed gives a relatively high yield, together with its close compositional similarities to

Table 4.45 The elements with their percentage values of the niger seed fatty acid composition

Elements	%
Palmitic acid ($C_{16}H_{32}O_2$)	5.0–8.4
Stearic acid ($C_{18}H_{36}O_2$)	2.0–4.9
Oleic acid ($C_{18}H_{34}O_2$)	31.1–38.9
Linoleic acid ($C_{18}H_{32}O_2$)	51.6–54.3
Myristic acid ($CH_3(CH_2)_{12}COOH$)	1.7–3.4

Source: From Duke (1983).

Table 4.46 The chemical elements of niger seed

Carbon	58.67%
Hydrogen	8.06%
Nitrogen	6.04%
Oxygen	27.19%
Sulfur	~0.04%

Source: Author.

Table 4.47 A comparison of the amount of oil obtained in cold pressed extraction from one kilogram of seed

Species	Oil Extracted (mL)	Percent Extracted (First Press)
Canola	350	83.3
Niger seed	310	86.0
Camelina	300	81.1
Linseed	275	88.0
Cramble (in hull)	225	72.5

Source: Adapted in part from Francis and Campbell (2003).

sunflower oil, which make niger seed oil a good source for the production of bio-fuel. To illustrate the above, during cold pressed oil extraction, niger seed came out at the top in comparison to camelina, linseed, and cramble (with the exception of canola). In fact the percentage of extraction for niger seed is even higher than canola. The oil itself was obtained from 1 kg during the first press only (Table 4.47) (Francis and Campbell, 2003).

Unlike sunflower seeds, niger seed has less potential when it comes to its use as a bio-fuel, diesel substitute or as part of a co-firing method. The reason for this can be attributed to the production process in general, that is, niger seed yields globally are too low to be able to meet the possibility of higher demand. Present day production cannot meet the commercial demands for the production of bio-fuels (Duke, 1983). Competition also exists on the

Table 4.48 Niger seed calorific value: dray matter with 7% moisture

Sample	Energy (H) J/g	Moisture %
Niger seed	25,918	7

Source: Author.

Box 4.11

ENERGY CROPS UK

In order to assess the potential and constraints of growing perennial crops for energy production in the UK we must undertake a full literature review of the social, environmental and economic barriers and opportunities. This should include data collated from scientific analysis in previous studies on energy crops, including impacts on:

a) Social and environmental:
   i. Food production
   ii. Biodiversity
   iii. Water use
   iv. GHG emissions
   v. Educational
b) Economic
c) Legislative
d) Technical

Scenarios of likely or potential uptake can then be developed on this basis.

NNFCC (2012)

international market for niger seed to be sold as feed for livestock. The present niger seed market is mostly related to birdseed. Consequently, unless new, much higher levels of production takes place, the commercial development of niger seed for the purpose of using it as a fuel is still far from certain, despite the high potential the seeds have in regard to their chemical composition (Table 4.45) and energy output (Table 4.48) (Box 4.11).

4.3.13.5 Niger Seed Ash Composition. Figure 4.18 is an illustration of niger seed ash composition.

The elements of niger seed ash (the plot obtained via FEI Quanta 600 scanning electron microscope). The x-axis in Figure 4.18 is energy in kev (thousands electron volts) and the y-axis is accumulated counts in kcnt (thousand counts).

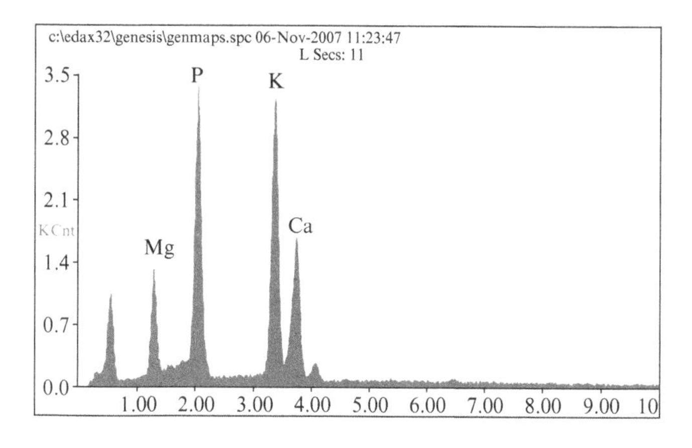

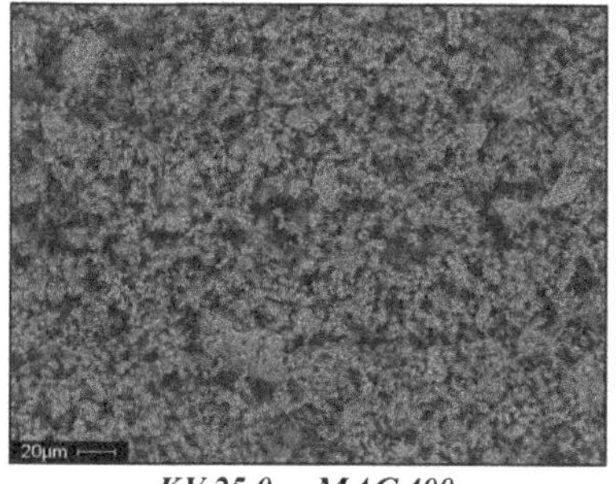

Element	*Wt%*	*At%*
MgO	16.26	32.01
P2O5	42.57	23.81
K2O	24.63	20.76
CaO	16.54	23.42

Figure 4.18 Niger seed ash elements plus a magnified image (400×) of the ash. *Source*: Author.

4.3.14 Rapeseed (*Brassica napus*)

4.3.14.1 General Introduction. The period in which rapeseed was used by humans is not accurately recorded. There are no reliable data by historians or archaeologists that may provide information related to the use of the seeds and/or the oil extracted from them. However, rapeseed was first mentioned around 2000 BC. The rapeseed oil itself was used for illuminations, as archaeologists in India discovered (Cyberlipid Centre, 2008).

Rapeseed is an annual winter or spring herbaceous crop, largely harvested for the production of oil and meal. This type of crop differs from a number of edible vegetable oil source plants in that rapeseed originates from various species that belong to the mustard family (cabbage, broccoli, cauliflower, and turnip). Examples of these species are *Brassica. campestris* or turnip rape, *Brassica. juncea*, and brown mustard or *Brassica napu*. What this means is that the name rapeseed is usually used to refer to a number of species, which may differ in their chemical makeup but can be similar in their appearance (Erickson and Bassin, undated). During the 1940s, rapeseed was widely grown for its industrial oil. In the 1960s, there were breeding efforts that eventually led to the removal of two compounds, erucic acid ($C_{22}H_{42}O_2$) and glucosinolates,

Figure 4.19 Dry rapeseed (seeds only) used during laboratory tests. *Source*: Author.

consequently changing the plant to an edible oilseed named canola (Raymer, 2002). Rapeseed can be referred to by other names as well, such as rape, summer turnip, oilseed rape, and rapa. It germinates within a period of approximately 7 days, and the plant grows between 3 and 5 ft high, producing bright yellow flowers with four petals. After self-fecundation, the fruits of the plant (or siliques) form shortly afterward. The seeds are small, round, black or brown-black, and can measure from 1/32 to 3/32 of an inch in diameter (90,000–150,000/lb) (Fig. 4.19). The plant roots form a deep taproot. In addition to this main root, a root system close to the soil surface grows, which can form protection against soil erosions (Boland, 2008).

From the 1950s to the year 2006, there was a clear and noticeable increase in the production of rapeseed, worldwide. The most noticeable period is from 1985 to 1995, where output was high, that is, from 19.2 to 34.2 MMT (Table 4.49).

The top three world rapeseed producers up to March 2013 are China (~13.50 MMT) followed by Canada (~13.31 MMT), and India (~6.80 MMT) (Table 4.50).

According to the HGCA Knowledge Centre, rapeseed accounts roughly for 1/8th of total world production of oilseed. Rapeseed contains around 40% oil, while the residual percentage is made up of rape meal and waste. The same source also mentioned that 25 members of the European Union make up around 32% of world production. China is close to 26%, Canada 20%, and India 14%. The percentage of rapeseed oil in relation to the world's total

Table 4.49 Worldwide rapeseed production from 1950s to 2006/2007

Year	Million Metric Tons
1950s	3.5
1965	5.2
1975	8.8
1985	19.2
1995	34.2
2006/7	47.6

Source: Adapted from Soyatech (2008) and FAO (2013b).

Table 4.50 Estimated figures for nine rapeseed producers, 2013

Country	Million Metric Tons
China	13.50
Canada	13.31
India	6.80
Germany	4.84
France	5.43
United Kingdom	2.60
Poland	1.88
Australia	3.09
World Total	60.63

Source: Data from USDA (2013e).

vegetable oil need is around 18%. Rapeseed oil used globally during 2006/2007 was about 17.9 Mt (HGCA Knowledge Centre, 2008).

4.3.14.2 Characteristics. Rapeseed is relatively tolerant to saline soil and low temperature. It does best when planted in a drained soil with its ideal pH being somewhere between 5.5 and 8.3 (Sattell et al., 1998). As mentioned earlier, rapeseed can be widely grown in various parts of the world. Presently, Canada and Europe are home to most of the world's rapeseed cultivation.

Rapeseed can be classified into two types with relation to its fatty acid composition:

A. High erucic acid rapeseed oil (HEAR) approximately 40% to 50% erucic acid.
B. Low low erucic acid rapeseed oil (LEAR) approximately 0–2% erucic acid.

The term "industrial rapeseed" usually refers to a type of rapeseed with high level of erucic acid, which can be 40% or higher of the total seed content.

The oil from high erucic acid rapeseed (HEAR) is mostly used for industrial purposes as a lubricant. By reducing the erucic acid and glucosinolates to a very low level, the canola oil type product can be obtained.

In order to produce 1 lb of canola oil, 110,000–140,000 seeds would be required (Buntin et al., 2007).

At approximately 40% moisture content, the seeds are considered to be physically mature. Ripening can happen within a short period of time, which means that the timing of the rapeseed harvest is very important in order to obtain the highest yield possible. Rapeseed is not suitable for continuous cropping; consequently, it would be beneficial to rotate it with other types of crops, such as potatoes and carrots (The Tokyo Foundation, 2008).

The scoring for rapeseed according to ideal biomass energy crop table (Table 4.51) is 5, 10, 13, and 14 points.

Finally, rapeseed can help to reduce the growth of weeds, and the root system can help loosen the soil for plowing, improve soil tilth, and produce large amounts of biomass (Sattell et al., 1998).

4.3.14.3 Composition.

At an average production of around 17 MT a year, rapeseed is considered to be the third largest source of oil, where only soybean oil and palm oil production are reportedly higher (The Lipids Library, 2008). As worldwide demand for rapeseed rises, reflected in the form of continuous higher production since the 1950s (Table 4.49 and Table 4.50), its commercial

Table 4.51 Characteristics of an ideal biomass energy crop: rapeseed

Crop Characteristics	Score
1. Photosynthesis pathway	C3 –
2. Long canopy duration	
3. Perennial (no need for annual tillage or planting)	Annual –
4. No known pests or diseases	
5. Rapid growth in spring to out compete weeds	+
6. Sterile; prevent "escape"	
7. Stores carbon in the soil (soil restoration and carbon sequestration tool)	
8. Partitions nutrients back to roots in fall (low fertilizer requirement).	
9. Low nutrient content, that is, <200 mg/MJ nitrogen and sulfur (clean burning)	
10. High water use efficiency	+
11. Dry down in the field ("zero" drying costs)	
12. Good winter standing (harvest when needed; "zero" storage costs)	
13. Utilises existing farm equipment	+
14. Alternative markets (high quality paper, building materials, and fermentation).	+

Source: Data from various sources and adapted in part from Long (1999) and Heaton et al. (2003).

Table 4.52 Structural analysis of rapeseed

Alcohol/benzene extractives %	Holocellulose %	Lignin %	α-Cellulose %	Gross CV (MJ/kg)
16.2	50.3	27.7	13.6	19.4

Source: Adapted from Haykiri-Acma and Yaman (2008).

Table 4.53 Percentage values for the composition of rapeseed lipid

Lipid	%
Palmitic acid (C_{16} H_{32} O_2)	4
Stearic acid (C_{18} H_{36} O_2)	2
Oleic acid (C_{18} H_{34} O_2)	62
Linoleic acid (C_{18} H_{32} O_2)	22
Linolenic acid ($C_{18}H_{30}O_2$)	10

Source: Adapted from the Lipids Library (2008).

potential will continue to grow upward. For this reason, an accurate look at the composition of rapeseed and rapeseed oil would make a difference in the choice of usage, together with the possible necessary development that may be needed for efficient and wider use at a lower cost.

Structural analysis of rapeseed has shown that more than 50% is composed of holocellulose (Table 4.52), and the gross calorific value, according to this test, is 19.4 MJ/kg (Haykiri-Acma and Yaman, 2008).

Rapeseed oil has less saturated acid (palmitic acid) than other commodity of oils. Oleic acid is counted at 62%, which is higher than sunflower and niger seed oil (Table 4.53). This type of composition makes niger seed oil very attractive, not just for its industrial applications, but also for a variety of food uses as well.

Fatty acid composition is vital when it comes to commercial and industrial use. As mentioned in Section 6.3, oleic acid (as well as erucic acid) is an important factor when it comes to resistance to oxidation in relation to fuel, lipochemistry and lubrication. A comparison between diesel, biodiesel, and canola oil shows that density and calorific values are close to each other for the three types of oils.

The quality of their ignition is also close as indicated by their cetane number. The only obvious difference is when it comes to oil density, where canola oil, similar to many other types of vegetable oil, differs with a much higher percentage (Table 4.54).

Finally, laboratory tests during the work on this book showed that the total percentage of carbon and nitrogen in a single seed is 53% and 4%, respectively (Table 4.55).

Table 4.54 Various factors comparison between diesel, biodiesel, and canola oil

Factors	Diesel	Biodiesel	Canola Oil
Density kg/L at 15.5°C	0.84	0.88	0.92
Calorific value MJ/L	38.3	33–40	36.9
Viscosity mm^2/s at 20°C	4–5	4–6	70
Viscosity mm^2/s at 40°C	4–5	4–6	37
Viscosity mm^2/s at 70°C	–	–	10
Cetane number	45	45–65	40–50

Source: Adapted from Journey to Forever (2008).

Table 4.55 The chemical elements of rapeseed

Carbon	53.28%
Hydrogen	8.02%
Nitrogen	4.16%
Oxygen	34.49%
Sulfur	~0.05%

Source: Author.

4.3.14.4 Suitability as a Fuel. It is very well known that rapeseed can be a good source of fuel for the production of biodiesel (Peterson et al., undated). In addition, it can be used in co-firing as a crushed rapeseed cake (rapeseed meal) with coal for the generation of electricity.

Industrial rapeseed and canola are both suitable for this process.

Rapeseed produces more oil per unit of land area compared with other types of oil sources, such as soybeans. Reportedly, this is one of the reasons that makes rapeseed more popular as a renewable source of energy.

Presently, Europe is one of the top world producers of biodiesel from rapeseed oil (Oilgae, 2008). Using a process called transesterifaction, converting rapeseed oil into a suitable biodiesel fuel can be a simple process. Transesterifaction is defined as adding a catalyst for example, lye (sodium hydroxide, NaOH) to any type of alcohol, such as ethanol, which has already been heated before adding rapeseed oil (or other types of vegetable oil). In relation to the rapeseed oil viscosity, the composition of many vegetable oils is usually made up of around 50% esters of mono, as well as polyunsaturated fatty acids (the organic ester chemical formula is represented in the form RCO_2R', where R is the carboxylic acid and R′ is alcohol. Both are part of the hydrocarbon) (Bojanowska, 2006). The viscosity of the fuel is a result of the makeup of the earlier-mentioned acids. The type of acids contained in rapeseed oil, such as monounsaturated acids and oleic acid, make rapeseed oil one of the best bio-oils for use as a fuel due to its higher stability in comparison with other types of vegetable oils. However, in RME molecules (RME: rapeseed methyl ester *biodiesel*), where multiple bonds of unsaturated acids exist, these can speed up oxidation, which makes it unsuitable for long-term basis storage. It is possible for a derivative fuel (from crude oil) to be mixed with rapeseed fuel because both of them have certain similarities in that they contain long carbon chains (Bojanowska, 2006).

A ratio from 2% to 20% mixture of rapeseed oil with fossil fuel, respectively, can be made for the production of biodiesel, but this may make the final price of the product on the market more expensive than the use of diesel fuel on its own. This is simply because of the cost involved related to the production, in particular the processing cost of the rapeseed oil. However, this process depends largely on the makeup of the fatty acid in the rapeseed oil itself, that is, depending on the type of rapeseed and the environment the plant has grown in, the composition of the rapeseed oil fatty acid will vary accordingly, which means the rapeseed oil can have different uniformities. With such different uniformities, the processing of rapeseed oil for the production of biodiesel may produce various types of bio-diesel oil with different qualities (Bahadir et al., undated). This kind of result may add to the original cost if the above is not analyzed accordingly prior to the production stage, for example a high quality of rapeseed oil that may be used as a fuel for a combustion engine.

One final note is worth mentioning in relation to energy content. The high energy rapeseed oil contains, as laboratory tests have shown during the work on this book, an energy value (excluding the effect of the moisture content) almost synonymous with the same level of energy obtained from the South African bituminous coal (Table 4.56).

4.3.14.5 Rapeseed Ash Composition. Figure 4.20 is an illustration of rapeseed ash composition Fig. 4.20.

The elements of rapeseed ash (the plot obtained via FEI Quanta 600 scanning electron microscope). The x-axis in Figure 4.20 is energy in kev (thousands electron volts) and the y-axis is accumulated counts in kcnt (thousand counts).

4.4 CONCLUSION

A variety of main and reference samples have been presented and discussed in this chapter, mainly for the purpose of building a concept and an outline

Table 4.56 Rapeseed calorific value: dry matter with 8.3% moisture compared with South African bituminous coal

Sample	Energy (H) J/g	Moisture %
Rapeseed	26,387	8.3
Coal	26,819	1.6

Source: Author.

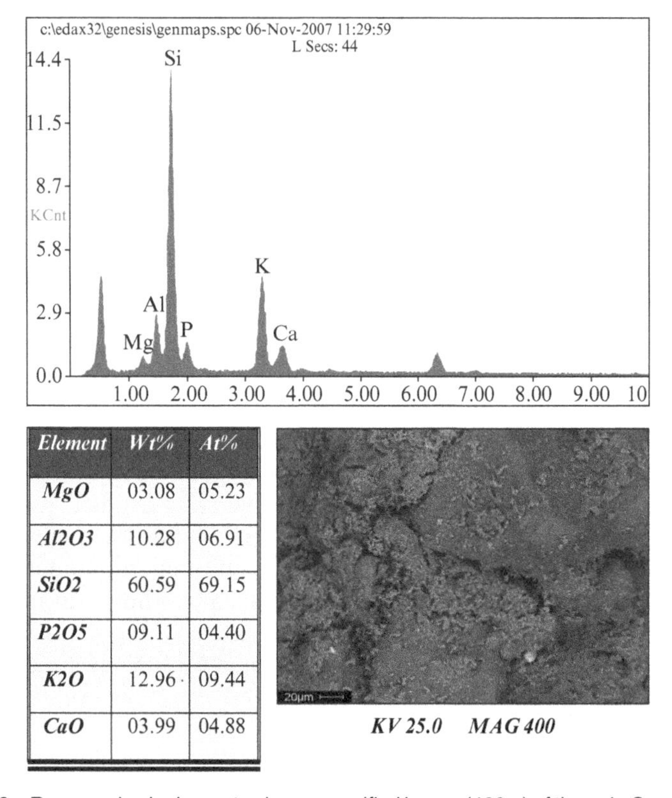

Element	*Wt%*	*At%*
MgO	03.08	05.23
Al2O3	10.28	06.91
SiO2	60.59	69.15
P2O5	09.11	04.40
K2O	12.96	09.44
CaO	03.99	04.88

Figure 4.20 Rapeseed ash elements plus a magnified image (400×) of the ash. *Source*: Author.

prior to the construction of REA1 methodology. This is how various parts of this book were provisionally designed and developed.

There are 18 samples used in this book. Three reference samples have already been looked at and examined briefly. Eight main samples (out of 15 biomass samples—Table 4.57) have then been discussed in detail.

4.4.1 Samples Selection

The selection process for the samples seemed an apparently simple and easy procedure, in the first instance. Working on the details involved, whether as part of the literature review or as part of the observation, selection, or general initial testing procedures, proved to be a huge task to undertake and a very time-consuming endeavor. The method used to look at each sample was in some sense connected to the general need for a selection process to help in choosing the most suitable biomass sample (or samples) for use as a source of energy. For this reason, the discussion and examination of each sample is listed under four different headlines. These headlines are general introduction, characteristics, composition, and suitability as a fuel.

Table 4.57 Biomass samples data: summary

Sample	Energy (H) J/g	Moisture %	Absolute Density g/cm^3	Packing Density g/cm^3	Space %	Ash %	Ignition Temp. °C	Volatile Matter %	Nitrogen Emission %	Carbon Emission %	Hydrogen Emission %	Sulfur Emission %
Corn	17,334	12.5	1.3911	0.7157	48	2.2	280	78	3.76	44.83	6.39	~0
Wheat	15,128	14	1.4712	0.8338	43	1	283	80	1.46	39.67	6.20	~0
Miscanthus	16,847	9.6	1.4511	0.2989	79	2	253	72	2.86	46.24	5.45	~0
Rice	15,188	13	1.4690	0.8966	39	0.4	289	78	2.97	39.45	5.78	~0
Pot barley	15,740	11	1.4186	0.8637	39	1	286	75	2.59	41.81	6.37	~0
Sunflower Black Seed	24,711	6.4	1.0848	0.4813	56	2.3	269	88	5.60	64.68	9.33	~0
Sunflower Striped Seed	27,099	7	1.0826	0.4857	55	2.3	269	88	6.80	67.31	9.48	~0
Niger seed	25,918	7	1.1273	0.6812	35	4	268	82	6.04	58.67	8.06	~0
Rapeseed	26,387	8.3	1.0978	0.7467	32	5.2	261	83	4.16	53.28	8.02	~0
Rapeseed meal	17,943	11	1.3530	0.6203	54	6	221	70	7.41	43.34	5.94	~0
Switch grass	17,138	7.8	1.3317	0.1397	89	3.3	271	91	2.91	48.80	5.27	~0
Reed canary grass	17,035	8	1.3166	0.1690	87	5	266	78	3.36	48.27	5.22	~0
Distilled dried corn	18,680	12.4	1.3565	0.4629	66	4.5	230	73	6.60	49.92	6.38	~0
Straw pellets	16,465	9.6	1.4825	0.5315	64	6.4	257	70	3.08	43.34	5.90	~0
Apple pruning	16,971	7	1.4236	0.3352	76	2.4	256	68	2.82	48.98	5.54	~0

Source: Author.

4.4.2 The Next Step

Having a good knowledge and understanding of all the samples selected prior to the forming and application of REA1 methodology is a necessary step that has helped greatly in the following stages of writing this book. Details about each sample are easily accessible and available for references.

This chapter is the launchpad for the main and most important parts of this work, that is, the methodology and results contained in the following chapters (Box 4.12 and Box 4.13).

Box 4.12

FOOD CROPS (FC) AND SHORT ROTATIONAL CROPS (SRC)

There is a historical and long connection related to the experience and knowledge associated with energy crops, such as food crops, which can be readily used as a source for generating electricity on a commercial scale. Unlike short rotational crops (SRC), which are advocated to be used for dedicated fuel supply systems, experience and knowledge in this field is still lagging behind compared with FC. This is especially so in large commercial farming, transportation, storage, environmental issues, and market commercial aspects (local, national and international).

Box 4.13

WOOD (APPLE TREE WOOD)

Using wood taken from pruned trees is one of the best ways for obtaining energy, while at the same time, encouraging the tree's new growth. In this way, there is a possibility of maintaining the level of CO_2 in the atmosphere.

Most of the apple tree wood is hard with close grain and generally heavy. Fresh wood obtained from apple trees usually contains a high percentage of moisture. In fact, most of the wood weight is actually water. By cutting the wood into small pieces and leaving them exposed to air, the moisture in the wood can be reduced after a period of time (a year or more) to around 20%. Like most of the other types of biomass materials, when burning starts, wood gives up (in the first instant) its moisture. As the temperature rises, volatile matters within the wood begin to separate and consequently ignition takes place.

The chemical elements of apple wood are the following:

Carbon	48.98%
Hydrogen	5.54%
Nitrogen	2.82%
Oxygen	42.60%
Sulfur	~0.06%

Source: Author.

REFERENCES

Aggarwal D (2003) Use of rice husks as fuel in process steam boilers. Tata Energy Research Institute, Asia-Pacific Environmental Innovation Strategies (APEIS), India. http://www.iges.or.jp/APEIS/RISPO/inventory/db/pdf/0004.pdf (last accessed December 30, 2008).

Agri Publication (2008) Sunflower production: a concise guide. KwaZulu-Natal, Department of agriculture and environmental affairs. http://agriculture.kzntl.gov.za/portal/AgricPublications/LooknDo/SunflowerProduction/tabid/134/Default.aspx (last accessed January 15, 2009).

Appalachian Blacksmiths Association (2008) Coal types. http://www.appaltree.net/aba/coaltypes.htm (last accessed December 27, 2008).

Babcock B (2008) Wheat, corn and ethanol fight for acres. On line interview by washingtonpost.com with the director of the centre for agriculture and rural development at Iowa State University. 30.4.2008. http://www.washingtonpost.com/wp-dyn/content/discussion/2008/04/25/DI2008042502592.html (last accessed October 31, 2008).

Bahadir M, Krahl J, Ondruschka B, Ralle B (undated) Biodiesel. http://www.rsc.org/education/teachers/learnnet/green/docs/biodiesel.pdf (last accessed February 2, 2009).

Baik B-K, Ullrich ES (2008) Barley for food: characteristics, improvement, and renewed interest. Journal of Cereal Science 48(2):233–242.

Baldwin SB, Holmberg BK (2006) Growth characteristics important for field establishment of giant miscanthus. Dept. of plant and soil sciences, Mississippi State University. http://ms-biomass.org/conference/2006/presentations/track3_presentation4.pdf (last accessed November 19, 2008).

Barba C (2002) Synthesis of carbomethyl cellulose from annual plants pulps. PhD dissertation. Roviral i Virgili University.

Barnwal BK, Sharma MP (2005) Prospects of biodiesel production from vegetable oils in India. Renewable and Sustainable Energy Reviews 9(4):363–378.

BBC News (August 10, 2012) US biofuel production should be suspended, UN says. http://www.bbc.co.uk/news/business-19206199 (last accessed January 13, 2014).

Bojanowska M (2006) Fatty acid composition as a criterion for rapeseed application for fuel production. Electronic Journal of Polish Agricultural Universities 9(4):52.

Boland M (2008) Agriculture marketing resource centre: Rapeseed. Ag Marketing Resource Center, Kansas State University. http://www.agmrc.org/commodities__products/grains__oilseeds/rapeseed.cfm (last accessed January 30, 2009).

Bondioli P, Folegatti L (1996) Evaluating the oxidation stability of biodiesel: an experimental contribution. La Rivista Italiana Delle Sostanze Grasse LXXIII – August:349–353.

Buntin D, Grey T, Harris HG, Phillips D, Prostko E, Raymer P, Smith N, Sumner P, Woodruff J (2007) Canola production in Georgia. Bulletin 1331, cooperative extension, the University of Georgia. http://pubs.caes.uga.edu/caespubs/pubcd/B1331/B1331.htm (last accessed February 1, 2009).

Cereal Knowledge Bank (2008) Quality characteristics of milled rice. http://www.knowledgebank.irri.org/riceQuality/Qly03.htm (last accessed December 27, 2008).

Clifton-Brown J, Valentine J (2007) Asian elephant grass (Miscanthus) for bioenergy. http://www.aber.ac.uk/en/media/07ch4.pdf (last accessed October 5, 2009).

Clifton-Brown JC, Lewandowski I, Andersson B, Basch G, Christian D, Kjeldsen J, Jørgensen U, Mortensen J, Riche A, Schwarz K, Tayebi K, Teixeira F (2001) Performance of 15 miscanthus geneotypes at five sites in Europe. Agronomy Journal 93:1013–1019.

Commodity Online (2008) Sunflower oil. http://www.commodityonline.com (last accessed July 20, 2008).

Cooperative Extension Service (2000) Grain characteristics of rice varieties. Rice Information No. 146, University of Arkansas, Division of Agriculture, U.S. Department of Agriculture and County Governments Cooperating.

Cyberlipid Centre (2008) Main world sources of oil: rapeseed. http://www.cyberlipid.org/glycer/glyc0051.htm (last accessed January 30, 2009).

Defra (2007) Best Practice Guidelines: planting and growing miscanthus. For applicants to Defra's energy crops scheme. http://www.defra.gov.uk/erdp/pdfs/ecs/miscanthus-guide.pdf (last accessed February 5, 2008).

Demirbas A (2007) Importance of biodiesel as transportation fuel. Energy Policy 35(9):4661–4670.

Department of Agriculture, Sri Lanka (2008) Sunflower: *Helianthus annuus* L. http://www.agridept.gov.lk/Techinformations/Oilseed/Sflower.htm (last accessed January 12, 2008).

Dubcovsky J, Dvorak J (2007) Genome plasticity a key factor in the success of polyploid wheat under domestication. Science 316(5833):1862.

Duke AJ (1983) Handbook of energy crops: *Guizotia abyssinica* (L.f.) Cass. Electronic publication only (web). http://www.hort.purdue.edu/newcrop/duke_energy/duke index.html; http://www.hort.purdue.edu/newcrop/duke_energy/Guizotia_abyssinica.html (last accessed January 5, 2009).

Earth Policy Institute (2013) Grain harvest: global grain stocks drop dangerously low as 2012 consumption exceeded production. http://www.earth-policy.org/indicators/C54/grain_2013 (last accessed March 14, 2013).

Ehleringer JR, Monson RK (1993) Evolutionary and ecological aspects of photosynthetic pathway variation. Annual Review of Ecology and Systematics 24:411–439.

Encyclopaedia Britannica (2008) Bituminous coal. http://www.britannica.com/eb/article-9015437/bituminous-coal (last accessed February 25, 2008).

Energy Information Administration (2009) Petroleum. Official energy statistics from the U.S. government. http://www.eia.doe.gov/oil_gas/petroleum/info_glance/petroleum.html (last accessed April 4, 2009).

Energy Information Administration (2012) Natural gas. Official energy statistics from the U.S. government. http://www.eia.gov/naturalgas/annual/ (last accessed August 17, 2012).

Erickson BD, Bassin P (undated) Rapeseed and crambe: alternative crops with potential industrial uses. Bulletin 656, Agricultural Experiment Station, Kansas State University, Manhattan. http://www.oznet.k-state.edu/library/crpsl2/sections/sb656_a.pdf (last accessed January 29, 2009).

EUBIA (2007) About Biomass/Biomass characteristics. European Biomass Industry Association. http://p9719.typo3server.info/115.0.html (last accessed June 16, 2008).

ExxonMobil (2013) The Outlook for energy: a view to 2040. http://www.exxonmobil.co.uk/Corporate/Files/news_pub_eo2013.pdf (last accessed July 22, 2013).

Farm Direct (2001) Modern wheat: wheat: characteristics and uses. http://www.farm-direct.co.uk/farming/stockcrop/wheat/wheatcurr.html (last accessed February 6, 2009).

Farm Direct (2008) Barley, the crop. http://www.farm-direct.co.uk/farming/stockcrop/barley/crop.html (last accessed November 10, 2008).

FAO (1990) Food balance sheets 1984–86. Food and Agricultural Organisation, Rome.

FAO (1992) Corporate document repository, 1992. http://www.fao.org/documents/en/docrep.jsp;jsessionid=269A5A69D375D4712269D99BB9FFD01E (last accessed January 10, 2009).

FAO (2013a) The International Rice Commission. http://www.fao.org/agriculture/crops/core-themes/theme/treaties/irc/en/ (last accessed August 30, 2013).

FAO (2013b) World Food and Agriculture. FAO Statistical Year Book http://issuu.com/faooftheun/docs/syb2013issuu (last accessed January 22, 2014).

FAO, Forestry Department (2004) Unified Bioenergy Terminology (UBET): parameters, units and conversion factors. Wood Energy Programme, Food and Agriculture Organisation of the United Nations. http://www.fao.org/docrep/007/j4504E/j4504e08.htm (last accessed January 8, 2009).

Flagella Z, Di Caterina R, Monteleone M, Giuzio L, Pompa M, Tarantino E, Rotunno T (2006) Potentials for sunflower cultivation for fuel production in southern Italy. HELIA 29(45):81–88.

Fløjgaard Kristensen E, Kristensen JK, Nikolaisen L, Bjerrum M, Nørgaard Jensen T (1999) Alternative biofuels in combustion appliances from 20 to 250 kW Report for Energistyrelsen, J.nr. 51161/96-0033 and 731327/97-0134, Energy Division, Danish Technological Institute, Kongsvang Allé 29, DK-8000 Århus C, Denmark.

Fowler PA, McLauchlin AR, Hall LM (2003) The potential industrial uses of forage grasses including miscanthus. BioComposites Centre, University of Wales, Bangor, Gwynedd, LL57 2UW, UK.

Francis MC, Campbell CM (2003) New high quality oil seed crops for temperate and tropical Australia. A report for the rural industries research and development corporation. http://www.rirdc.gov.au/reports/NPP/03-045.pdf (last accessed January 27, 2009).

Gattamaneni RNL, Subramani S, Santhanam S, Kuderu R (2008) Combustion and emission characteristics of diesel engine fuelled with rice bran oil methyl ester and its diesel blends. Thermal Science 12(1):139–150.

Global Warming (2008) Global warming: the rise of CO_2 and warming. http://earthguide.ucsd.edu/globalchange/global_warming/03.html (last accessed February 2, 2009).

Gmelin L (1817) Handbuch der theoretischen Chemie. Jubilee Facsimile Reprint of the First Edition of 1817, in celebration of the 150th anniversary of Gmelin's Handbook of Inorganic Chemistry, Verlag Chemie, Weinheim, 1967.

Gopalan C, Ram Sastri BV, Balasubramanian SC (1982) Nutritive values of Indian foods. National Institute of Nutrition (ICMR). Hyderabad, India.

Gramene.org (2009) Hordeum introduction. http://www.gramene.org/species/hordeum/barley_intro.html (last accessed January 4, 2009).

Hallgren AL, Oskarsson J (1998) Minimization of sintering tendencies in fluidized-bed gasification of energy crop fuels. In Biomass for Energy and Industry, Proceedings of the 10th European Biomass Conference, Würzburg, Germany: June 1998. Rimpar, Germany: C.A.R.M.E.N. Publishers, pp. 1700–1703.

Haykiri-Acma H, Yaman S (2008) Thermal reactivity of rapeseed (Brassica napus L.) under different gas atmospheres. Bioresource Technology 99(2):237–242.

Heaton AE, Clifton-Brown J, Voigt BT, Michael B, Jones MB, Long PS (2003) Miscanthus for renewable energy generation: European Union experience and projections for Illinois. Crop Sciences and Plant Biology, University of Illinois. http://www.springerlink.com/content/j0610uv467x11501/fulltext.pdf (last accessed December 25, 2008).

HGCA Knowledge Centre (2008) The main oilseeds: rapeseed. http://www.hgca.com/minisite_manager.output/2006/2006/Knowledge%20Centre/Overview%20of%20the%20World%20Oilseeds%20Market/The%20Main%20Oilseeds.mspx?minisiteId=11 (last accessed January 31, 2009).

Hicks BK, Flores AR, Taylor F, McAloon JA, Moreau AR, David B, Johnston BD, Senske EG, Brooks SW, Griffey AC (2005) Current and potential use of barley in fuel ethanol production. 2005 EWW/SSGW Conference. http://209.85.229.132/search?q=cache:MeaoWaA0hY8J:www.uky.edu/Ag/Wheat/wheat_breeding/EWW_SSGW/documents/kevin_hicks.doc+Barley+for+fuel+come&hl=en&ct=clnk&cd=1&gl=uk (last accessed January 7, 2009).

Himmel ME, Ding SY, Johnson DK, Adney WS, Nimlos MR, Brady JW, Foust TD (2007) Biomass recalcitrance: engineering plants and enzymes for biofuels production. Science 315:804–807.

Hollander JM (1992) The Energy-Environment Connection. Washington, DC: Island Press.

IEA (2012) Electricity Information 2012. http://www.iea.org/media/training/presentations/statisticsmarch/ElectricityInformation.pdf (last accessed January 17, 2014).

IEA Bioenergy (2013) Biogas from energy crop digestion. http://www.biogas.org.nz/Publications/Resources/biogas-from-energy-crop-digestion.pdf (last accessed December 12, 2013).

Ienica.net (2003) Sunflower. http://www.ienica.net/crops/sunflower.pdf (last accessed January 18, 2009).

Ingredients101.com (2008) Barley grain. http://www.ingredients101.com/barleyg.htm (last accessed November 22, 2008).

International Grain Council (2013) Supply & demand: all rice. http://www.igc.int/en/grainsupdate/sd.aspx?crop=Rice (last accessed March 11, 2013).

International Rice Research Institute (2002) Rice Almanac, 3rd ed. Oxon, UK: CABI Publishing.

IPCC (2007) Chapter 9: Forestry. In M Apps, E Calvo, eds., Climate Change 2007: Working Group III: Mitigation of Climate Change. Cambridge: Cambridge University Press, p. 545. IPCC Fourth Assessment Report: Climate Change 2007. http://www.ipcc.ch/publications_and_data/ar4/wg3/en/ch9.html (last accessed August 30, 2013).

ISI (International Starch Institute) (2001) Maize (corn). Science Park Aarhus, Denmark, 2001. http://www.starch.dk/isi/starch/maize.htm (last accessed July 1, 2008).

Jadhav SJ, Lutz SE, Ghorpade VM, Salunkhe DK (1998) Barley: chemistry and value-added processing. Critical Reviews in Food Science and Nutrition 38:123–171.

Jones M, Walsh M (2001) Miscanthus for Energy and Fibre. London: James & James. Google Books Search.

Journey to Forever (2008) Oil yields and characteristics. http://journeytoforever.org/biodiesel_yield2.html (last accessed February 1, 2009).

Juliano OB (1993) Rice in Human Nutrition. Rome: FAO. Published with the collaboration of the International Rice Research Institute and Food and Agriculture Organization of the United Nation (FAO).

Karvy Comtrade Ltd. (2008) Barley short term report: 24th June 2008. http://www.karvycomtrade.com/downloads/karvySpecialReports/karvysSpecialReports_20080624_01.pdf (last accessed January 5, 2009).

Kent LK, Evers DA (1992) Technology of Cereals, 4th ed. Pergamon, p. 93.

Kikuchi K, Takatsuji I, Tokuda M, Miyake K (2006) Properties and uses of horny and floury endosperms of corn. Journal of Food Science, 47(5):1687–1692.

Kulp K, Ponte GJ (2000) Handbook of Cereal Science and Technology, 2nd ed. New York: Marcel Dekker.

Kurkovál M, Klikal Z, Martinec P, Pěgřimočová J (2003) Composition of bituminous coal in dependence on environment and temperature of alteration. Bulletin of the Czech Geological Survey 78(1):23–34.

Leibtag E (2008) Corn prices near record high, but what about food costs? Amber Waves. http://www.ers.usda.gov/AmberWaves/February08/Features/CornPrices.htm (last accessed February 4, 2009).

Lewandowski I, Kahnt G (1993) Development of a tissue culture system with unemerged inflorescences of *Miscanthus* "giganteus" for the induction and regeneration of somatic embryoids. Beitrage zur Biologie der Pflanzen 67:439–451.

Lewandowski I, Clifton-Brown B, Andersson B, Baschd G, Christiane DG, Jørgensenf U, Jones MB, Richee AB, Schwarzf KU, Tayebid K, Teixeirad F, et al. (2003) Environment and harvest time affects the combustion qualities of Miscanthus genotypes. Agron. J., 95:1274–1280.

Lin L, Ying D, Chaitep S, Vittayapadung S (2008) Biodiesel production from crude rice bran oil and properties as fuel. Applied Energy 86(5):681–688.

The Lipids Library (2008) Oils and fats in the market place: commodity oils and fats Rapeseed (Canola) oil. http://www.lipidlibrary.co.uk/market/rapeseed.htm (last accessed January 31, 2009).

Long SP (1999) Environmental responses. In R Sage and RK Monson, eds., C4 Plant Biology. San Diego, CA: Academic Press, pp. 215–242.

Lorenz JK, Kulp K, eds. (1990) Handbook of Cereal Science and Technology. New York: Marcel Dekker, ch. 3.

Mangelsdorf PC (2008) Corn: Its Origin, Evolution and Improvement. Cambridge, MA: Harvard University Press.

Marvey BB (2008) Sunflower-based feedstocks in nonfood applications: Perspectives from olefin metathesis. Review, International Journal of Molecular Sciences 9:1393–1406.

Marvey BB, Segakweng CK, Vosloo HCM (2008) Ruthenium carbene mediated metathesis of oleate-type fatty compounds. International Journal of Molecular Sciences 9:615–625.

McGee H (1984) On Food and Cooking. New York: Charles Scribner's Sons.

Melvin A (1988) Natural Gas: Basic Science and Technology. Bristol, UK: IOP Publishing.

Michniewicz J, Biliaderis CG, Bushuk W (1990) Water-insoluble pentosans of wheat: composition and some physical properties. Cereal Chemistry 67(5):434–439.

Microsoft Encarta (2008) Corn. http://encarta.msn.com/text_761559467__1/corn.html (last accessed February 2, 2009).

Moilanen A, Nieminen M, Sipila K, Kurkela E (1996) Ash behavior in thermal fluidizedbed conversion processes of woody and herbaceous biomass. In Biomass for Energy and the Environment, Proceedings of the 9th European Bioenergy Conference, Copenhagen, Denmark. June 1996, Pergamon/Elsevier Publishers, pp. 1227–1232.

Moore MJ (1997) No_x emission control in gas turbines for combined cycle gas turbine plant. Proceedings of the Institution of Mechanical Engineers, Part A: Journal of Power and Energy 211(1):43–52.

Mosier N, Wyman C, Elander R, Lee YY, Hotzapple M, Ladisch M (2005) Features of promising technologies for pretreatment of lignocellulosic biomass. Bioresource Technology 96:673–686.

mt.gov (2006) Wheat varieties. http://wbc.agr.mt.gov/Producers/Variety_releases/wheatvarietiesformt06.pdf (last accessed December 14, 2013).

Muzafaroy DC, Mazhidov KK (1997) Chemical composition of husked and polished rice. Chemistry of Natural Compounds 33(5). UDC 641.5.002.237.

National Sunflower Association (2008) All about sunflower. http://www.sunflowernsa.com/all-about/default.asp?contentID=41 (last accessed January 10, 2009).

National Sunflower Association (2013) Sunflower statistics > world supply & disappearance. http://www.sunflowernsa.com/stats/world-supply/ (last accessed March 1, 2013).

NDSU (North Dakota State University) (2007) Sunflower production. http://www.ag.ndsu.nodak.edu/aginfo/entomology/entupdates/Sunflower/a1331sunflowerhandbook.pdf (last accessed March 17, 2013).

Nelson T, Langdale JA (1992) Developmental genetics of C4 photosynthesis. Annu. Rev. Plant Physiol. Plant Mol. Biol. 43:25–47.

NIFES Consulting Group (2008) Renewable energy: biomass projects: wood-fuel learning visit to Sweden, March 20–22, 2006. http://www.nifes.co.uk/ (last accessed January 8, 2009).

NNFCC (2012) Report title: domestic energy crops; potential and constraints review. Project Number: 12-021. https://www.gov.uk/government/uploads/system/uploads/attachment_data/file/48342/5138-domestic-energy-crops-potential-and-constraints-r.PDF (last accessed May 15, 2013).

Nomura M, Katayama K, Nishimura A, Ishida Y, Ohta S, Komari T, Miyao-Tokutomi M, Tajima S, Matsuoka M (2000) The promoter of rbcS in a C3 plant (rice) directs organ-specific, light-dependent expression in a C4 plant (maize), but does not confer bundle sheath cell-specific expression. Plant Molecular Biology 44:99–106.

Oilgae (2008) Biodiesel from Rapeseed, Rapeseed oil as bio-diesel, biofuel: reference and resources. http://www.oilgae.com/energy/sou/ae/re/be/bd/po/rap/rap.html (last accessed February 2, 2009).

Olsson M (2006) Wheat straw and peat for fuel pellets: organic compounds from combustion. Biomass and Bioenergy 30(6):555–564.

Optima Energy Group (2007) About crude oil. http://www.optimaenergygroup.com/content/abt_crude/about_crudeoil.html (last accessed February 26, 2008).

Oryza.com (2008) Japan: fuel made with rice as power alternative. http://oryza.com/Asia-Pacific/Japan-Market/japan-alternative-rice.html (last accessed January 2, 2009).

Panoutsou C, Elbersen B, Böttcher H (2011) Biomass future: energy crops in the European context. http://www.biomassfutures.eu/public_docs/final_deliverables/WP8/D8.4%20Energy%20crops%20in%20the%20European%20context%20(contribution%20to%20FNR%20workshop).pdf (last accessed May 14, 2013).

Paredes-Lopez O (2000) 2: The wheat grain. Plant Foods for Human Nutrition 55(1):15–20.

PCC (Pima Community College) (2008) Plants: type of photosynthesis. Arizona, USA. http://wc.pima.edu/~bfiero/tucsonecology/plants/plants_photosynthesis.htm (last accessed December 25, 2008).

Peterson CL, Brown J, Guerra D, Drown DC, Withers RV (undated) Rapeseed oil as diesel fuel an overview. pp. 916–921. http://www.biodiesel.org/resources/reportsdatabase/reports/gen/19920101_gen-302.pdf (last accessed February 1, 2009).

Plants for a Future: Database Search Results (2008) *Guizotia abyssinica*. http://www.ibiblio.org/pfaf/cgi-bin/arr_html?Guizotia+abyssinica (last accessed January 25, 2009).

Quinn J, Myers LR (2002) Nigerseed: specialty grain opportunity for Midwestern US. p. 174–182. http://www.hort.purdue.edu/newcrop/ncnu02/v5-174.html (last accessed January 25, 2009).

Raymer PL (2002) Canola: an emerging oilseed crop. p. 122–126. http://www.hort.purdue.edu/newcrop/ncnu02/v5-122.html (last accessed January 30, 2009).

Reinhold V (2002) Agronomy and environmental impact of high oleic-sunflower production in Germany. Twelfth European Biomass Conference, Amsterdam vol. 1 (2002) ISBM 88-900442-5-x.

The Rice Association (2008) Varieties. http://www.riceassociation.org.uk/About/varieties.htm (last accessed February 6, 2008).

Roecklein CJ, Leung P (1987) A Profile of Economic Plants. New Brunswick: Transaction Books.

Ruiz HA, Silva DP, Ruzene DS, Lima LF, Vicente AA (2012) Bioethanol production from hydrothermal pre treated wheat straw by a flocculating *Saccharomyces cerevisiae* strain-effect of process conditions. Fuel 95:528–536.

Sandhu KS, Singh N, Kaur M (2004) Characteristics of the different corn types and their grain fractions: physicochemical, thermal, morphological, and rheological properties of starches. Journal of Food Engineering 64(1):119–127.

Satake Corporation U.K. Division (2008) Maize milling: the origin, cultivation and types of maize http://www.satake.co.uk/cereal_milling/maize_origin.htm (last accessed October 29, 2008).

Sattell R, Dick R, Ingham R, Karow R, Kaufman D, McGrath D (1998) Oregon cover crops: rapeseed. EM 8700, extension service, OSU. http://extension.oregonstate.edu/catalog/html/em/em8700/ (last accessed February 1, 2009).

Scurlock J (1999) Miscanthus: a review of European experience with a novel energy crop. Environmental Science Division, Oak Ridge National Laboratory, prepared for US Dept. of Energy. http://bioenergy.ornl.gov/reports/miscanthus/toc.html (last accessed November 1, 2008).

Shapouri H, Duffield JA, Wang M (2002) The energy balance of corn ethanol: an update. Agric. Econ. Rept. 813. USDA/OCE, Washington, DC. http://www.transportation.anl.gov/pdfs/AF/265.pdf (last accessed February 5, 2009).

Shon K-J, Lim S-T, Yoo B (2005) Rheological properties of rice starch dispersions in dimethyl sulfoxide. Starch 57(8):363–369.

Singh VJ, Rausch KD, Yang P, Shapouri H, Belyea RL, Tumbleson ME (2001) Modified dry grind ethanol process. Public. No. 2001–7021. Univ. of Illinois at Urbana-Champaign, Urbana, IL. http://abe-research.illinois.edu/pubs/k_rausch/Singh_etal_Modified_Dry_Grind.pdf (last accessed February 4, 2009).

Skhonde MP, Herod AA, van der Walt TJ, Tastsi WL, Mokoena K (2006) The effect of thermal treatment on the compositional structure of humic acids extracted from South African bituminous coal. International Journal of Mineral Processing 81(1):51–57.

Smith L (2007) Rapeseed biofuel produces more greenhouse gas than oil or petrol. *The Times* (September 22, 2007).

Sohail S (2005) Coal's price is rising, but can it clean up? The Guardian, August 2005. http://www.guardian.co.uk/environment/2005/aug/22/oilandpetrol.climatechange (last accessed February 26, 2008).

Southeast Exotic Pest Plant Council (2008) Southeast exotic pest plant council invasive plant manual. Website article, University of Georgia, College of Agricultural and Environmental Sciences. http://www.se-eppc.org/manual/MISI.html (last accessed October 17, 2008).

Soyatech (2008) Sunflower facts. http://www.soyatech.com (last accessed January 18, 2009).

Sullivan W (1982) Cross between corn and a wild relative yields a perennial crop. New York Times, February 16, 1982.

Sun R, Sun XF, Liu GQ, Fowler P, Tomkinson J (2002) Structural and physicochemical characterization of hemicelluloses isolated by alkaline peroxide from barley straw. Polymer International 51(2):117–124.

Sustainable Energy Ireland (SEI) (undated) Quality assurance for rape-seed oil as vehicle fuel. http://www.sei.ie/uploadedfiles/InfoCentre/RapeseedOilfinal.pdf (last accessed February 3, 2009).

Teagasc, Agri-Food and Bioscience Institute (2011) Miscanthus Best Practice guidelines. http://www.afbini.gov.uk/miscanthus-best-practice-guidelines.pdf (last accessed December 12, 2012).

Thompson P (2008) Alternate energy source: natural gas. The Beginner's Guide to Peak Oil, The Wolf at the Door (website). http://wolf.readinglitho.co.uk/subpages/natgas.html (last accessed February 27, 2008).

The Tokyo Foundation (2008) Vol. 8: rapeseed oil. http://www.tokyofoundation.org/en/articles/vol.-8-rapeseed-oil (last accessed February 1, 2009).

Triandafyllis J, Katopodis SP, Vosniakos F, Grammatikis V, Kalafatis E, Mantzaris D, Mallas CH (2003) Use of sunflower oil in direct injection engines. Journal of Environmental Protection and Ecology 4(3):509–517. http://www.gen.teithe.gr/~bena/VOL4NO3_2003/Paper1.pdf (last accessed January 19, 2009).

U.K. Agriculture (2008) Barley farming in the UK. http://www.ukagriculture.com/crops/barley_uk.cfm4.12.2009 (last accessed February 17, 2009).

UN Data (2008) Trade of goods, US$, HS 1992, 12 Oil seed, oleagic fruits, grain, seed, fruit, etc. http://data.un.org/Data.aspx?d=ComTrade&f=_l1Code%3A13 (last accessed January 18, 2009).

UNCTAD (2008) Rice. Info comm, market information in the commodities area, United Nation. http://r0.unctad.org/infocomm/anglais/rice/characteristics.htm#desc (last accessed December 11, 2008).

USDA (2013a) Corn. Economic research service. http://www.ers.usda.gov/topics/crops/corn/background.aspx (last accessed July 5, 2013).

USDA (2013b) World agricultural supply and demand estimates. http://www.usda.gov/oce/commodity/wasde/latest.pdf (last accessed March 2, 2013).

USDA (2013c) PSD online (production, supply and distribution online). http://www.fas.usda.gov/psdonline/psdQuery.aspx (last accessed March 3, 2013).

USDA (2013d) World barley production, consumption, and stocks. http://www.fas.usda.gov/psdonline/psdreport.aspx?hidReportRetrievalName=BVS&hidReportRetrievalID=378&hidReportRetrievalTemplateID=7 (last accessed March 12, 2013).

USDA (2013e) Rapeseed area, yield, and production. http://www.fas.usda.gov/psdonline/psdreport.aspx?hidReportRetrievalName=BVS&hidReportRetrievalID=922&hidReportRetrievalTemplateID=1 (last accessed March 12, 2013).

USDA (2013f) Wheat area, yield, and production. http://www.fas.usda.gov/psdonline/psdreport.aspx?hidReportRetrievalName=BVS&hidReportRetrievalID=448&hidReportRetrievalTemplateID=1 (last accessed March 14, 2013).

USDA (2014) World Agricultural Supply and Demand Estimates – Corn (February 10, 2014). WASDE – 526. http://www.usda.gov/oce/commodity/wasde/latest.pdf (last accessed February 16, 2014).

USDE (2013) Energy efficiency and renewable energy: vehicle technologies program. http://www.afdc.energy.gov/pdfs/47414.pdf (last accessed May 15, 2013).

U.S. Department of Energy Office of Science (2008) Genomic: GTL: system biology for energy and environment: understanding biomass: plant cell walls. genomics. energy.gov. http://genomicsgtl.energy.gov/biofuels/placemat2.shtm (last accessed December 3, 2008).

van der Vossen HAM, Mkamilo GS (2006) Plant Resources of Tropical Africa 14: Vegetable Oils. Wageningen, Netherlands: PROTA Foundation/Backhuys Publishers/CTA.

Velasquez JA, Ferrando F, Farriol X, Salvado J (2003) Binderless fiberboard from steam exploded *Miscanthus sinensis*. Wood Science and Technology 37:269–278.

Wang M (2003) Effect of pentosans on gluten formation and properties. Thesis, Department of Agrotechnology and Food Sciences, Wageningen University, The Netherlands.

Watson SA, Ramstad PE (1991) Structure and composition. In Corn-Chemistry and Technology. St. Paul, MN: AACC, pp. 53–82.

Wayne's Word (2000) Sunflower family (Asteraceae): the largest plant family on earth. Volume 9, number 3 http://waynesword.palomar.edu/ww0903a.htm (last accessed January 12, 2009).

WCC (Western Coordinating Committee) (1999) Enhanced use of barley for feed and food. Petition. http://www.lgu.umd.edu/lgu_v2/pages/attachs/410_Petition.PDF (last accessed January 5, 2009).

Winteringham FPW (1992) Energy use and the environment. Boca Raton, FL: Lewis Publishers.

World Coal Association (2013) Coal statistics: top ten coal producers (2011e). http://www.worldcoal.org/resources/coal-statistics/ (last accessed March 8, 2013).

World Coal Institute (2008) Coal facts 2007. http://www.worldcoal.org/pages/content/index.asp?PageID=188 (last accessed October 7, 2008).

5

METHODOLOGY: PART 1

5.1 METHODOLOGY APPROACH

5.1.1 Introduction

The process of selecting a workable commercial biomass fuel is nothing but gradual steps of testing and refining procedures, similar to those used when designing and producing a new hardware product. Part of the procedure involved is the creation and implementation of a new methodology. The methodology is a tool that will play an important part during and after the selection of biomass samples required to formulate a new bio-fuel. Any approach in building a new methodology has the tendency to fluctuate between original ideas, which in the case of this book are the principles of scientific and technical (S&T)/business factors (BF) (Fig. 5.1) and their percentage values, and unforeseen issues, challenges, and inquiries. The methodology's main focus, therefore, is to find biomass materials with the highest output of energy that are readily available at a low cost, that is, the correct selection process is one of the main parts of the methodology. This approach is always kept in mind during each stage of the process.

The first issue is how to select the required biomass samples. Basic requirement for the biomass sample should be taken into account in relation to the types of biomass materials needed. Some of these requirements can be summarized in the following points:

The Selection Process of Biomass Materials for the Production of Bio-fuels and Co-firing, First Edition. Najib Altawell.

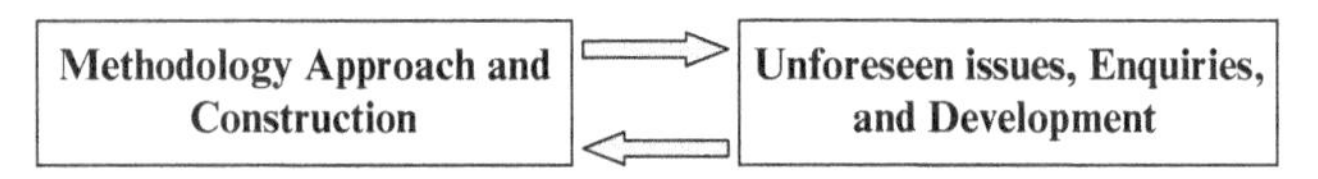

Figure 5.1 The balance between a new methodology creation and issues of further developments with unexpected problems. *Source*: Author.

1. Low cost
2. Available throughout most of the year
3. High energy content
4. Simple to process
5. Fewer unwanted by-products.

5.2 THE PYRAMID

To illustrate the creation of a successful commercial RE/biomass business, or any business for that matter, a pyramid is used to explain the principles symbolically.

There are three layers or steps within the pyramid. The first step is the base of the pyramid, which represents "environmental and alternative sources of energy." The second step is the "scientific and technical issues," and the third step is the "business issues." Without the base, factors in relation to a renewable commercial source of energy will not exist. If it is first possible to establish sound base, that is, "environmental and alternative sources of energy," then the work can start on the next step (i.e., "S&T") without any delay.

The base is made up of three different factors. The first factor concerns choosing which type of renewable energy is the most applicable to the proposed business needs. The second factor is associated with government regulations. The third factor concerns human health, environmental aspects, and their regulations.

"Scientific and technical" aspects will build on the regulations and rules already established by the "environmental and alternative sources of energy" base.

Overall, 10 general factors are taken into consideration, 80% of which form part of the S&T section, that is, three factors from "environmental and alternative source of energy," and seven main factors from S&T. The total number of factors for S&T when subdivided is 10. However, the original total number for S&T was 11, when the "space"* factor was included. The space factor was dropped from the final calculation as there was no need for it during the

* The internal space within the structure of biomass materials.

application of the methodology for the example discussed in this book; however, this factor can be added if and when an energy business requires it during their calculation.

The last step, "business," can be established *if* the basics of the previous two steps are acceptable and can be achieved successfully.

There are nine main BF at this step (31 factors when subdivided), all of them form part of the BF section that completes the pyramid model.

The completed pyramid structure (Fig. 5.2) signals that there is a successful commercial environment for this kind of work, but only as and when the business section gradually grows from the previous two steps and is in harmony with them.

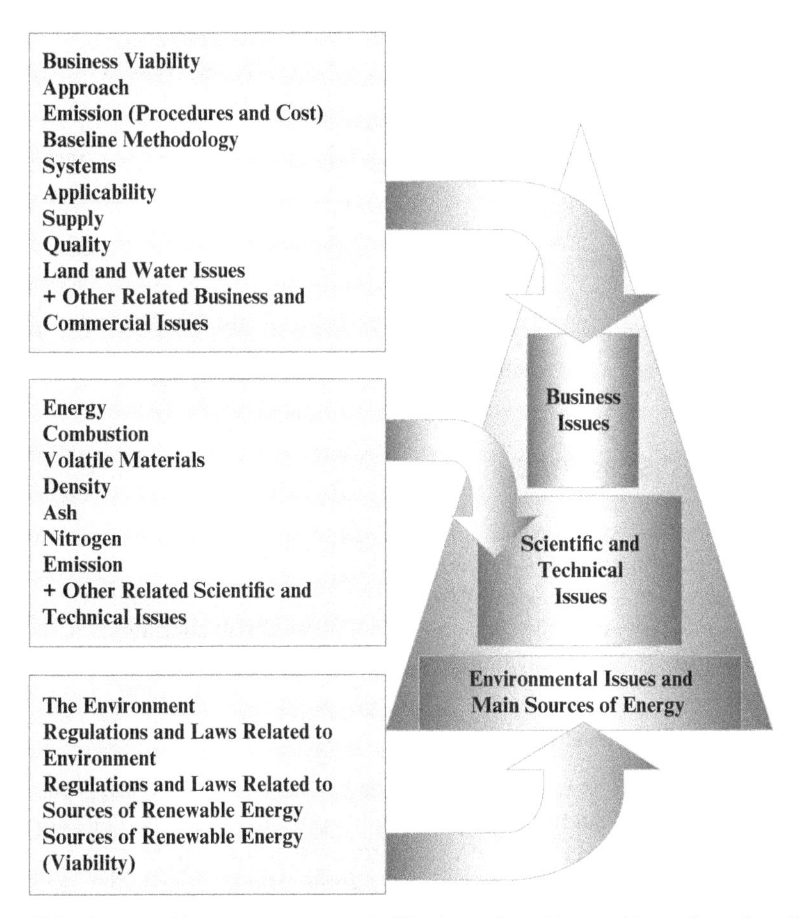

Figure 5.2 A renewable energy source project/business (e.g., biomass) is made up from three important parts represented here in the shape of a pyramid/cone. *Source*: Adapted from Altawell (2012).

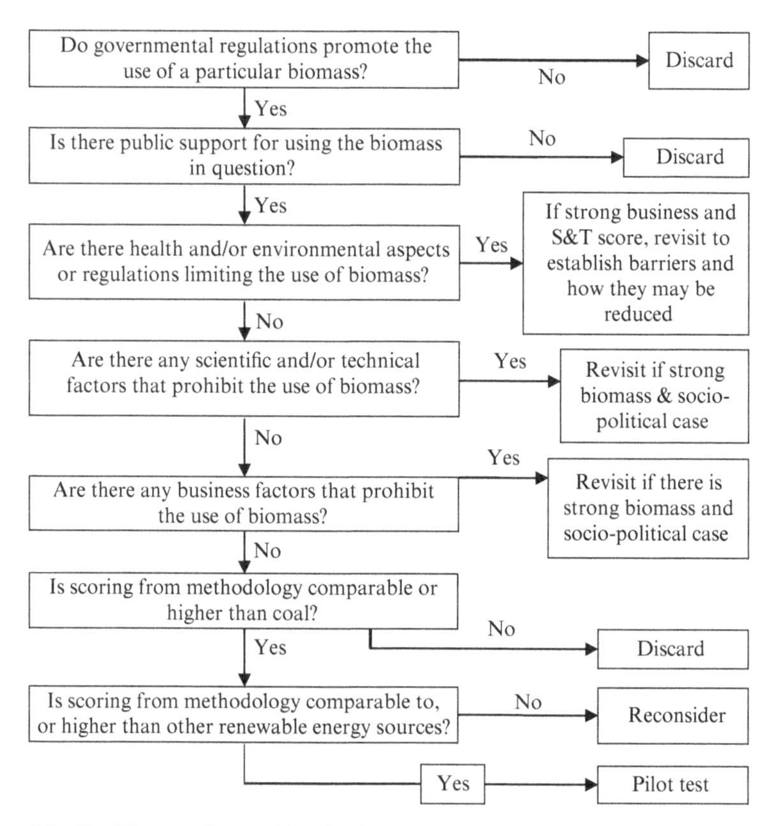

Figure 5.3 Decision tree for reaching the final selected biomass sample(s). *Source*: Author.

5.3 THE DECISION TREE

To make the process approach simpler within the technical and commercial factors, a decision tree was employed (Fig. 5.3). This kind of approach can make the first decisions easier by simply checking each area (and/or relevant field) with an answer of "yes" or "no." This process will shed light on areas of biomass energy prospects, which may be missed with other approaches.

5.3.1 Steps for the Biomass Fuel

As illustrated in the general approach for biomass fuel selection—under the "Decision Tree" in Section 5.3, a guiding approach is used. This approach shows the various steps in the selection process for the final biomass sample(s) that the proposed methodology produces later on. In particular, the basic preparations and steps needed for the production of a new bio-fuel. This test can be repeated several times before the concluding decision is arrived at in choosing the final selected samples, that is, between step 4 and 5 (Fig. 5.4). In

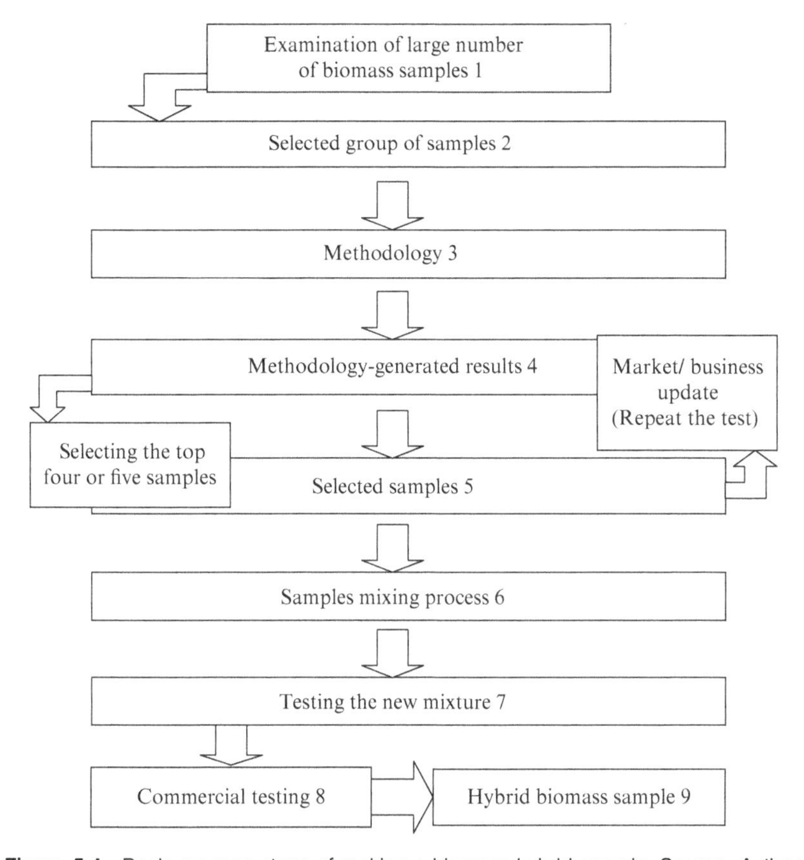

Figure 5.4 Basic process steps of making a biomass hybrid sample. *Source*: Author.

order to fulfill this successfully, four important guidelines have been examined during every step throughout the construction of the methodology. One of these guidelines is that a balance should be made between resources needed and the expected usage and/or demand for the new fuel.

The four vital points which should be considered as part of the methodology's aim and objectives can be summarized as follows (Ayoub et al., undated):

1. Cost minimization
2. Emission minimization
3. Energy consumption minimization
4. Balance between resources and expected demand.

5.3.2 Three Numbers

To provide a scoring mechanism to help in the selection process of biomass materials, a simple numerical scale has been designed. The scale is used for comparison between different materials within each factor of the methodology. The decision is made, therefore, to employ three numbers (1, 0, –1) as the basic scoring values.

Different approaches were also tested during the methodology construction. However, the same or similar results were obtained. Since the book is aimed at a wider scale of readers, with or without formal education, a simple numerical approach to the problem is adopted, rather than any other type of mathematical equation(s)*. Using the above approach is an easy and effective way of obtaining accurate and direct results.

5.4 METHODOLOGY TERMS AND DEFINITION FOR BF AND S&T

During the application of the methodology by a power generating company, the allocation of the percentage value for each factor and subfactor should be provided from their own business working procedures and requirements.

5.4.1 BF

The "BF" as one of the main factors is created from a long list of factors and subfactors (e.g., market, legal, environmental, business, and similar related factors) for the success, or otherwise, of a commercial biomass energy enterprise. These factors be grouped within nine different sections (Fig. 5.5, Box 5.1).

5.4.2 S&T

The scientific element is the result of laboratory tests completed on each sample. These tests were divided into the requirements needed for the selection of the final biomass samples. As illustrated in Figure 5.6, S&T factors have been grouped within seven different sections.

5.5 BF AND S&T DATA

5.5.1 Why Are Data for the BF and S&T Needed?

The methodology can only deal with data in the form of numerical values. For this reason, a scoring mechanism is needed to represent the values of BF and S&T with numbers (i.e., –1, 0, +1) as a representation of the original values for these two factors. All samples, including the reference samples, require

* To help in the calculation and to avoid human errors, as well as for the purpose of converting the methodology into software, a formula has been designed for this purpose. See Section 9.2.1.4.

Factor	Definition
Approach	Project approaches to establish power generating business in biomass/fossil fuel
Present Prices	Today's market prices
Prices' Tendency	Possible price increase or decrease
Harvest/Exploration/Mining	Present production
Available Acres/Reserve	Possible future production
Applicability	Business risks, policies, and adjustment required
List of Risks	All risks related to the new business in this field
Baseline Methodology	Business data, ideas, experience, and fuel preparation
Fuel Preparation	Process and method of preparation
Knowledge	Present knowledge related to the biomass or fossil fuel
New Products	New product emerging from the biomass or fossil fuel
Innovations	New ideas and invention to support biomass or fossil fuel
Business Viability	Present and future available business factors
Government Regulations	Central and local government laws
Investment	Private and governmental project/business investment
Method	Methods related to the general dealing of the business from within and without including marketing and advertising
Energy	The energy used to produce the product (input energy) compared with energy the same product producing(output energy)
Technological	Present advancement in technology/engineering to produce the energy from biomass/fossil fuel
Emission	CO_2 and SO_x NO_x and other gases emissions

Figure 5.5 Definition of business factors (BF). *Source*: Author.

Land Issues and Water	Matters related to the use, cost and purchasing of lands for commercial use related to biomass and fossil fuel. Also, the availability, cost, and use of water for the same purpose.
Supply	Overall supply of biomass/fossil fuel needed for the business locally and internationally
Systems	Systems to generate power from biomass/fossil fuel
Existing Systems	Systems being used presently to produce energy from biomass/fossil fuel
Emerging Systems	New systems and/or systems being developed to be used in the near future
Technology Issues	Life cycle, maintenance and engineering issues
By-products	Deposit (slag)/ash and various types of by-products from biomass and fossil fuel
Quality	Business issues related to quality assurance and quality control

Figure 5.5 (*Continued*)

Box 5.1

DEFINITION

The term biomass was first introduced by Congress in the Powerplant and Industrial Fuel Use Act of 1978 (P.L. 95-620) as a type of alternate fuel. Biomass was first defined in the Energy Security Act of 1980 (P.L. 96-294), in Title II, Biomass Energy and Alcohol Fuels, as "any organic matter which is available on a renewable basis, including agricultural crops and agricultural wastes and residues, wood and wood wastes and residues, animal wastes, municipal wastes, and aquatic plants."

Congressional Research Service (2012)

complete data in order to obtain the fitness value of the biomass sample, that is, in comparison with the value of the reference sample.

5.5.2 How Are Data for the BF Obtained?

The value of each BF depends on changeable factors, some of which cannot be estimated with numerical figures, such as "today's market," the "emerging

Factor	Definition
Energy Content	The total number of joules/gm in the sample
Moisture	The percentage of moisture in relation to the total mass
Combustion Index	The main factor related to the process of combustion, i.e., the process of combustion in a specific period of time
Ignition	Sample's temperature ignition
Burning Period	Sample's temperature during the length of combustion
Ash	The type and amount of ash deposited after 3 days in the oven of 1000°C
Ash Quality	The types of minerals that make up the ash
Ash Quantity	The size and weight of the ash
Volatile Matter	The percentage of VM in relation to the total mass of the sample
Nitrogen Emission	The percentage of nitrogen emitted in relation to the total mass of the sample
Density	The density of the sample for both packing and absolute
Packing Density	The number of storage units per length or area of a storage device
Absolute Density	The absolute density of a material is the weight of a given quantity of the material divided by the sum of the volumes of the particles contained in the same quantity

Figure 5.6 Definitions of scientific and technical (S&T) factors. *Source*: Author.

market,"* and "government regulations." These factors will receive the representational values of either 1 or 0 or −1, according to the present information and data available. To give this kind of immediate representative value to BF, there is the need for good background knowledge of these factors. Up-to-date information on the energy crops market and related variables associated with it and what is going on within the energy commercial sector as a whole are all vital sources in getting accurate data as an input for the BF section of the methodology.

* Concerning biomass materials prices and availability in today's market as well as the future market, these are constantly changeable values that can be compared with what has been termed as a "stochastic process," that is, a random process where variables are constantly changing in their values or level.

5.5.3 How Are Data for the S&T Obtained?

The data for S&T were obtained through a number of technical and laboratory tests, as described in Chapter 3 (further details can be found in Chapter 10). The laboratory tests performed on the selected samples were rigorous and performed in triplicate on each sample. A list of conditions was made before and after laboratory tests had been completed. The list contained technical and scientific aspects related to biomass materials and coal. As discussed in Chapter 3, the technical aspects and their influence provide a platform on how to deal with S&T data. Basic facts are obtained by examining the quality of the sample, the weight of the sample, the physical state of the sample (i.e., powder, seeds, or pellets), and the conditions for and accuracy of the machine being used to perform the test (prior to and after the actual tests). The same data can be obtained repeatedly if similar conditions are implemented as those observed above.

5.6 SCORING SYSTEM

5.6.1 The Method

As mentioned in the previous sections, the scoring method for the methodology (REA1) is to involve the use of three basic numbers. With these three numbers (+1, 0, −1) calculation for the final scoring of the sample (both for the biomass sample and the reference sample) can easily be obtained. The methodology uses coal (crude oil, or crude oil derivatives and natural gas can be used as well) as a reference to compare with the biomass samples, although any fuel may be used in practice as previously discussed. The reference sample* may score higher in comparison with some biomass samples, while other biomass samples may score higher than the reference sample itself. The principle of the methodology is to find the best sample that can be used to generate electricity and/or be used for transportation, heating or cooling systems. This can only happen if the sample scores highly in both BF and S&T sections, regardless of the scoring of the reference sample.

A standard percentage boundary value should be set up in order to mark possible changes in the values of each factor related to S&T and BF, that is, a standard percentage boundary where above or below significant changes may occur.† A value of 3% has been allocated to the calculation. This 3% boundary is used in this book as an example during the application of the methodology,

* Any biomass materials or fossil fuel samples can be used as a reference sample.

† The approach is similar to MCDM hierarchy system. The method is a mechanism for analyzing a problem with multiple conflicting objectives in a scenario where a number of parameters have directly or indirectly certain influence on the overall final outcome.

that is, this percentage can be changed according to the judgment and inclination of how high or low a percentage value each commercial biomass business should use in accordance to their needs. However, the 3% boundary line for the complete working methodology has no effect as it stands. This is because there are already two tables monitoring the boundary level for S&T, as well as the basis of the methodology, which provides a percentage value to every factor, including BF. What this means is that there is no boundary "gap" step, there is only a boundary "line" step, that is, there is no "space" to put any figure/number within the boundary line. The 3% figure allocated to take the scoring from one level into another is purely a "separation" boundary, that is, for distinguishing one level from another. Therefore, the 3% is nothing but a "spare part" of the methodology to be referred to (or used) when and if the need arises for it (further details are in Chapter 8).

To explain the methodology scoring approach, Figure 5.7 illustrates the REA1 mechanism. Additional explanations are also available in Sections 5.6.2 and 5.6.3 in the following pages.

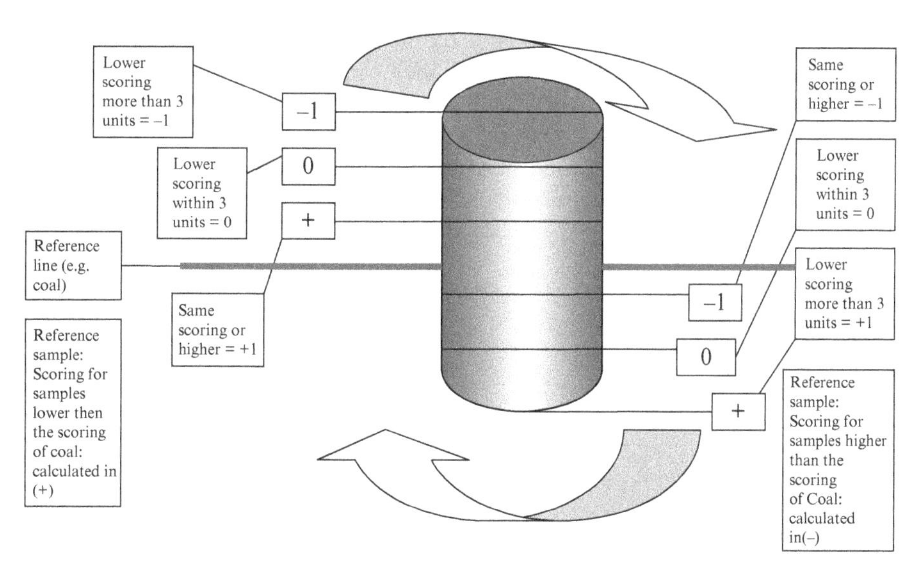

Figure 5.7 The cycle of scoring within the BF and S&T factors in REA1 methodology. 1. Simple and flexible scoring method. 2. Criteria percentage boundaries. 3. The values can be chosen freely, according to the need of the user. 4. Changeable parameters are not critical, as the value chosen represent "today" or future situation. 5. There is no need for complicated mathematical equations. 6. The reference sample in essence is the fulfilling points for the objectives and aims of the proposed project. 7. No limit on the number of factors/variables need to be considered. 8. The final result will be representing the fact on the ground on "today" or "future" terms, that is, according to the choice made in valuing the factors prior to the final scoring/calculation of the overall values. *Source*: Author.

5.6.2 Calculating the Score When the Reference Sample Is Set in a Positive Mode

If the score is the same as the reference sample, or higher, then a score of +1 will be given to the sample and +1 for the standard sample, that is, the reference sample. If the scoring is lower than the reference sample, within three units (3%), then the score applied will be 0 for the sample and +1 for the reference sample. If the sample scores more than three units (>3%) below the reference sample, then the final scoring will be −1, but the reference sample will still score +1 (Altawell, 2012).

5.6.3 Calculating the Score When the Reference Sample Is Set in a Negative Mode

If the scoring of the biomass sample is higher than the reference sample by more than three units (>3%), then a score of +1 will be given to the sample and the reference sample will receive −1. If the scoring is within three units (3%) higher than the reference sample, then the sample will be given +1, and the reference sample will score 0. If the biomass sample is the same as the reference sample, then the sample will receive +1, and the reference sample +1 likewise. When the scoring of the biomass sample becomes equal to or lower than that of the reference sample, then the process will return to Point A and the scoring process will begin again. This is simply because all the scoring has been checked, and therefore, the process has to go back to Point A starting a new scoring for another sample (Fig. 5.7 and Fig. 5.8) (Altawell, 2012).

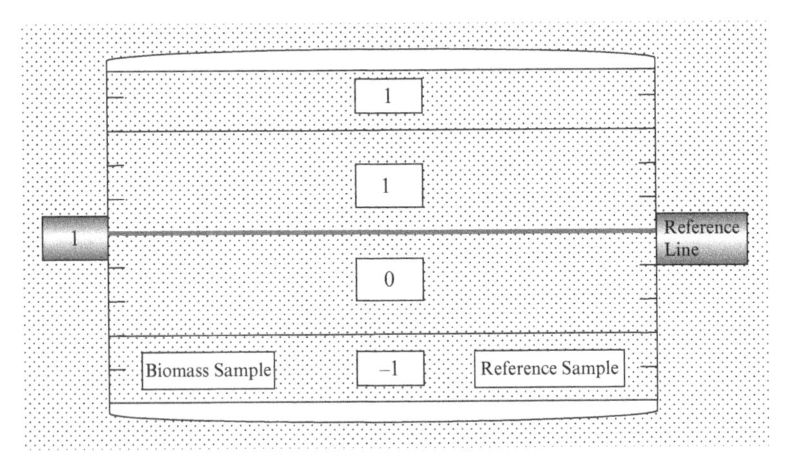

Figure 5.8 Scoring scale for reference and biomass samples. *Source*: Adapted from Altawell (2012).

Table 5.1 S&T factors with higher or lower needed function value

Number	Factor	Function	Abbreviation
1	Energy	F1	E
2	Volatile matter	F2	VM
3	Packing density	F1	PD
4	Absolute density	F1	AD
5	Moisture	F2	M
6	Ash quantity	F2	AQn
7	Ash quality	F2	AQl
8	Nitrogen emission	F2	N
9	Ignition	F1	Ig
10	Burning period	F1	BP

Source: Adapted from Altawell (2012).

The following page provides a simple mathematical equation regarding the functions of S&T factors.

By including the main and subfactors of S&T, the overall total is 10 factors (Table 5.1), from which each sample can be analyzed according to various biomass and fossil fuel characteristics. A simple mathematical model can be made for both BF and S&T factors. This type of modeling can be expressed for the S&T factors in the following mathematical terms. Let F denote any of the 10 factors valued below, $F1$ denote the higher value needed for the factor, and $F2$ denote the lower value needed for the factor. Let A denote any of the samples. A function can be defined as: $F[A]$.

The function $Q[F, B, K]$ is the comparison function for a biomass sample with a coal sample (Altawell, 2012), where Q = function, B = biomass fuel, K = Fossil fuel, f = any of the 10 S&T factors. Using $F1$ and $F2$ functions for higher and lower values (in relation to the required need for each of the 10 factors), the mathematical equations can be written as follows:

$$\mathrm{Q1}[F1, B, K] = \frac{F1[B] - F1[K]}{F1[K]} \times 100 \quad \text{for a higher value of the factor} \tag{5.1}$$

$$\mathrm{Q2}[F2, B, K] = \frac{F2[K] - F2[B]}{F2[K]} \times 100 \quad \text{for a lower value of the factor.} \tag{5.2}$$

The score function is represented by the letter S. $S[F, B, K]$ is as shown in the following equations:

$$S[F, B, K] = \begin{cases} 1 & \text{if } Q[F, B, K] \geq 0 & (5.3) \\ 0 & \text{if } -3 \leq Q[F, B, K] < 0 & (5.4) \\ -1 & \text{if } Q[F, B, K] < -3. & (5.5) \end{cases}$$

5.6.4 Boundaries for S&T

The comparison method for S&T factors has been designed in a more specific way in that there are specific values for each S&T factor, which is not the case with some of the BF.* This is because each factor has an upper or lower limit, whether in connection with fossil S&T factors or biomass factors (Altawell, 2012).

5.6.5 Boundaries for BF

There are no boundaries for BF for value of any named factor in this category.* This means that BF can be represented directly within the three values of REA1 methodology, that is, +1, 0, or −1, in accordance with the properties of the fuel in general, and in accordance to the co-firing process and/or the makeup of the new bio-fuel (Altawell, 2012).

5.6.6 Reference Sample Boundaries

The actual values used in the scoring of the reference sample (e.g., coal†) obviously differ from those of biomass samples (Altawell, 2012). For example, the moisture parameter for coal will score 1 if the values range from 2% to 0%, while for biomass, it will score the same value if the moisture is between 10% and 0%. Biomass possesses a naturally higher percentage of moisture, as well as the capacity to absorb more from its surrounding environment than coal.

For "ash quantity," the scoring for coal means that if it is higher than 20% it will be given −1, while from 20% to 10%, it will score 0. If it is lower than 10%, it will score 1.

Another example of scoring for coal is "energy content." The scoring in this particular case means that lower than 16,000 J/g will be given −1, while from 16,000 to 22,000 J/g, it will score 0. On the other hand, if it is higher than 22,000 J/g, then it will score 1. These values are introduced as a general guideline after a discussion with a representative from a power generating company and with experts in the field of biomass and fossil fuels, including from those who took part in the survey (Chapter 8).

The values of these factors and ranges are changeable if the power generating companies and/or any other type of energy-related business decided to use

* Exact specifications and boundaries are needed within the S&T factors. This is possible to obtain; however, BF, unlike S&T, represents different factors that may not be the same from one day to the next.

† Anthracite coal has the highest percentage of carbon, approximately 86–98%, that is, it is high in energy, and therefore has the highest rank of coal. Bituminous coal ranks second after anthracite coal with carbon content of approximately 46–86%. The lowest is lignite, or brown coal with carbon content of approximately 46–60%, that is, lower in energy compared with other types of coal.

Table 5.2 S&T scoring data for the reference sample (coal)

	Fossil Fuel (e.g., Coal)	S&T	
	−1	0	+1
Moisture	>10%	2–10%	2%–0
Ash quantity	>20%	20–10%	<10%
Ash quality (SiO_2 and Al_2O_3)	<50%	50–90%	>90%
Combustion index (ignition)	<300°C and >550°C	300–350°C and 450–550°C	350–450°C
(burning period)	<200 and >400°C	200–250°C and 350–400°C	250–350°C
Absolute density	<1.3 and > 2.00	1.3–1.4 and 1.8–2.00	1.4–1.8
Packing density	<0.5 and > 2.00	0.5–0.9 and 1.8–2.00	0.9–1.8
Nitrogen emission	>3	1–3	<1
Energy content (samples as received)	<16,000	16,000–22,000	>22,000
VM	<20% >60%	20–30% 50–60%	30–50%

Source: Altawell (2012).

it for their co-firing or bio-fuel production. These businesses can easily introduce their own values and ranges to apply within the methodology. Table 5.2 and Table 5.3 are the result of expert input for both coal and biomass S&T boundaries.

5.6.7 Biomass Boundaries

Biomass boundaries are represented according to the "qualities" that power generating companies are seeking as minimum requirements. The contents of Table 5.3 can be explained as follows (Altawell, 2012):

For "packing density," values between <0.5 and >2.00 g/cm^3 will be −1, while from 0.5 to 0.9 g/cm^3 will score 0. On the other hand, for values between 0.9 to 1.8 g/cm^3, then the score will be 1. For volatile matter (VM), the scoring system means that scores of lower than 30% and larger than 80% will be given −1, while 30–40% and 70–80% will score 0. On the other hand, 40–70% will accrue a score of 1. Other factors, such as the "combustion index," are divided into two parts: part 1 is related to the ignition temperature, and part 2 to the burning period. If the ignition temperature ranges from <200 to >400°C, then the scoring is −1. When the range is between 200 and 250°C or between 350 and 400°C, then the scoring is 0. On the other hand, when the range is 250–350°C, then the scoring is +1. For the "burning period," if the range is >400°C smaller than 150°C, then the scoring will be −1. However, if the range is

Table 5.3 S&T scoring data for the biomass samples

	Biomass (e.g., Energy Crops)	S&T	
	−1	0	+1
Moisture	>30%	30%–10%	<10%
Ash quantity	>5%	5%–2%	<2%
Ash quality (CaO, KO_2, and Mg)	>60%	60%–30%	<30%
Combustion index (ignition)	<200°C and >400°C	200–250°C and 350–400°C	250–350°C
(burning period)	>400 and <150°C	150–200°C and 300–400°C	200–300°C
Absolute density	<1.3 and > 2.00	1.3–1.4 and 1.8–2.00	1.4–1.8
Packing density	<0.5 and > 2.00	0.5–0.9 and 1.8–2.00	0.9–1.8
Nitrogen emission	> 3	1–3	<1
Energy content (samples as received)	<16,000	16,000–22,000	>22,000
VM	<30% >80%	30%–40% 70%–80%	40%–70%

Source: Altawell (2012).

150–200°C and between 300 and 400°C, then the score will be 0. Finally, when the range is 200–300°C, then the score will be +1.

The definition of the "burning period" in this context is the time scale from the point of ignition to the complete extinction of flame and combustion (external and internal).

5.6.8 Scoring Plan for BF

Under the methodology pertaining to the business aspect of biomass energy production, factors and results can be "weighted" according to present business situations and perceived trends. The weighing mechanism applied is explained in the following points using some of the BF methodology factors as an example (Altawell, 2012).

1. Each "value" for each "factor weight" or percentage value as described in the BF scoring, represents what is going on in the current market/situation for that particular factor. This means that if we consider the factor "adjustment" in coal, then it is clear that the present established successful coal businesses does not need commercial adjustment, to the extent that, for example, the straw pellets business does in order to establish itself on the market. The scoring for the coal therefore is "1," as its situation in business terms is a positive one (at the time of writing this

section of the book). The straw pellets business meanwhile is still in the process of trying to establish itself on the market and needs further adjustment to its business approach in order to match the coal business in this respect. For this reason, the straw pellets when it comes to adjustment factor would score −1.

2. The same thing can be said for the other remaining factors. For example, "CO_2 emission" factor for coal would be given −1, as there are many emerging markets in the field of energy that will compete with the present business of coal on the market when it comes to CO_2 emission. That in itself is bad news for the coal business in general, if the biomass energy businesses were to get off the ground successfully.
3. The emerging market for straw pellets, in the form of a new biomass energy business, can be considered as part of an emerging market trying to compete with the traditional established fossil fuel energy market, such as coal, oil, or natural gas. This is positive, regardless of how big the challenge may be for this type of energy business. This is an emerging market and consequently the scoring in this instance for straw pellets is 1.

5.7 METHODOLOGY SURVEY

For the sake of obtaining an example to help in the creation of the priority and value percentage tables, a survey method was specifically constructed. However, the result(s) obtained from the survey method may or may not have value for the power generating companies and for those who work in the biomass energy sector. This is because power generating companies and energy businesses have their own priority tables and their own value percentages, as discussed on a number of occasions in the previous chapters. Each business is completely unique, even when there are similarities and/or when they are producing energy from the same sources, that is, businesses are considered to be "unique" regardless of whether they are operating in the same field or not. This means that different businesses may have different approaches, different managements, different suppliers, different locations, different skills, a different workforce, different aims and objectives, different budgets, different hardware/software, different risks, a different market and so on. A question can be asked, therefore: if the survey method is not important, why bother researching it at all? The answer is simple, there is the need to have priority tables and value percentages for the methodology factors in order to provide a realistic example as close as possible to the need of biomass energy business. This is similar to a "standard format" that is needed for certain projects and methodologies such as REA1.

Finally, the biomass energy businesses do not need a survey method of their own as they already have their own data needed for the application of REA1 methodology.

5.8 THE SURVEY METHOD

Introducing priority listing tables and percentage values for both BF and S&T factors using a survey method needs preparation and careful research. For this reason, a survey to find out the respondents' opinions was thought to be useful in applying factors and data to the methodology. The survey itself is based on three different groups, all of them are working in the field of biomass energy. Group 1 is university staff researchers, group 2 is engineers and technicians, and group 3 is businessmen/economists. These groups originate from various parts of the world.

The main point of the survey was to find out what were/are the methods of listing for the experts in the field of biomass energy when it comes to the S&T factors and their counterpart, the BF.

5.8.1 Aim

The aim of the research is to find out via a survey method, the level of importance regarding BF, and S&T factors among people with a solid connection to biomass energy research and industries. Biomass energy's commercial aspects and CO_2 emissions are the main focus of the survey. Emphasis on large commercial power generating companies where the earlier two factors are related is the main driver in achieving the aim of this survey, that is, to obtain expert input from well-established energy companies dealing fully or partly with biomass energy.

5.8.2 Objective

The survey objective is to establish certain facts related to the makeup of biomass energy methodology factors and to find out the effect of individuals, groups, and overall factors in the selection process of biomass materials.

5.8.3 What Is the Survey Looking For?

The survey is looking at the overall general response toward biomass energy in technical and commercial terms. The questions in the survey are within the field of: energy, combustion, volatile matter, density, moisture, ash, nitrogen emission, systems, production and market, fuel preparation, business viability (short and long terms), business risks, land and water issues, supply, quality, and emission.

5.8.4 Survey Methodology

5.8.4.1 Research Design. The survey used a cross-sectional method (researchers, engineers, technicians, businessmen, and economists). Why cross-sectional? The cross-sectional survey is ideal for a "descriptive analysis." Facts are defined as they are seen, experienced, and learned by people with different backgrounds, working within one central field that is, biomass energy. These are the people that the survey covered that is, the target.

5.8.5 Mode

The survey comes in the form of a written questionnaire. This method has been selected simply because it is less time consuming and cheaper to carry out than many other methods (Henry and Valliant, 2009). In addition, the method can reach all of the people targeted in this survey.

Biomass energy is a vast subject; for this reason, the questionnaire contained additional explanations in order to clarify any difficult expressions/words used. Personal and direct explanations were offered to each individual who volunteered to complete the S&T and BF forms whenever this was applicable and possible.

5.8.6 Mode Effect

After completing the survey, a number of questions were raised and discussed with a number of people who helped in completing this work. Some of these questions have been summarized as follows:

What are the possible negative aspects of this mode? Might the questionnaires not have reached the people intended? Might not all the questions be answered, or maybe questions misunderstood? Could they be answered incorrectly? (Research Methods Knowledge Base, 2009). In addition, respondents may not have given back all the completed forms, or given back incomplete answers to a question or the questionnaire.

These questions explore some of the possible problems with a survey method, which could happen in any large-scale sampling approach.

5.8.7 Questionnaire Design

The questionnaire contained questions about factors mostly related to biomass energy. The questions requested the choice of a number (from 1 to 7 for S&T and from 1 to 9 for BF) plus a value percentage. There were no closed questions, that is, questions with the option of "yes" or "no"; neither were there questions with the option of choosing one or more ready-made answers.

5.8.8 Sample Design

There were 16 questions in the questionnaire itself. The hard copies were printed in two parts: part 1 deals with S&T and part 2 deals with BF. The two parts are on two separate sheets. There are general questions that deal with the type of business, institution, or project. In addition, there is a question related to the type of biomass materials regularly/irregularly used. A space for comment(s) and clarification has also been included (see Chapter 6, Fig. 6.1, and Chapter 7, Fig. 7.2).

5.8.9 Sample Size

More than 1500 copies were printed and around 1000 people were contacted and asked to complete the questionnaire. Some of the people contacted requested two or more of the questionnaire forms. Out of this handout, 475 completed (or partially completed) forms were received back. Out of these 475 forms, there were 107 forms from research staff from various institutes and universities, 188 from people connected to the biomass business energy, that is, businessmen, dealers, and economists, 83 engineers, and 97 in the form of technicians and workers in the biomass industries. That means that out of 1000 handouts, slightly less than 50% responded.

As the survey method used was a written questionnaire, a sampling error would not be relevant to this particular research, especially given that the number of people involved was less than 500.

Reportedly, an ideal number in any large sampling survey is 1000 (De Vaus, 2002). This magic number is considered by those specialized in survey methods as the top of the scale in getting optimum results, that is, a sample of this size should produce a good representation in a large population survey.

5.8.10 Pretesting and Piloting

A pretest was done at the School of Chemical and Environmental Engineering, Nottingham University, UK. The first design of the questionnaire was sent to all the staff and postgraduate students. This was to find out about the way the questions were answered and whether any adjustment might be needed to the questions, as well as to the form, as a first guideline before launching the main survey.

5.8.11 Reducing and Dealing with Nonresponse

Regarding the completion of the forms, the majority of contact made with subjects was face to face. When this approach was not possible, telephone calls and e-mails were used for this purpose. The survey took place at various conferences (in the United Kingdom and the United States), power generating stations, companies, manufacturers, small businesses, and with office workers connected directly or indirectly to the field of biomass energy. In addition to this, announcements were made during the opening of various conferences emphasizing the importance of answering all the questions. There was also a brief explanation on how to answer the questions. The above procedure has clearly reduced non-responses from a number of people who were not sure, or possibly hesitant, about completing the questionnaire.

5.9 CONCLUSION

In order to arrive at the original aim of this book, that is, finding an environmentally friendly biomass fuel and consequently helping to balance CO_2 in the atmosphere, a number of approaches took place in the first stage of the methodology construction. These approaches (e.g., approaches to areas of investigation related to biomass materials, methodology, and bio-fuels) can be compared with larger and smaller scale research. The first area of investigation is made up of a very wide and general selection of biomass materials and related energy aspects. The focus is slightly more specific within the second area, and consequently, the field of investigation became smaller. Progress continued in this fashion, that is, from a larger field of research progressing into a smaller one, until the final target reached successfully, fulfilling the aim and objectives related to this section of the book. This means that obtaining commercially viable fuels from biomass is possible by following a similar process. By trying to build a successful selection process for the production of biomass fuels, a number field was progressively examined and considered. These are the following:

1. A general outlook at biomass materials and possible drivers for their use as a source of energy
2. The definition of all methodology factors and their role
3. The design of a simple but effective scoring mechanism for both BF and S&T factors
4. Agreeing/selecting a standard (reference sample) for S&T tables.

The "pyramid approach" helped in dissecting various factors into their appropriate steps, making the development of a commercial business specialized in biomass fuel much easier. The findings about S&T and BF priority listing and percentage values produced a good response from a large number of participants. Although the total number of those participants was less than 500 (475 out of 1000), that is, a 47.5% response, the final result was better than expected because of the high return to the organizer. The survey method is probably accurate and suggests that the majority of people who were questioned have an in depth knowledge and/or experience in the field of biomass energy. The survey section has been completed within the guidelines discussed in this chapter. This was possible because of the support provided from people and organizations during the launching of the survey. The figures obtained, therefore, have successfully met the objectives specified in this book. This chapter has illustrated the first part of the methodology. Table 5.4 provides a summary. The following two chapters will examine S&T and BF sections in detail (Box 5.2).

Table 5.4 REA1 methodology basic approach: summary

REA1	Factors, Reference, Data and Scoring Values											
BF	Approach	Present prices	Prices' tendency	Harvest/ exploration/ mining	Available acres/ reserve	Applicability	List of risks	Baseline methodology	Fuel preparation	Knowledge	New products	Innovations
	Business viability	Government regulations	Investment	Method	Energy	Technological	Emission	Land issues and water	Supply	Existing systems	Emerging systems	Technology issues
	By-Products	Quality	→ Possible additional factors									
	BF boundaries											
S&T	Energy content	Moisture	Combustion index	Ash	Volatile matter	Density	→ Possible additional factors					
	S&T boundaries											
Reference sample	Coal/crude oil derivatives/natural gas/biomass											
	Reference sample boundaries											
Data sources	Survey, laboratory tests, and commercial environment											
Scoring values	1, 0, –1											

Source: Author.

Box 5.2

UK DEPARTMENT OF ENERGY AND CLIMATE CHANGE (DECC)

Survey—Renewables: Key Finding

- More than three quarters of people (79%) support the United Kingdom's use of renewable energy to generate electricity, fuel, and heat, with 30% strongly supporting it. Just 4% of people are opposed.
- While overall support for the UK relying on a range of renewable energy sources remained high, solar energy was found to have the highest levels of support (82%), followed by offshore wind (72%), and wave and tidal (71%). Onshore wind was opposed by 13% of respondents.
- More than a third of people (37%) support the use of nuclear energy to generate electricity in the UK, whilst a quarter (25%) are opposed. Over a third (36%) neither support nor oppose using nuclear energy to generate electricity.

DECC (2013)

REFERENCES

Altawell N (2012) Energy crops optimisation selection process for the commercial production of bio-fuels. GSTF Journal of Engineering Technology 1(1):25–30.

Ayoub N, Seki H, Naka Y (undated) A methodology for designing and evaluating biomass utilization networks. Slide presentation, process systems engineering division, Tokyo Institute of Technology, Japan.

Congressional Research Service (2012) Powerplant and Industrial Fuel Use Act of 1978 (P.L. 95–620). Energy Security Act of 1980 (P.L. 96–294), Title II, Biomass Energy and Alcohol Fuels. http://www.loc.gov/crsinfo/research/ (last accessed August 4, 2012).

De Vaus D (2002) Surveys in social research, 5th ed. Oxon, UK: Routledge.

DECC (2013) Public attitudes tracker: wave 4: summary of key findings. https://www.gov.uk/government/uploads/system/uploads/attachment_data/file/73107/Key_findings_wave_4.pdf (last accessed December 3, 2013).

Henry K, Valliant R (2009) Comparing sampling and estimation strategies in establishment populations. Survey research methods. European Survey Research Association 3(1):27–44.

Research Methods Knowledge Base (2009) Selecting the survey method. Web Centre for Social Research Method. http://www.socialresearchmethods.net/kb/survsel.php (last accessed December 3, 2013).

6

METHODOLOGY: PART 2

6.1 INTRODUCTION

More than 90% of the total biomass of the earth is made up of plants (Sengbusch, 2003), which simply means that the selection process for locating energy sources from biomass materials that originate from plants is a vast and complex procedure. A random selection of crops alone is clearly a huge task and certainly a time-consuming one.

As mentioned previously, biomass materials can be used for co-firing or on their own, for example, for heating/cooling systems, or developed further to function as a fuel for transportation. In order to achieve this successfully, "scientific and technical" factors (S&T) should be part of the selection process.

The chosen biomass samples should be subjected to laboratory tests using factors from the S&T section. These factors have been designed to help in meeting the standard for a new biomass fuel, as required by power generating companies.

The percentage factors have been left open so that a power generating company, which has the best understanding of the importance of each factor, can choose the appropriate percentage which is best suited to its needs. Nevertheless, the technical and commercial environment within this business can

The Selection Process of Biomass Materials for the Production of Bio-fuels and Co-firing, First Edition. Najib Altawell.

change from time to time, hence the values of S&T factors will change as well. That means at any time in the future power generating companies can change their percentage values regardless of the time scale. This is usually associated with cost, supply, market, and regulations.

In this chapter, the methodology section of S&T factors has been examined and analyzed in order to test the accuracy of their values.

6.1.1 Biomass Samples and Methodology

The method of co-firing using selected types of biomass materials with fossil fuel (e.g., REA1 methodology) is important part for present and future applications. Various materials are considered to be viable as a source of energy; among them are the selected 15 biomass samples.

In order to improve present co-firing methods, a methodology has to be specifically designed for this purpose. As discussed in the previous chapter, the new methodology is divided into two main parts. This chapter examines in detail the deals with the S&T factors, while Chapter 7 examines the business factors (BF, which is the content of the following chapter.

Building on the principles of simple and complex methodologies (Amari, 2009), the baseline methodology assumes that the generation of electricity is viable, from a business and environmental point of view, merely by using selected biomass materials that apply the S&T and BF approach. A question arises concerning the method used to extract energy from biomass sources and whether the optimum result obtained by using the co-firing method is the best way (i.e., biomass with coal). Perhaps using 100% biomass materials would be as good, if not better. To assess either of the above, accurate data are needed, that is, data related to the effect of various S&T factors involved in a co-firing method or for a method using energy crops on their own. For this reason, the question above would be used as part of the investigative tools for the purpose of constructing the methodology in general, and S&T factors in particular.

There were 10 different types of laboratory tests (or S&T factors as termed in the methodology) performed on the samples. The very fact that a priority listing table for S&T created is a reflection of how highly important each factor is, in particular with relation to the other factors within S&T. The decision to choose certain percentage values for these factors depends on the scientific and technical facts related to them.

6.2 S&T VALUES ANALYSIS

One of the purposes of this book is to find a biomass sample (or samples) that contain the highest possible scoring of S&T factors. These factors form the backbone of the S&T section of REA1 methodology. They are important aspects of the selection process and the outcome of the final results. As mentioned previously, the methodology has left the option open for power generating companies to give their own percentage value for each priority factor in S&T, as well as within the "BF" in the other section of the methodology. Each power generating company has a different approach and different priorities in the selection process for S&T factors. Consequently, the power generating companies should be able to decide upon the appropriate values which can be given to each factor. In this way, realistic results will be obtained concerning co-firing with biomass and/or the use of biomass materials on their own for electricity generation and/or for the production of biofuels for transportation and heating/cooling systems. Questions may be asked as to why a certain percentage is allocated to each S&T factor and how. The answer is within the structure of a list for S&T factors (Table 6.13). This list presents all the factors according to their order of priority in relation to the S&T section of the methodology. This prioritized listing table has been designed for use before percentage values are allocated. The priority table was created according to the results obtained from the survey (Fig. 6.1 and Box 6.1).

6.3 S&T FACTOR EVALUATIONS

6.3.1 Energy Factor (EF)

Biomass samples contain different amounts of energy, but in general, dried biomass materials have a heating value of 5000 Btu (5.27 MJ) to 8000 Btu/lb (8.44 MJ) (Cuff and Young, 1980). The standard measure of the energy content in a fuel is its heating value, sometimes called the calorific value (CV) or the heat of combustion. The "heating value" is usually referred to as the "standard way" for measuring the energy content of a fuel (Livingston and Babcock, 2006).

Two terms for lower and higher values of energy have been formulated, higher heating value (HHV) and lower heating value (LHV). The reason a division (or difference in the values) is made under these two titles for CV is because of the water evaporation heat formed from the moisture and the hydrogen in the material itself. The difference in CV between HHV and LHV depends mainly on the chemical composition of the fuel. During the complete oxidation of a fuel, the maximum stored energy will be released and the HHV will be obtained.

LHV is a term invented during the late nineteenth century when it was discovered that heat below 150°C had no practical value as it was impossible for the flue gases of sulfur-rich coal to condense (European Biomass Industry Association [EUBIA], 2007). LHV is the total amount of heat released through combustion of a known quantity of materials initially at 25C°, rising to 150C°.

The ratio of hydrogen to carbon can provide the CV for any fuel.

Date:		Name (optional):							Form Number:
In considering the importance of factors in biomass energy production at your business *(or research)*, is there any priority order? If so, then please rate them in order of priority *(in the Priority Table)*.	**Factor**	Highest	**Priority Table** (please tick a number)				Lowest	Is there a percentage value of importance for these factors? If this is the case, then please give a percentage value to each factor *(in the Percentage Value column)*.	Percentage Value
	Energy	1	2	3	4	5	6	7	
	Combustion	1	2	3	4	5	6	7	
	Volatile Materials	1	2	3	4	5	6	7	
	Density	1	2	3	4	5	6	7	
	Moisture	1	2	3	4	5	6	7	
	Ash	1	2	3	4	5	6	7	
	Nitrogen Emission	1	2	3	4	5	6	7	
	Other Factors:	Factor Name				Value			
									Total 100%
Company/Project:					**Type of biomass materials:**				
Comments:									

Figure 6.1 Part of a survey sample form showing the method of questionnaires for biomass scientific and technical factors (S&T). *Source*: Author.

Box 6.1

THE SURVEY METHOD ADVANTAGES

1. Any size of sample representation is possible.
2. Lower cost than other methods of collecting data on a large scale.
3. Simple to implement.
4. Important statistical outcome.
5. Large-scale surveys can eliminate subjectivity.
6. By taking into consideration possible percentage error, accurate results can be obtained.

DISADVANTAGES

1. May not be suitable for every subject/situation.
2. Question design may not be suitable for some participants.
3. May represent strict rules and approaches, therefore it is inflexible method.

Process and heat combustion (Soo, 2006) are illustrated by the following equations:

$$2C + O_2 \rightarrow 2CO\,(2440\ \text{kcal/kg}) \tag{6.1}$$

$$C + O_2 \rightarrow CO_{2.}\,(8100\ \text{kcal/kg}) \tag{6.2}$$

$$2H_2 + O_2 \rightarrow 2H_2O\,(33{,}910\ \text{kcal/kg}) \tag{6.3}$$

$$S + O_2 \rightarrow SO_2\,(2210\ \text{kcal/kg}). \tag{6.4}$$

The aim is to find a biomass sample that contains the highest percentage of energy. Energy, or the level of energy a biomass sample contains, is one of the most important factors within the construction of the methodology. This means that the priority related to energy is high, and consequently the percentage allocated for S&T factors is also high, when compared with the factors included in the priority listing table (Table 6.13). But how high should it be? The answer might be found in examining other biomass methodologies and researching the importance of the energy value within power generating companies, that is, in relation to other S&T factors.

6.3.1.1 EF Priority and Percentage Allocation. By checking the present priorities in the selection process of biomass materials, whether from the point of view of power generating companies, researchers, industrialists or businessmen, the energy percentage comes close to about a third of the total value in

Table 6.1 Priority and factor weight for energy factor in S&T

S&T Main Factor	Priority Listing	Factor Weight
Energy	1	30%

Source: Author.

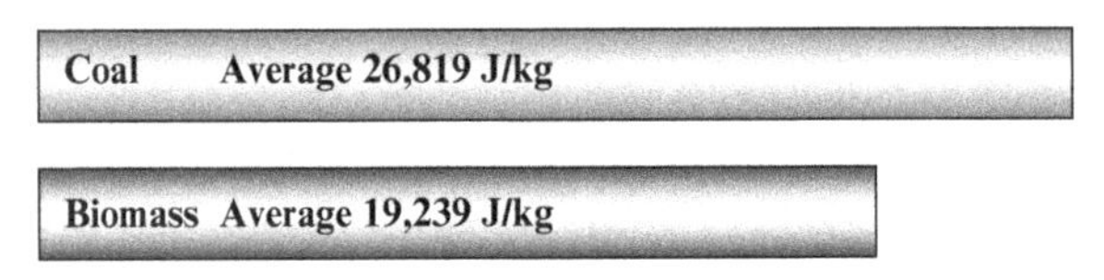

Figure 6.2 Coal and biomass energy: dry heating value (DHV). *Source*: Author.

comparison with other S&T factors. This means it was the top S&T factor (survey, Chapter 8). One might ask why the energy factor is 30% and on what basis the decision was made? Is it accurate to give 30% within the S&T section of the methodology?

All these questions are irrelevant in the actual commercial or noncommercial use of this methodology. The reason is that the percentage value has been left open for the company to decide, as explained previously. Second, the survey produced a result close to this value. That means the percentage value of energy is allocated at slightly less than a third, that is, 30%. Third, the feedback (see Chapter 8, Sections 8.3.1 and 8.3.2) from various staff researchers and those who are connected and/or working within the energy field of bioenergy, from the United Kingdom and various other parts of the world, states that the majority of them marked the energy factor at the top of their list (Fig. 6.1 and Table 6.1). Testing the percentage value of 30% for energy on REA1 is simply to check the viability of the 30% allocated for the energy factor as an example in this book.

The average energy content is represented by the total average values of biomass samples compared with the standard sample (coal) (Fig. 6.2).

6.3.2 Combustion Index Factor (CIF)

A combustible fuel can be defined as a substance that readily burns and as a consequence releases significant amounts of energy. The combustion system (or index) is made up of two different subfactors. These are the following:

1. Ignition
2. Burning period.

The task is to find a biomass sample for which the factor of ignition can easily be achieved. At the same time, the burning period must last longer in comparison to the other biomass materials.

Biomass materials contain higher percentages of volatile matter than a variety of fossil fuels (Mitchell et al., 2004), which means ignition for most of biomass materials can take place easily, if there is not a high percentage of moisture within them. Having said that, a high percentage of volatile matter (VM) can make the burning period of the biomass materials much shorter than that required for a good quality fuel, plus particulates (ACEA/AAMA/JAMA/EMA, 2006).

During combustion, oxygen is used from the air. Each kilogram of oxygen is mixed with around 3.76 kg of nitrogen. The result of the combustion is flue gas, which contains nitrogen. The following equation illustrates:

$$CH_4 + 2O_2 + 7.52N_2 \rightarrow CO_2 + 2H_2O + 7.52N_2 + \text{Heat}. \quad (6.5)$$

For each ton of oil or coal fuel burned at a power plant, the flue gas contains 3–3.5 tons of carbon dioxide.

The amount of carbon in a fossil fuel is much higher than within many biomass materials (Chapters 4 and 9). This is one of the reasons why it is possible to get higher energy, in many cases, from fossil fuels than from certain types of biomass materials. Apart from high or low volatile matter (VM) presence, the length of the burning period can be affected by other factors. These can be related to the amount of carbon within the fuel, and the density of the materials being burned. There are additional factors as well, mostly related to hardware design (e.g., differences in biomass materials are not taken into account when used in some boilers at power stations), flame stability, and hardware maintenance.

Ignition temperature is measured by using two tangents: where they meet indicates what temperature the materials start to ignite at (Fig. 6.3). Since

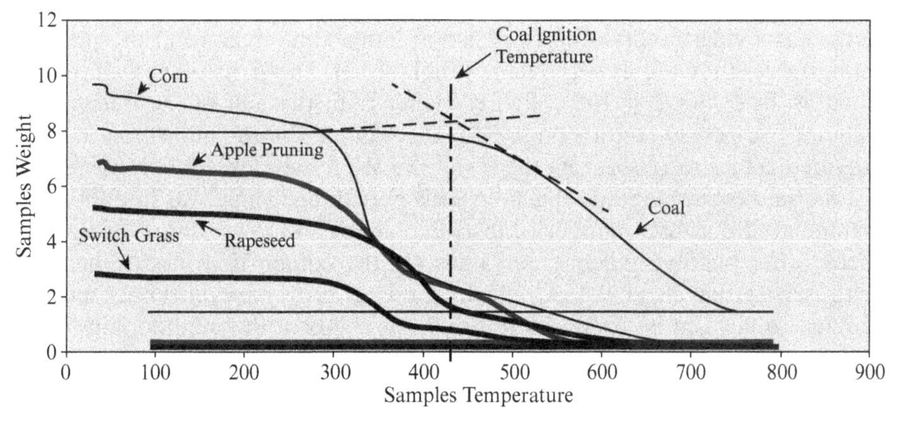

Figure 6.3 Ignition temperature for coal, apple pruning, corn, rapeseed, and switch grass (the ignition temperature for coal is measured by using tangents, where the two tangents cross, the ignition temperature of coal begins). *Source*: Author.

Table 6.2 Priority and factor weight for combustion index in S&T

S&T Main Factor	Priority Listing	Factor Weight
Combustion index	2	16%

Source: Author.

approximately 70% of a biomass substance consists of voids, in many cases, the burning period will be much shorter than some fossil fuels. To help in the assessment of the value of the combustion index factor (CIF) in relation to S&T factors, the following points should be considered:

1. The importance of the combustion index is in its usage of biomass as a fuel for power generating companies.
2. A noticeable difference exists between one biomass sample and another within the biomass samples tested when it comes to the combustion index, for example, ignition and burning period (see Chapter 9, Section 9.3). For this reason, the Combustion Index has additional importance in selecting the "required" ignition temperature and the "required" burning period of the sample itself. This will be the most efficient and economical value for the combustion index, possibly close to those of the fossil fuel values, such as coal (since there are no agreed International Biomass Fuels Standard).
3. Good combustion depends on flame stability (DTI, 2007), even though this may largely be related to the boiler hardware design and the quality of the mix of biomass and coal. Another variable may be the size of particles being injected.

6.3.2.1 CIF Priority and Percentage Allocation. According to the result of the surveying method, the combustion index came second in the priority listing table after the energy factor (Table 6.13). The percentage value allocated as an example is 16%, slightly higher than the half of the percentage value of the energy factor (Table 6.2). The example of the allocation of 16% provided actual results during the testing of REA1 compared with numerous of random research data in this field. This percentage value was intentionally manipulated during a number of tests for higher and lower values, simply to check other possible percentage values for the combustion index. The final test result settled close to 16%, which was used as an *example*. Different percentage values can be used according to the needs of the methodology user. The biomass ignition point is an important factor for power generating companies and for a variety of energy hardware systems, as it is part of the energy being used and consequently part of the overall cost (Fig. 6.3). Each biomass sample has its own ignition point, although the difference between the

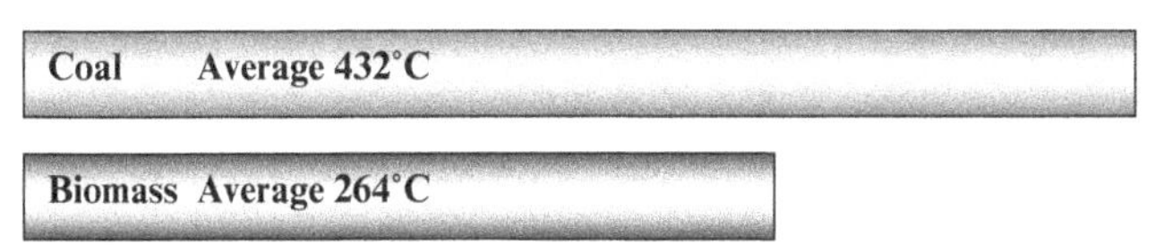

Figure 6.4 South African bituminous coal (Kleinkopje) and biomass average ignition temperature (dry matter). *Source*: Author.

ignition temperatures of the 15 samples is relatively small (see results section). The average ignition temperature of the 15 biomass sample is 264°C, while the ignition temperature for South African bituminous coal (Kleinkopje) is 432°C (Fig. 6.4). Such a large difference in ignition temperatures between coal and biomass materials makes any co-firing process in a single boiler an inefficient and costly procedure. The use of dry matter (average moisture from 12% to 15%) requires a stable ignition point within the boiler in order to be cost-effective. This is one of the reasons that the combustion index came second in the priority list (Table 6.2), but only when it comes to biomass being used to generate electricity. It would have come third on the list if the biomass was used as a liquid fuel (e.g., for transportation, see Chapter 4, Section 4.3.4), as viscosity would be more important than the combustion index.

Finally, below is an example of gas production (for percentages of gas used during a burning stage) throughout the "burning period," according to Lobert et al. (1991).

Element	Flaming	Smoldering
CO_2	63	37
CO	16	84
CH_4	27	73
NMHCs	33	67
----------	---------	----------

Element	Flaming	Smoldering
NH_3	15	85
HCN	33	67
CH_3Cl	28	72
NOx	66	34
----------	---------	----------

6.3.3 Volatile Matter Factor (VMF)

This factor is listed in third place within the priority listing table in relation to the importance attached to it within the S&T methodology section. This importance of the volatile matter factor (VMF) to the biomass fuel, whether with co-firing or without it, is high on the list used to assess a good quality fuel. Too little or too much of volatile matter is not suitable for any type of quality regulated fuel. A balance must be found in which the "right" amount of volatile material forms an important part of the new biomass fuel. In a sense, the standard sample (i.e., coal) has the closest possible level of volatile material make-up within an "ideal" biomass sample needed. Out of the 54% valuation left in the remaining other eight factors, the percentage value for the

Table 6.3 Priority and factor weight for volatile matter in S&T

S&T Main Factor	Priority Listing	Factor Weight
Volatile matter	3	15%

Source: Author.

VM, for the purpose of applying it to the biomass methodology as an example, is decided upon the following factors:

1. The importance of the VMF in relation to the combustion index is a vital chemical element factor (as long as it is close to the required standard). This importance came third on the list (according to survey).
2. Some differences in value exist between one biomass sample and another (see Chapter 9) among the biomass samples tested for VMF. The solution for all these biomass samples is in the reduction of the volatile material to make it closer in value to that within the standard sample. One way of achieving this is through a mixing recipe designed for the final four or five samples in order to produce a new hybrid biomass sample (see Chapter 10, Sections 10.5 and 10.6).
3. The right amount of volatile matter, such as in the standard sample coal, should provide continuous flame stability (Pronobis, 2005).
4. From the test done on the 16 samples, including coal, the difference of VM between these biomass samples and the standard sample can be expressed in the following average value in Figure 6.5.

6.3.3.1 VMF Priority and Percentage Allocation. According to the survey method, the VMF's acceptable percentage came to slightly less than half of the value of the energy percentage: 15% (Table 6.3). The VMF is important in that the percentage value should be reflected, accordingly, in the methodology calculation. The energy and the combustion index play an important role (Yang et al., 2005), and will always precede in value and importance to the VMF when used as a fuel to generate electricity. The decision to place VMF in third place in the priority table, as the results show in the survey method, means that the percentage value would also be less than the previous factors already discussed. It would also be very close to the level of the combustion index, that is, 16% for the combustion index and 15% for the VMF.

The survey showed that there is strong competition, if this word is appropriate here, for position 3 on the listing table between moisture and VMF. In some sense, the response showed that there is an almost equal reason for allocating this position to moisture or VMF. The final decision was to locate VMF above MF, simply because there is a very small percentage on the side of VMF (see Section 6.3.4.1).

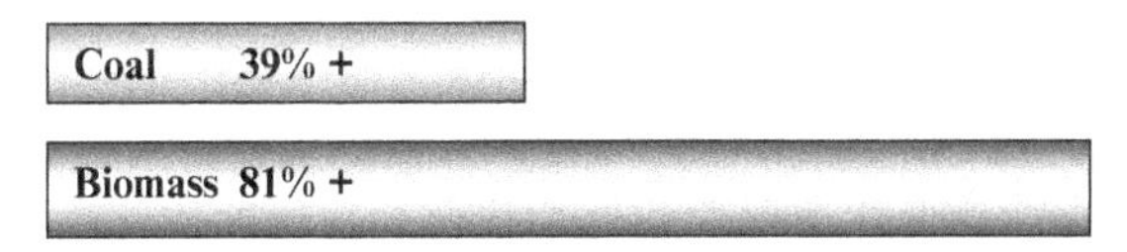

Figure 6.5 VM comparison (between coal and biomass materials). *Source*: Author.

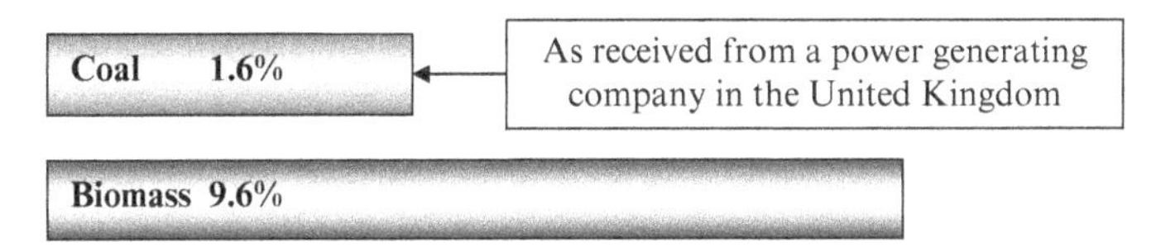

Figure 6.6 Moisture comparison (between coal and biomass materials). *Source*: Author.

6.3.4 Moisture Factor (MF)

Fuel that contains a higher percentage of moisture has less heating value than a dry weight of the same material (Fig. 6.6).

The moisture content in biomass material has a significant effect on various process stages during and after the processing of the final biomass fuel, as well as during its use within the hardware system. A simple question related to biomass for energy use is "why is it important to minimize the content of moisture in biomass materials?" Moisture in biomass materials used as a fuel can be the cause of the following negative effects (Arumugam, 2004):

a. Wastage of unnecessary energy during the initial stage of combustion.
b. Reduction in the system's efficiency.
c. Reduction in the combustion temperature to below the optimum.
d. Incomplete combustion of the fuel, which is usually associated with a higher percentage of creosote and tars. These materials can be the cause of fire or blockage.
e. Condensation of water in the flue, which is the main cause of corrosion.
f. If moisture is high, then a complete shutdown of the system may take place as these systems have been designed and programmed to a standard required for biomass and/or fossil fuels.
g. Storage of biomass materials with a high percentage of moisture can be the cause of composting, damaging the biomass materials and increasing the risk of fire. This means biomass with less energy must be stored within a bigger space.
h. Transporting biomass materials in these conditions means transporting a heavier weight, which yields less energy due to the water within.

Figure 6.7 Moisture effect on flame temperature in biomass boilers. *Source*: Adapted from Mather and Freeman (2008).

Table 6.4 Priority and factor weight for moisture in S&T

S&T Main Factor	Priority Listing	Factor Weight
Moisture	4	12%

Source: Author.

These are some of the main reasons why moisture is an important factor for the power generating companies, especially when associated with the use and storage of biomass materials. Biomass materials tend to absorb far more moisture than coal. This makes them unsuitable for storage under similar conditions to storing coal incorporated into the present design of power generating stations. The average moisture content in the 15 biomass samples used is 9.6%, while the average moisture contained within South African bituminous coal (Kleinkopje) is 1.6% (as received from one of the power generating companies in the United Kingdom), that is, in this case, around eight times more moisture (average value) than the standard sample (Fig. 6.6 and Fig. 6.7).

6.3.4.1 MF Priority and Percentage Allocation. MF is listed fourth on the priority table, according to the survey method (Table 6.4). However, 49% of those in the survey method argued that moisture should be third on the list on the priority table, while 49% believe it should be fourth on the list, 2% mentioned that it could be either third or fourth on the list. What this means is that MF is almost at the same level as VM level, that is, third place on the priority table. According to the need and priority related to the power generating companies, MF can change places in the priority listing table, accordingly. Fortunately, S&T factors within the methodology are flexible enough to allow for any changes needed when it comes to the priority listing of S&T factors and their percentage values. According to the survey method, the MF can be seen almost in the middle of the priority table (Section 6.3.3.1).

6.3.5 Ash Factor (AF)

Ashes from biomass material are chemically different to those from coal. Biomass ash is mostly made up of a mixture of inorganic compounds/elements such as Si, Ca, K, and P (Mitchell et al., 2004). However, ash from coal usually occurs as a material with an aluminoslicate structure (Raask, 1984).

The ash factor has two subfactors. These are the following:

1. Ash quantity
2. Ash quality.

Waste product and by-product are terms that describe biomass and/or fossil fuel ash. However, the amount of ash leftover from both process differs considerably.

Ash quantity is categorized as the amount and weight of ash left after the complete combustion of biomass material or coal in a boiler/furnace.

The ash quality is a consequence of the type of minerals that ash is made up of and may reflect upon any detrimental effects to the boilers used. Another factor is whether or not the ash can be a saleable by-product or not.

If we look at coal ash, then it is clear that this type of by-product has become part of the commercial market, mainly for the cement manufacturing industries. It has been used as a backfilling material, as well as for roadbed material for more than 60 years.

Coal produces much higher quantities of ash than other types of biomass materials (see Chapter 9). The characteristics for the makeup of ash from coal differ from that of a biomass ash sample (Chapter 4 provides analysis for the basic elements of ash, while the results section in Chapter 9 provides an illustration of a comparison, e.g., rapeseed and coal ashes).

The importance of the ash quantity subfactor is related to the type of business that a power generating company may run. If a power generating company sells the ash to other companies and/or farmers as part of another source of income, then the importance here could be high. On the other hand, if the ash produced is nothing but unwanted extra work with no benefit to the business, then the percentage factor accredited to ash quantity can be low. Regardless of whether the ash quantity factor is a positive or a negative one, the percentage value still has to be taken into consideration, even if the value is very small.

When it comes to ash chemical composition and fusion behavior, according to Bryers (1996), there are three types of biomass ash:

1. Low fusion temperature with high silica/high potassium/low calcium
2. High fusion temperature with low silica/low potassium/high calcium
3. Low fusion temperature with a high calcium/high phosphorus.

The ash quality factor is the makeup of the ash and how this may affect the hardware system (such as corrosion) and other types of problems associated with ash (e.g., slagging and fouling), then the ash quality factor would be very important indeed (Pronobis, 2005). Examining the content of biomass ash, various chemical elements can be found (e.g., Fe, Al, Ca, Mg, K, Si, P, and Na), which have originally come from air, water, and soil. The percentage of ash content in biomass differs from that of coal (Fig. 6.8). At the same time, the

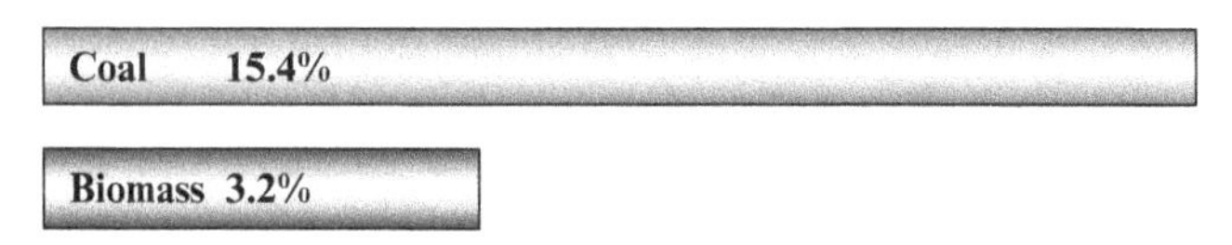

Figure 6.8 Percentage of ash generated after combustion of coal and biomass (ash comparison between coal and biomass). *Source:* Author.

melting point for ash is lower than the melting point of ash obtained from coal (Hills Emergency Forum, 2007). Different ash melting points can be a good indicator of the amount of slagging and fouling in the combustion boiler, as and when one type of biomass material is used in comparison with another (Livingston and Babcock, 2006).

The percentage of ash from most biomass is small, for example, the content of ash from biomass plants can be as little as 1%, or possibly less than this. However, the percentage in certain plants can go up to 12% (in the case of straw and sugar cane residue, referred to sometimes as Bagasse ($C_{3.7}$ $H_{6.4}$ O_3) (Yang et al., 2005). As a result, long-term complex problems occur when combusting these kinds of biomass materials. The problems can be resolved by manipulating the existing chemical elements within the biomass as they are the main causes of slagging and fouling. A new boiler designed specifically for biomass materials should be part of the new design of the biomass hardware system as it has an important part to play in the overall solution to these problems. The percentage value of biomass used during a co-firing method is calculated specifically in order to producc the minimum by-product while keeping the same benefits needed from this process. The first option, that is, manipulating the existing chemical elements within the biomass, is possible to achieve by using calculated mixing percentages from the final four or five samples as one combined biomass sample. Ash quality, as described earlier, is far more important for a power generating company than the amount of ash produced to be sold as a second income (Livingston and Babcock, 2006). This is because inefficiency in heat production (burning of fuels) caused by the ash quality type deposited in the boiler can lead to a higher cost of maintenance during the life cycle of the hardware system.

6.3.5.1 AF Priority and Percentage Allocation. In the priority-listing table, AF has been allocated to number 5, which means that energy, combustion index, VM, and moisture are all allocated at a higher level, and consequently, can be allocated a higher percentage value as well (Table 6.5). The allocation of 11% (as an example) for AF has been selected according to the conclusion made during the work on this book in addition to the result obtained via the surveying method. The two subfactors, ash quality and ash quantity, have been allocated two different values according to their importance to power generating companies. Ash quality, as explained in Section 6.3.5, has a higher priority than the ash quantity factor. The value of 8% is allocated to the ash quality

Table 6.5 Priority and factor weight for AF in S&T

S&T Main Factor	Priority Listing	Factor Weight
Ash	5	11%

Source: Author.

Table 6.6 Priority and factor weight listing for the AF subfactor ash quality in S&T

S&T Subfactor	Subpriority Listing	Subfactor Weight
Ash quality	1	8%

Source: Author.

Table 6.7 Priority and factor weight for the AF subfactor ash quantity in S&T

S&T Subfactor	Subpriority Listing	Subfactor Weight
Ash quantity	2	3%

Source: Author.

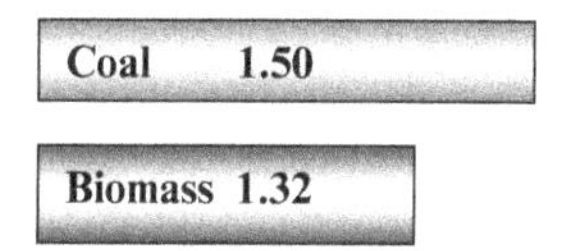

Figure 6.9 Coal and biomass average AD compared for the purpose of illustration as a mean in assisting S&T valuation. *Source*: Author.

factor, while the remaining 3% is given to ash quantity. This is in accordance to the work concluded in this section and from the results obtained via the survey (Table 6.6 and Table 6.7).

6.3.6 Density Factor (DF)

Sixth in the list of priorities is density. Density is made up of two different subfactors. These are the following:

1. Packing density
2. Absolute density.

"Packing density" (PD) is the number of storage units per length or area of a storage device. The "absolute density" (AD) of a material is the weight of a given quantity divided by the sum of the volume of the particles contained within (Fig. 6.9). The importance of the PD subfactor is mainly related to

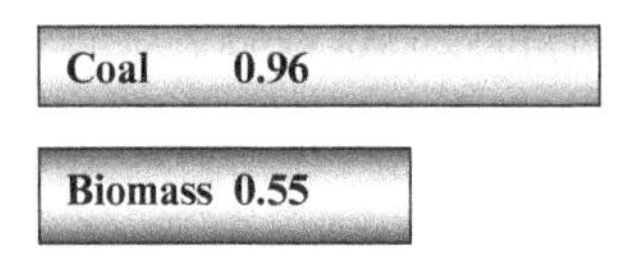

Figure 6.10 Coal and biomass average PD compared for the purpose of illustration as a mean in assisting S&T valuation. *Source*: Author.

transportation and storage, while the AD subfactor relates to the amount of matter within the sample.

Biomass materials, as mentioned previously, are mostly void, that is, around 70% of the total volume is space. In examining PD for biomass materials, a number of factors should be taken into consideration such as "ash content and composition," "moisture content," and "flow characteristics" (Wigley et al., 2007) (Fig. 6.10).

It is important that moisture is always kept to an absolute minimum. When it comes to "ash content and composition," biomass materials have less ash deposit than fossil fuels but with a higher percentage of alkaline minerals (Pronobis, 2005). "Flow characteristics" mean that the granular part of the biomass materials should be made uniform to make the flow easier in relation to silos and bunkers. The percentage value for this factor should reflect its importance in accordance with the priority listing table. Before a percentage value was given to these two S&T subfactors, that is, the AD and the PD, they were first compared with each other. The comparison was performed by changing the percentage values for each subfactor and inputted them into the REA1 methodology. The results obtained were compared with tested values of established data (NREL, 2006; U.S. Department of Energy, 2006).

6.3.6.1 *DF Priority and Percentage Allocation.* When calculating "energy," "combustion index," "volatile matter," and "absolute density," the latter, even though it is part of the main density, can be a little smaller in value than the "packing density." This is because AD, in relation to this work, refers to the amount of carbon per sample. The higher the amount, the higher the energy. Therefore, as the energy has been already calculated, in the form of J/g, the AD has less relevance here. This is one aspect of AD; the other aspect has a connection to the "burning period" under the "combustion index." This factor can have a positive or negative effect on the "burning period" (Yang et al., 2005). The density factor has been allocated the value of 9% (in the example applied in this methodology) (Table 6.8). The following points explain the reason behind this decision:

1. The importance of the DF for the purpose of milling, transportation and storage, and for the amount of carbon within. This came sixth on the priority list.

Table 6.8 Priority and factor weight for density in S&T

S&T Main Factor	Priority Listing	Factor Weight
Density	6	9%

Source: Author.

Table 6.9 Priority and factor weight for the subfactors of packing density in S&T

Subfactors	Subpriority Listing	Subfactor Weight
Packing density	1	(6%)

Source: Author.

Table 6.10 Priority and factor weight for the subfactors of absolute density in S&T

Subfactors	Subpriority Listing	Subfactor Weight
Absolute density	2	(3%)

Source: Author.

2. PD and AD related to biomass materials (results contained in Chapter 9) are close in value to each other. This means that the makeup of the SFS sample will be easy to adjust to the required characteristics of the new SFS biomass fuel.
3. PD is more important as it has more influence on the factors mentioned previously than the factors related to AD. The application of the percentage values for the S&T factors in this example are given as 6% (Table 6.9) for PD and the remaining 3% for AD (Table 6.10). The reason for this kind of percentage allocation is that PD has three important aspects (flow, transportation and storage) ahead of AD, which has only one aspect here (percentage of space).

6.3.7 Nitrogen Emission (N_x) Factor (NEF)

Various environmental agencies around the world accept that emissions of N_x from coal and biomass materials have a dangerous impact on health and on the environment (EPA Report on the Environment, 2008). Some of these concerns have been summarized in the following points (DTI, 2007):

a. The generation of ground level ozone
b. One of the causes of acid rain
c. One of the causes of global warming
d. One of the causes of toxic chemicals and fine particles in the atmosphere

Table 6.11 An example of power plant fuel composition

Component	N_2	CO_2	H_2O	O_2	SO_2 and NO_x
Volume	72	13	12	3	<1 ppm

Source: Livingston and Babcock (2006).

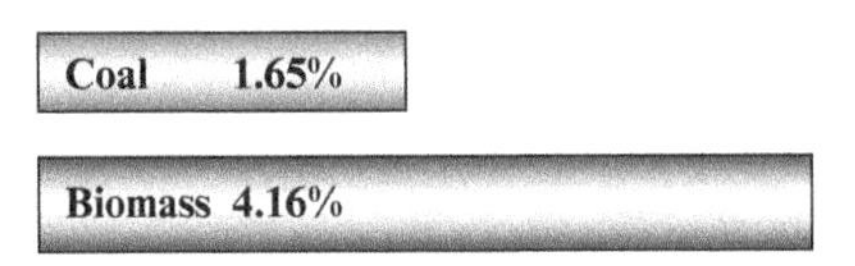

Figure 6.11 Nitrogen comparison (between coal and average content of biomass materials). *Source*: Author.

e. One of the causes of the deterioration of water quality and impairment of atmospheric visibility.

These health and environmental problems make the control of N_x emission from coal, as well as from biomass materials, a very important matter, especially if these fuels are regularly used (Table 6.11). This kind of usage, whether by the power generating companies or via transportation or heating/cooling systems, has to meet a national and international standard in relation to the amount of nitrogen compound allowed to be emitted. The good news is that co-firing can reduce the emission of a number of gases, including N_x.

There are many theories put forward regarding the formation of N_x within combustion systems. It is believed that the combustion of fuel and air may result in the conversion of nitrogen in the air (or within the fuel) into different types of oxides nitrogen, that is, NO_x (e.g., NO, nitric oxide; N_2O, nitrous oxide; NO_2, nitrogen oxide) (DTI, 2007). There are two methods to control NO_x:

1. Combustion control
2. Postcombustion control.

The first method tries to prevent the occurrence of conditions needed for the formation of NO_x during the combustion period. The second converts NO_x into N_2 using reagents. The percentage value for nitrogen emission (from burning energy crops) is relatively small (DTI, 2007) (Fig. 6.11) (Chapter 9, Section 9.3). This small emission can be reflected in the S&T percentage value factor.

The importance placed on the nitrogen factor depends on the efficiency of the hardware system, in particular within the combusting sector at a power generating company. Importance is also placed on the amount of effort and technology used in the postcombustion, in order to reduce the N_x emission. If

Table 6.12 Priority and factor weight for nitrogen in S&T

S&T Main Factor	Priority Listing	Factor Weight
Nitrogen	7	7%

Source: Author.

this is working efficiently as part of a controlling mechanism, then clearly the nitrogen emission will be low.

6.3.7.1 NEF Priority and Percentage Allocation. Despite NEF being an important factor, from both an environmental and health point of view, NEF listing and the value for the power generating companies is not at the top of their considerations. As long as governmental laws regarding the environment and health issues have been adhered to, including the emission of N_x, then the issue here is how efficiently and less costly it can be controlled. NEF from a technical perspective is the responsibility of the power generating companies to manage and control as the technical know-how in this field is already well established. The business factor, that is, the cost for NEF, is acceptable in commercial terms for generating electricity.

As this is the last allocation in the priority-listing table, the priority number is seventh on the list and, incidentally, the percentage value is seven as well (Table 6.12).

6.4 S&T ALLOCATION RESULTS

6.4.1 Introduction

In the literature, there are two types of allocation technique. The first is referred to as the "simple method" and the second one is the "complex method" (Falcone et al., undated). Deciding on which type of method to use depends mostly on data/information available about the system. The simple method is used when there are no sufficient data available about a factor's/component's characteristics and some of the data are obtained from subsystem factors. However, when there is already enough data related to the main and subfactors value and their characteristics, the complex method is a better choice. Examples of allocation methods with brief explanations (Amari, 2009):

A. *Equal (or Equal Apportionment).* A simple method that can be used when data are scarce concerning the system. A number of components or subfactors systems can be used if they are available.
B. *Base.* Assumes that failure rates related to subsystems and their relative difficulties in reducing them are already known.

C. *ARINC.* Assumes that subsystem's failure rates are already known by understanding the exiting failure data and/or average failure rate possibility.
D. *AGREE.* Looks into various factors, such as subsystem complexity and time factors. Possible failure is applied to all elements within the system, as well as the importance accorded to the subsystem of the overall operation.

These are a few examples of different allocation methods used in a variety of field systems, such as scientific, engineering, economic, business, and social science. Some of them may need statistical methods to build upon. Regardless of the application, the principles are the same, that is, the target in any methodology is simply to reduce error and achieve certain aims and objectives within the system. The S&T factors used both "simple" and "complex" approaches during the building of this section of the methodology. Examining a general approach to S&T factors in connection to how accurate the final result in this section of the methodology can be, as well as the time and cost aspect vital data obtained regarding the ten factors which added to the priority and percentage allocation to become an important part of the main methodology.

Main factors and subfactors (Fig. 6.12) make up the main block for S&T that can be later be built upon if an additional factor or factors are needed/or need to be removed from the methodology tree. The whole methodology, including the S&T section, is designed and built so that additional changes and future upgrades can be made without major adjustments or difficulties.

6.4.2 The Priority List

From the survey conducted during the writing of this book (Chapter 8), and from conclusions drawn from the work on biomass and fossil fuel samples, it is evident that the factor of "energy" is the number one priority in the majority of cases. This is regardless of whether it is related to electricity generation or for the production of liquid and gas fuels for combustion engines and related hardware.

Listing these factors in order of priority, the results obtained are in the following order shown in Table 6.13.

Having obtained the position of each factor's priority in the list, the second task is to allocate a percentage value to each factor. The mechanism for this is made on two fronts:

A. The first is feedback from those who are experts in the field of biomass energy (from scientific, technical, and engineering backgrounds, along with businesses operating in this field).
B. The second is the manipulation of different percentages within the methodology factors. These different percentages are checked by the comparisons between these results. This comparison will indicate what

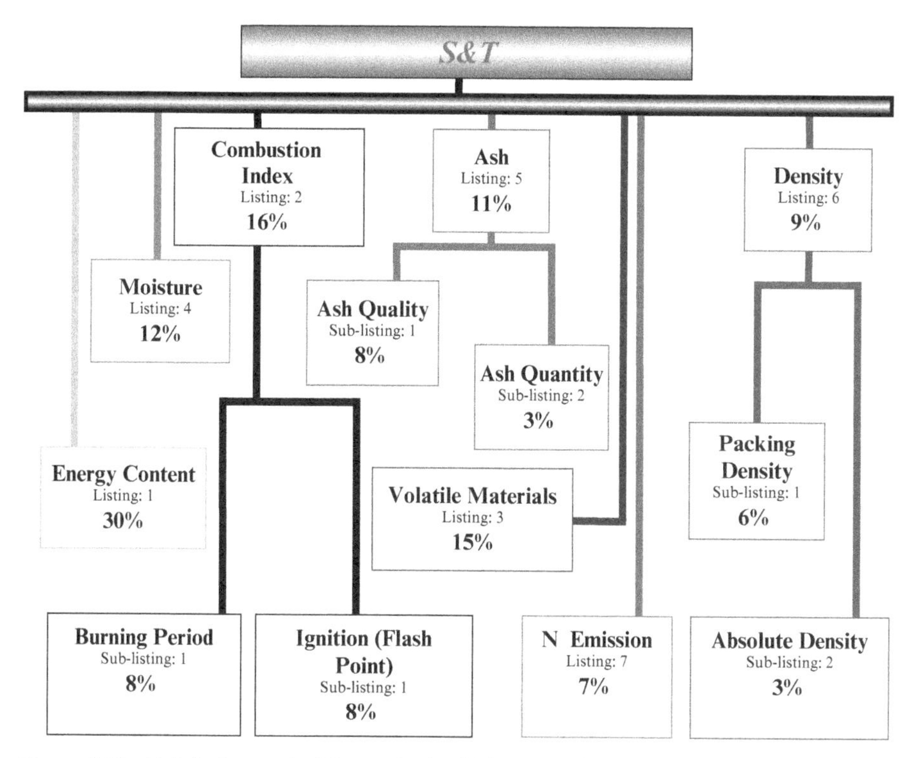

Figure 6.12 Main factors and subfactors for the S&T section of the methodology. *Source*: Author.

Table 6.13 Priority main factors listing table for S&T

S&T Main Factors	Priority Listing
Energy	1
Combustion index	2
Volatile materials	3
Moisture	4
Ash	5
Density	6
Nitrogen emission	7

Source: Author.

is the "right" percentage value needed for each S&T factor for a particular situation or example.

Table 6.14 is the predicted best outcome for allocating each percentage for the main S&T factors.

Table 6.15 shows the allocations of percentages to the subfactors of S&T. This of course can be changed should the need arise by those who may require a different percentage.

Table 6.14 Percentages of main factors (weight factors) for S&T factors

S&T Main Factors	Priority Listing (Weight Factor) (%_)
Energy	30
Combustion index	16
Volatile materials	15
Moisture	12
Ash	11
Density	9
Nitrogen emission	7

Source: Author.

Table 6.15 Priority subfactors listing table

Subfactors	Priority Listing (Weight Factor)
Ignition (combustion index)	1 (8%)
Burning period (combustion index)	1 (8%)
Ash quality (ash)	1 (8%)
Packing density (density)	2 (6%)
Absolute density (density)	3 (3%)
Ash quantity (ash)	3 (3%)

Source: Author.

Another testing approach used is the "comparison method." This kind of testing involves changing the percentage value for each factor and each subfactor using the flexibility of the methodology REA1. The input of various percentage values of the above factors and subfactors into REA1 creates several results. The results obtained are always compared with the values of data from similar work. When the data matche or are close to other researched data, then it is clear that the percentage value given to that particular factor (or subfactor) is correct. The best way to ascertain how accurate the S&T methodology section data are would be to compare it with available published data.

Finally, a percentage error has been allowed for the system to range from <1% to <9% (Chapters 9 and 11).

6.5 CONCLUSION

When using biomass in a co-firing system (for the purpose of generating electricity) and for the commercial production of bio-fuels (for the combustion engines and related hardware), a number of factors have to be considered before a decision can be made regarding the type of biomass material best suited for these procedures. The same thing will be applied should a power company decide to use 100% biomass materials for the purpose of generating electricity.

The conclusion of this chapter can be summarized in the following points:

1. The S&T factor's percentage value for the main methodology have been examined and checked using comparison methods through the application of the REA1 methodology. By using a different percentage value for each application, and for each factor and subfactor, a balance has been achieved in reaching a standard example.
2. In addition, other methods were used to verify the percentage values of the factors and subfactors, such as the results obtained by using the "surveying method" (see Chapter 8), and the consideration given to the level of importance of each factor to power generating companies.
3. Technical issues, such as biomass degradation, corrosion, transportation, storage, and by-products, have to be taken into consideration when examining S&T factors. (Technical and engineering issues have already been examined in detail in Chapter 3.)
4. The scientific test and the making of a new hybrid biomass sample is another way of improving the final fuel and keeping the cost of biomass materials relatively low.

A successful validation of S&T factors, allocations, and values is an essential step for the next stage in the mathematical pyramid model. Here, BF will be investigated and validated in a similar way, but from a commercial angle. The S&T analysis results will be kept in mind while BF are being validated. This is because BF are part of the same chain, which will be investigated in the third part of the methodology in the following chapter (Box 6.2).

Box 6.2

DISTILLERS DRIED CORN (DDC)

DDC is usually obtained from two sources:

1. Brewers
2. Ethanol plants.

The DDC can be a co-product or by-product within the food and drink industries. In the West, however, it is mostly related to spirit (e.g., whisky and beer-making). DDC is used mostly as animal feed. DDC can be obtained from grains (e.g., corn). The grains are crushed and then hot water is added. After the mixture cools, yeast is added, and the mixture is left for few days of fermentation. After fermentation is completed (around 7 days), the solid part that is left over is what has been termed distiller grains or distiller corn. The fact that two-thirds of corn are made up from starch means approximately 2.6 gal of ethanol, 17 lb of carbon dioxide, and a wet-mash will be produced from one bushel of corn. A number of steps are usually taken before a final DDC can be obtained, whether for the aim of producing animal feed or simply for the purpose of obtaining a by-product to be used as a fuel in a co-firing environment:

1. Centrifuges.
2. Evaporators.
3. Presses to produce soluble (liquid) and distiller grains (semi-dry).

Regarding animal feed, the soluble/liquid and distiller grains are blended and dried to produce approximately 17 lb of animal feed from one bushel of corn; alternatively, the distiller grains can be dried and used as a fuel.

The chemical elements of DDC are the following:

Carbon	49.92%
Hydrogen	6.38%
Nitrogen	6.60%
Oxygen	37.08%
Sulfur	0.02%

Source: Author.

REFERENCES

ACEA/AAMA/JAMA/EMA (2006) World Wide Fuel Charter, European Automobile Manufacturers' Association/American Automobile Manufacturers' Association/Japan Automobile Manufacturers' Association/Engine Manufacturers' Association, September 2006.

Amari S (2009) Choosing a reliability allocation method. Reliability articles, relex. http://www.relex.com/resources/art/art_allocat3.asp (last accessed December 3, 2013).

Arumugam S (2004) Nitrogen oxides emission control through reburning with biomass in coal-fired power plants. Thesis, Texas A&M University.

Bryers R (1996) Fireside slagging, fouling and high temperature corrosion of heat transfer surfaces due to impurities in steam raising fuels. Prog. Energy Combust. Sci. 2229–120.

Cuff J, Young WJ (1980) U.S. Energy Atlas. New York: Free Press/Macmillan.

DTI (2007) Engineering Report. Guidance document on biomass co-firing on coal fired power stations. Project 324-2.

EPA Report on the Environment (2008) Nitrogen oxide emission. U.S. Environmental Protection Agency.

EUBIA (European Biomass Industry Association) (2007) About biomass. http://www.eubia.org/about_biomass.0.html (last accessed August 26, 2008).

Falcone D, Silvestri A, Di Bona G (undated) Integrated factors method for reliability allocation: application to an aerospace prototype project. Department of Industrial Engineering, University of Cassino. http://www.scs.org/scsarchive/getDoc.cfm?id=2107 (last accessed March 2, 2009).

Hills Emergency Forum (2007) Biomass Management: fuel removal and mulching, biomass management working paper.

Livingston B, Babcock M (2006) Workshop on ash related issues in biomass combustion. Arranged by Livingston B., Glasgow.

Lobert JM, Scharffe DH, Hao WM, Kuhlbusch TA, Seuwen R, Warnesk P, Crutzen PJ (1991) Experimental evaluation of biomass burning emissions: Nitrogen and carbon containing compounds. In JS Levine, ed., Global Biomass Burning: Atmospheric, Climatic, and Biospheric Implications. Cambridge, MA: MIT Press, pp. 289–304.

Mather MP, Freeman MC (2008) Moisture & char reactivity modeling in biomass cofiring boilers. Dinesh Gera, Fluent Inc., National Energy Technology Laboratory, Combustion and Environmental Research Facility. http://www.fluent.com/about/news/newsletters/01v10i1/s3.htm (last accessed September 11, 2008).

Mitchell RE, Campbell PA, Ma L, Sørum L (2004) Characterization of coal and biomass conversion behaviors in advanced energy systems. GCEP technical report, 202–209.

NREL (National Renewable Energy Laboratory) (2006) Biomass Research, USA.

Pronobis M (2005) Evaluation of the influence of biomass co-combustion on boiler furnace slagging by means of fusibility correlations, institute of power engineering and turbomachinery, Silesian University of Technology.

Raask E (1984) Mineral impurities in coal combustion: behavior, problems, and remedial measures. Washington, DC: Hemisphere,

Sengbusch PV (2003) The flow of energy in ecosystems—productivity, food chain, and trophic level. Botany on line: Ecosystems. http://www.biologie.uni-hamburg.de/b-online/e54/54c.htm (last accessed December 3, 2013).

Soo MI (2006) Technology options for reusing biomass waste for energy recovery. Regional Training Seminar on EE and RE for SMEs in the Greater Mekong Subregion of ASEAN.

U.S. Department of Energy (2006) Biomass energy data book. Energy Efficiency and Renewable Energy, USA. http://cta.ornl.gov/cta/Publications/Reports/ORNL_TM_2006_571.pdf (last accessed April 22, 2009).

Wigley F, Williamson J, Malmgren A, Riley G (2007) Ash deposition at higher levels of coal replacement by biomass. Sciencedirect, Fuel Processing Technology 88(11–12):1148–1154.

Yang BY, Ryu C, Khor A, Sharifi NV, Swithenbank J (2005) Fuel size effect on pinewood combustion in a packed bed. Sciencedirect, Fuel 84(16):2026–2038.

7

METHODOLOGY: PART 3

7.1 BF PERCENTAGE VALUE SELECTION

7.1.1 Introduction

Accurate data for business factors (BF) are more difficult to establish compared with the data obtained for the scientific and technical (S&T) section of the methodology. This is because some of the business factors (BF) values are constantly changing and therefore they are subject to the date on which they have been obtained. This means that data in this section are measured much more strictly in order to establish their accuracy at the time they have been gathered and the reliability of the source. In addition, subjectivity can be a hindering factor when commercial and business aspects are analyzed for their actual figurative values. In essence, the data gathered should be able to complete the equation together with S&T data for a successful establishment of a commercial biomass energy business via the overall implementation of Renewable Energy Analyser One (REA1).

This chapter examines in detail individual factors in the form of evaluations, subjectivity/objectivity, percentage allocation, priority listing, and other related commercial aspects.

The Selection Process of Biomass Materials for the Production of Bio-fuels and Co-firing, First Edition. Najib Altawell.

7.1.2 BF Subjective and Objective Factors

The most common question about BF is whether these factors are generally "subjective" or "objective." This part of the methodology will discuss this gray area and market forces actively engaged in manipulating prices and supply of biomass materials.

Unlike S&T factors, the BF valuation of 1, 0, or −1 is a much more straightforward approach. The fact that BF has changeable values (e.g., prices, supply, and quality) makes the selection more like day-to-day valuation rather than the usage of a standard value. The user approach, therefore, for BF is to apply a different but simpler way of obtaining the scoring for a particular BF factor.

The BF have been analyzed under various circumstances, that is, examined for their practicalities, tested by the REA1 methodology, and researched via the survey. These tests have been applied to BF (factors) in order to ascertain, as closely as possible, the reality of a business, in general, and for biomass energy BF values, in particular. Calculating the value of any business factor whose influence can determine the outcome successfully or otherwise can be difficult to carry out accurately. This is more marked when the factors relate to a new business, such as in the field of commercially proposed projects or renewable energy. For this reason, an accurate method has to be designed in which a future business scenario would be possible. Whether or not these proposed businesses are provided with a "green light" to indicate that it is safe and practical to establish and invest in them is what the BF section (Section 7.3) of the methodology is all about.

Part of the present biomass energy book's aim and objectives is to solve the dilemma of whether or not certain types of biomass materials can be used within a commercial environment purely for the purpose of generating energy. BF are a combination of a large number of various categories related to different areas and fields, such as "market," "legislation," "business," "products," and "land and water." The aim of having such a large selection of categories is to understand the influence of these factors on biomass energy businesses. The scoring of these factors will give a good indication as to how and why choosing one particular type of biomass material is better than choosing another. This basic but important process of selection can perform a useful "before and after" function in any biomass project/business that will be launched on a commercial scale.

The question often asked is: "How accurate is the scoring system and how close is the final result generated through the BF method to the actual reality of running a biomass businesses today?" The answer to this question is within the application of BF by the power generating and bio-fuels companies using their own business factor values.

7.1.3 Percentage Allocation for BF

BF form an important part of the methodology selection process of biomass materials for energy production. These factors can make up to 75% (or more) of the total scoring, depending on the importance of the BF to the project or to the business as a whole. Unlike S&T factors, the allocations of the factors in a priority listing-table and percentage values, is not needed during the use of the REA1 methodology. The valuation can be made using positive, neutral and negative factor influence that is, 1, 0 or −1, directly. Anything left from the above percentage allocations (i.e., the remaining 25%) can be set aside for the S&T factors section. BF scores are high because there are more factors which need to be considered (e.g., main and subfactors) than S&T factors. These factors are important as they make up the main basis of a commercial biomass business. However, if a project is mainly for scientific purposes or completely unrelated to commercial operation then the BF percentage values will shrink, and in a few very unique cases may even be canceled altogether.

In the first instance, BF are allocated according to their importance in relation to the biomass power generating companies with the sole purpose of generating electricity. However, BF can also be applied to any bio-fuel business.

7.1.4 BF Values and Headlines

BF values are not taken from a particular business, rather they are the result of a survey conducted on a variety of biomass energy businesses. The final average values (percentage values) refer to the final allocated results. Table 7.1 contains the main BF.

7.1.5 Biomass Energy Commercialization and BF

For any business to succeed, regardless of how noble and useful the idea behind it, "cost" or "market cost" is one of the most important factors in its success. The same principle is applied to the commercialization of biomass energy. The basic calculated cost of producing biomass energy crops is illustrated in Figure 7.1. This is plotted against the cost of the energy output, which has to be lower, not just for this process but also for other types of energy

Table 7.1 A list of main BF factors

Business viability
Approach
Emission
Baseline methodology
Systems
Applicability
Supply
Quality
Land and water issues

Source: Author.

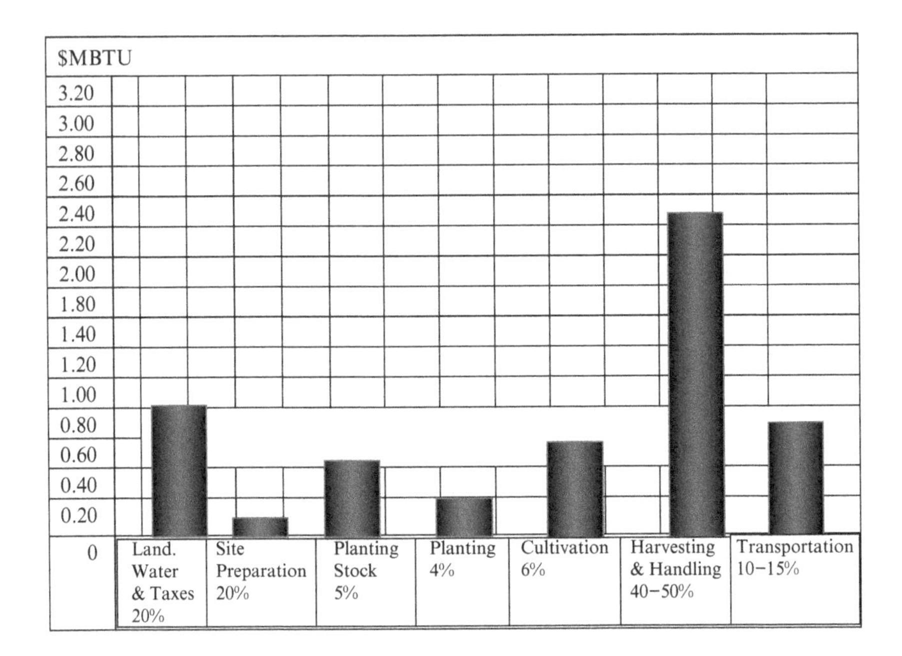

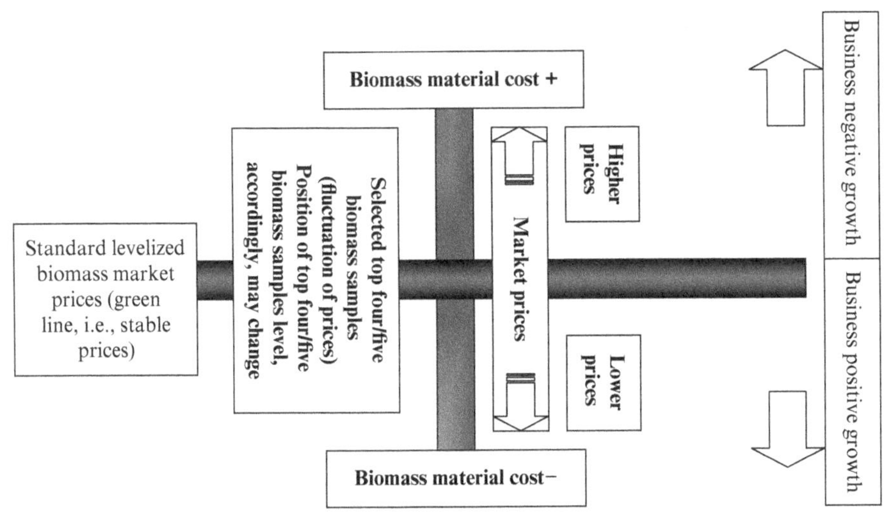

Figure 7.1 The top panel illustrates allocation cost for the production of energy crops (adapted from Turnbull, 1994, Electric Power Research Institute). The bottom panel illustrates the effect of market prices on the top selected biomass four/five samples and how their position level in the BF selection process may change, accordingly. *Source*: Author.

production. Figure 7.1 also illustrates how the market factor within BF may influence the top selected biomass samples position as prices (or other market factors) change.

The various stages related to the cost of producing fuel from biomass materials have been listed under the following points:

1. Biomass cost
   A. Cost of growing energy crops
   B. Transportation cost of biomass materials from farms
   C. Conversion facility delivery cost (if additional or differing from point B).
2. Energy conversion technology cost
   A. Capital cost
   B. Feedstock cost
   C. Operating cost
   D. Maintenance cost.

Considering these basic costs, energy produced from biomass should be examined in detail from various commercial points of view. This can only be done when the BF have been applied in full.

7.2 BF VALUES ANALYSIS

The purpose of this section is to find a biomass sample (or samples) that can meet the market, the business, the legal, and the economical needs for a new type of biomass fuel. The scoring system for BF, as mentioned earlier, is designed using a simple mechanism in the form of positive (1), neutral (0) or negative (–1) scoring for each factor. The methodology has left the option open for the power generating companies to give their own value to each factor.

Each biomass energy business has a different approach, and therefore different priorities in the selection process for business factors. Whether a power generating company wants to concentrate on one source of energy, or on a variety of sources, the business itself should be able to decide on what is the appropriate value given to each factor. In this way, realistic results can be obtained concerning co-firing with biomass and/or the use of biomass materials on their own, whatever the case for the power generating/bio-fuel company. All the questions that can be raised about why and how a certain percentage should be given to each individual factor are easily answered according to the BF priority list structure. This list should be compiled by the energy business itself. Thus, the business can allocate a value to each factor according to their order of "priorities." As an example for the BF section of the methodology, a priority listing table was designed before percentage values were given to the

Date:	Name (optional):	Form Number:

Is there a percentage value of importance for these factors? If this is the case, then please give a percentage value.	The importance of factors of your biomass energy production at your business *(or research)*, are there any priority order? If this is the case, then please rate them in order of priority.	**Factors**	Highest	**Priority** (tick a priority number that applies to you)							Lowest	**Percentage Value**
		System	1	2	3	4	5	6	7	8	9	%
		Production & Market	1	2	3	4	5	6	7	8	9	%
		Fuel Preparation	1	2	3	4	5	6	7	8	9	%
		Business Viability Long Terms	1	2	3	4	5	6	7	8	9	%
		Business Risks	1	2	3	4	5	6	7	8	9	%
		Land & Water Issue	1	2	3	4	5	6	7	8	9	%
		Supply	1	2	3	4	5	6	7	8	9	%
		Quality	1	2	3	4	5	6	7	8	9	%
		Emission	1	2	3	4	5	6	7	8	9	%
		Other Factors:	Factor Name					Value				%
												%
												Total 100%

Business/University/Project:	**Type of Biomass Materials:**

Comments:

Figure 7.2 Part of a survey sample form showing the method of questionnaires for biomass BF. *Source*: Author.

factors (in the same way as has been done in regard to S&T factors) with the help of results obtained from the survey method (Fig. 7.2).

7.3 BF EVALUATIONS

Evaluations factors for biomass businesses have been outlined and defined briefly in the following sections.

7.3.1 System Factor (SF)

The system factor is concerned with varieties of subfactors related to energy production systems within the biomass energy field. A general categorization can be made under the following titles:

1. Present systems
2. Emerging systems
3. Technology issues
4. By-products.

The examination of each of these factors is made from a commercial point view, even though the technical and engineering aspects are firmly part of these factors and therefore cannot be separated from them, especially in the final results. These technical and engineering issues have been discussed in detail in Chapter 3.

1. *Present Systems*. Many of the current biomass systems are still at the development stage and/or at an early stage of commercial trading. Across the world, commercial and noncommercial research and development in the field of energy production is developing fast. In fact, the development (apart from the Internet/PC and mobile phones) is faster than most of the other fields in recent human history. The need for a new source of renewable energy combined with the regulations related to global warming issues put work related to the field of energy higher on the agenda for governmental and non-governmental organizations.

A full list with details of various biomass energy production methods has been discussed in Chapter 2.

2. *Emerging Systems Factor (ESF)*. In order for most of the emerging biomass energy production systems to develop, a number of developments in related fields have to take place at a similar pace. Therefore, a system for the generation of electricity and/or for the production of bio-fuels from biomass may depend on a number of these other new fields (Andresen et al., undated):

a. New energy crops.
b. New type of oil seeds.
c. Developing biomass management.
d. New specialized farm machineries.
e. Higher energy crops plantation.
f. More refined biomass pretreatment (to include reduction in cost and time).
g. Improving supply chains.

h. Advances in combustion and co-combustion.

i. New biomass systems standard (national and international).

j. Better economical value for the end user.

Additional details concerning the above can be found in Chapter 2 and the following chapters.

3. *Technology Issues Factor (TIF)*. Present obstacles for the biomass energy industries are associated with a number of technical issues. These technical issues pose considerable problems related to maintenance cost, time, and energy waste. The TIF, as a subfactor of BF, requires a cost-effective solution in order to be part of a long-term successful commercial biomass energy business. Details related to technical and engineering issues have been discussed in Chapter 3.

4. *By-products Factor (BPF)*. By-products come in the form of ash and gases. Their effect on the hardware and on the environment is not part of this subfactor. Regarding ash, its BF importance will be decided by individual power generating companies on the basis of whether this type of by-product provides a negative or a positive aspect to the business. As has been mentioned previously, if ash provides a second income for the business, the scoring in the BF will be on the positive side. On the other hand, if the ash incurs additional cost to the business, then the score in the BF will be on the negative side. If the balance is equal, that is, no profit and no loss, then neutral scoring will be recorded. Other by-products, according to governmental regulations, will be treated in a similar way within the BF system.

Details concerning the ash factor and nitrogen emissions can be found in the previous chapter.

7.3.2 Approach Factor (AF)

Four different subfactors come under this factor. These are the following:

1. Present prices
2. Prices tendency
3. Harvest/exploration
4. Available acres/reserve.

The approach factor is related to production, reserves, and the commercial market. Production is connected to the harvest (energy crops) or exploration (for fossil fuel, that is, when BF are applied for a fossil fuel calculation) process. This factor is also concerned with the number of acres that may be available for energy crops commercial farming or the present known reserves for fossil fuels at a particular location.

The market factor is related to the present prices for biomass materials and/or fossil fuel prices. The market category in this instance is related to the examination of the actual prices and their tendency factors. The application of BF should be made and recorded on the same day that the market factor(s) are being observed (i.e., prices and their tendency).

7.3.3 Baseline Methodology Factor (BMF)

There are four subfactors pertaining to this factor:

1. Fuel preparation
2. Knowledge
3. Competition
4. Innovation.

Methods of assessment relate to fuel preparation and how competitive it is within the fossil fuel industries. The knowledge and experience gained is related to the field of biomass energy. Competition from new products in the field of energy, other than biomass energy, is also considered. Innovation and new ideas are provided by the research institutes and from those working within the biomass energy industries. These innovations and ideas can be applied to the practicality of the daily aspects of biomass energy production.

7.3.4 Business Viability Factor (BVF)

There are seven subfactors that come under business viability. These subfactors can be represented under the following titles:

1. Government regulations
2. Investment
3. Method
4. Emerging market
5. Today's market
6. Energy: (a) input; (b) output
7. Technological.

Factors included under this heading relate to biomass energy's technological development, that is, how the technology of biomass fuel can compete and/or reach the same level as fossil fuel technology. This subfactor should not be confused with the subfactor related to the technology issues under the heading System Factor. The first one (BF) is related only to the latest present engineering techniques used in the production side, while the second is related to the side effects and maintenance issues regarding the use of the present systems at various power generating companies (S&T).

Technological issues can be a barrier if the business is the first of its kind to be introduced to this type of application, which may raise technological concerns. New technology may reveal a shortage of skilled workers in this area, and consequently it may delay the work or even stop it completely. The issues from energy production stem from the cost of energy input versus the cost of energy output from biomass materials. The cost of energy input should be less than the cost of the energy output in order to make a biomass energy production business viable. The present market in general and the fuel market in particular, may affect the biomass energy business locally and internationally. The emerging market relates to other types of energy and sources of energy involved compared with biomass energy. The approach and method applied in dealing with various issues and planning for biomass energy businesses are another tipping point for the success of the business. The method used can be affected by a lack of experience or a lack of knowledge related to the use of biomass materials and biomass disposal practice. In addition to this, there is a lack of familiarity and availability of the latest technologies in this field. Questions can be related to investment (or lack of it) needed for the early stages of the business and development plus other issues related to investments in this field (such as loans and time scale of the loans, grants, interests on loans, and taxes). This can be a barrier if the return on equity is too low when compared with conventional projects. Also, real and/or perceived risk associated with new or unfamiliar technology or a particular business approach can be too high in the eyes of potential investors, that is, it may be difficult to attract investment. In addition to the above, funding may not be available for innovative projects or businesses within certain types of biomass applications and research. Finally, help or hindrance of various governmental regulations, directly or indirectly, can have influence on the outcome of any biomass energy business. Governmental regulations are important when it comes to the survival of the business and, consequently, this kind of importance is reflected within the BF system scoring.

Emission is listed as a main factor due to the importance it carries.

7.3.5 Applicability Factor (APF)

The applicability factor covers four subfactors:

1. List of risks
2. List of policies
3. Adjustments
4. Business limitation.

This factor deals with a list of risks for a project or a business within the biomass energy field.

7.3.5.1 List of Risks

a. *Commercial Risks.* There are a number of commercial risks associated with biomass energy production on a commercial scale. These risks can be summarized under the following points:

1. A secure supply of biomass materials and costs
2. Transportation costs
3. Fuel costs (production of bio-fuels cost vs. the selling price to the end user)
4. Competition (from fossil fuels and other types of bio-fuels) as well as cost competitiveness
5. Market conditions and access to market
6. Engineering know-how (development and additional staffing) plus technical risks related to energy conversion
7. The prospect of future energy demand
8. Local and national government support/approval
9. Present and future environmental aspects
10. Public and customers support.

Additional commercial risks also come in the form of international politics. International politics can play an important role in the success (or otherwise) of a newly developed international biomass energy enterprise.

b. *Environmental Risks.* By using biomass materials, instead of fossil fuels, it is possible to reduce waste and consequently, help in protecting the environment. However, apart from balancing CO_2 in the atmosphere via the use of energy crops, there are a number of environmental risks associated with biomass materials. Considerations for the overall importance and success for the proposed biomass energy business should be made in accordance with these risks. These risks can be summarized in the following points:

1. Methane emission is one of the main causes of global warming, especially if it is not utilized commercially as a source of energy. Also, uncontrolled methane emissions from decaying biomass materials can be the cause of explosion and fire, as well as the problem associated with its odor.
2. Manure from livestock is a source of environmental health risk as it contaminates water (both underground and surface) and be the breeding ground for *Salmonella*, coliforms, and *Brucella*.
3. Uncontrolled burning of biomass materials and forest fire are some of the causes of air pollution and air quality.
4. Unbalanced large-scale use of forestland as a source for biomass materials can cause soil erosion and may affect climate conditions.
5. Inefficient combustion of biomass materials from power generating stations can have a major effect on the environment and human health.

Reduction of emission, such as NOx and SO_2, is vital in protecting the environment.

In order to assess environmental risks of long-term commercial production of bio-fuels, a model is required. This model envisages all possible factors, which may have an influence on the environment in a negative way, directly and indirectly. Environmental risks for a biomass energy system should be considered following agreed guidelines (Department of Health and Ageing and enHealth Council, 2002) such as a *regulatory framework* for the project (current and anticipated national and regional laws, international standards, and best practice guidelines). An *environmental appraisal* of the project (assessing the environmental risk, determining mitigation measures, estimating the cost of risk management, and reporting the results) is necessary. Along with *assessing the environmental, social risks* and opportunities of the project/business (to provide an initial evaluation of the environmental risks and opportunities presented by a particular biomass project/ business, i.e., to provide the analyst with an estimate of the risk potential of the project/business with respect to a number of possible environmental issues).

c. *Human Health Risk.* In many ways, human health risk is associated with environmental risks. Using biomass to produce energy may reduce the risk to human health if the proper methods are in place. This may mean the "correct" disposal of biomass waste materials that would otherwise create an environmental and consequently human health risk. Efficient disposal methods are required that conform to local, national, and international regulations concerning the production of energy from biomass materials. Other similar aspects have been already discussed in the previous section; however, they are related to human health and therefore should be investigated before embarking on any type of biomass energy production system.

1. The danger to human health caused by gaseous emissions from biomass
2. Materials, such as methane and nitrogen oxide and unburned particles
3. Issues related to water contaminations
4. Danger related to waste from landfills
5. Health issues as a result of air pollution
6. Health issues related to acid rain and smog
7. Health issues related to forest fire and deforestation.

d. *Regulatory Risk.* This kind of risk is usually brought about by the fact that markets are created by policy mechanisms (Agnolucci, 2005), that is, from central and local governments as policy priorities usually mean the old rules are replaced with new ones. This kind of change in regulations may not benefit new contracts, or the renegotiation of old contracts where development in a new area is already taking place. Investors may respond, or predict that when the government makes changes to the law/regulations, for example, renewable obligation (RO), it increases the likelihood of further changes later in the future. This perception of increased regulatory risk can translate into higher risk premiums.

Various bodies are becoming increasingly aware of the ROC price, and as they do so, there is an increasing tendency among them to confuse the ROC price with the cost of renewable energy generation. Therefore, it should be made very clear to any policymaker that anything that increases regulatory risk, and consequently ROC prices, will increase the cost of the renewable energy generation capacity.

Higher ROC prices and increased regulatory risk are both negative aspects for the majority of renewable energy schemes.

7.3.5.2 List of Policies. A list of policies (internal and external) and their implications on the biomass business for present and future development is another important issue for which a BF score is needed. Policies related to the method of energy production, training, health and safety, marketing, recruitment, and various other internal and external procedures (which the business embarks upon under the umbrella of company policies) are all influencing factors for the success/failure of the business.

7.3.5.3 Adjustments. From time to time, adjustments may be needed to avoid or minimize business and commercial activity risks and/or to increase profitability and encourage expansion. The adjustment is part of the survival tools and the evolutionary process of the business as a whole. These kinds of "adjustments" are needed for the purpose of working towards success in terms of protection and gradual establishment within the local, national, and international market.

7.3.5.4 Business Limitations. This concerns the examination of various business limitations and their influence on biomass energy production levels, as well as business development. These limitations should be examined at various levels, connected directly or indirectly to the business. Examples of these limitations can be market penetration, expansion, or moving into other sources of renewable energy, new local or national and international regulations, output, supply, and competition.

7.3.6 Land and Water Issues Factor (LWIF)

> Land use causes various effects on biodiversity, the main being a reduction of biota in comparison with pristine forests. The extent to which biodiversity is affected depends on many and varying factors e.g. agricultural practices, type of crops, harvesting methods and forests type. (FAO, 1999)

The land issues factor deals with the availability and cost of using land for the purpose of energy crop plantations. Furthermore, there is the issue of water availability and how easy and/or costly it is to obtain, on both a short- and long-term basis. In some cases, the size of land needed depends on the amount of electricity a power plant is required to generate. For example, planting a short rotation coppice (depending on farming and harvesting methods used) for a 50 MW power plant operating at 30% overall efficiency, will require between 40,000 and 70,000 ha of land (Business Insights, 2007). In the United States, it is estimated by the Department of Agriculture that by allocating between 8 and 17 million ha of land for energy crops, 20,000 MW of electricity from biomass can be generated. Reportedly, economic models have shown that agriculture for the production of biomass energy will compete with agriculture for food in terms of the land areas they require (Field et al., 2008). By considering U.S. greenhouse regulations, such as carbon tax ($20/ton), land use for the purpose of energy production could equal agriculture areas used for the production of food in United States (Field et al., 2008). Land and water issues are vital in countries (or local areas) where either of them is in demand or short supply. Consideration, therefore, should be made in conjunction with cost and the effect on the local population, as well as for possible future development and business expansion. It is important to examine any historic map evidence of previous land use (as well as conducting a detailed soil analysis) before considering any project/business concerning a commercial energy crops plantation.

7.3.7 Supply Factor (SUF)

To succeed in any business, regular and secure supply of raw materials—as well as other related and relevant parts connected to the business—is an important factor for the business's survival. Large commercial businesses usually rely on two types of supplies:

1. Local (includes national)
2. International.

The supply factor for a biomass energy business within the BF may score negatively or positively, depending on the varieties of factors. These may include economic and political situations (directly or indirectly connected to the business), harvest outcomes, regulations, taxes and level of demands from individual customers. The supply of biomass materials is an important subject and should therefore be looked at in more detail on commercial, business and technical basis. There are barriers to using agricultural residues for biomass fuel, such as the following:

Availability. The need to ensure there is no bottleneck in supply that leads to poor utilization.

Competition. An acceptable agricultural residue may already have some other nonfuel use, and to direct it toward a source of fuel may stimulate a demand that increases the price.

Technical. Collecting agricultural residues in an efficient and cost-effective manner.

There are three areas that need to be examined when considering the supply factor; these are the following:

A. General
   1. How secure is the supply of biomass materials?
   2. How many suppliers are available/contracted?
   3. What is the extent of cooperation between various power generating companies (e.g., within a country and outside it) especially when it comes to the supply of biomass materials?
   4. Is there a shortage or surplus supply of biomass materials?
   5. What are the monitoring aspects related to the supply of biomass materials within the power generating company?

B. Supplies from within the country
   1. Availability and sustainability
   2. Extraction, processing and transport costs
   3. Consistency of quality (moisture and heat value)
   4. Drying and handling issues.

C. Supplies from abroad
   1. Significant and sustainable volumes
   2. Guaranteed quality (low moisture, consistent heat value)
   3. Transport CO_2/cost offset by less drying/handling risks
   4. Can the cost be competitive?

7.3.8 Quality Factor (QF)

For a higher quality biomass fuel and a reduction of the overall cost, a number of factors should be considered. These factors include the quality of the biomass materials being used, the water quality, soil quality, land use, slope, cropping history, and weather conditions (Andrew, 2006). These are few basic examples; however, there are other factors that should be considered in addition to these, such as storage time and moisture content at the time of harvesting.

There are two subfactors related to QF:

1. Quality assurance (QA)
2. Quality control (QC).

QA and QC are important parts of continuous process or processes for a variety of energy businesses. These processes are related directly and indirectly to the production of energy, and to the end-user of the final biomass energy product. Furthermore, they are connected to how well businesses are able to comply with the European (or equivalent) fuel quality standards.

Power generating companies require regular quality assurances that the biomass they are buying and using meets their specifications and boiler requirements. It is necessary for the power generating companies and suppliers to evaluate biomass feedstock and develop simple and cost-effective methods to ensure that the quality of the biomass materials meets an agreed standard. The standard should be clearly identified and publicized for end users.

The extent of QC and QA applied for various procedures should be monitored regularly, that is, from the first farming stage to the final stage of generating electricity and/or the production of bio-fuels. The creation of a database before and after the commencement of work can greatly assist in monitoring and analyzing variable data accumulated on a daily basis. If the company already holds commercial data related to the variables, the old data can be used as a comparison for the new data.

Scoring can be measured on the quality of biomass supply, energy output level from power generating companies and the end user's requirements/satisfaction.

7.3.9 Emission Factor

This factor deals with two types of emission:

1. CO_2 (ROC)
2. Other types of emissions such as SO_x and NO_x.

The cost and technology involved in controlling the above two subfactors should be taken into account when it comes to the scoring mechanism. A benefit is achieved through the ROC certificate and CO_2 overall balance/reduction from the power station. A biomass energy business should be able to reduce the CO_2 emissions from fossil fuel burning. Biomass materials emit methane (CH_4) due to the decay or burning of biomass materials and generate nitrous oxide (N_2O). In general, gas emission occurs during the combustion and the transportation* of biomass materials to a power station.

* Production and availability of biomass materials, including storage facilities, should be very close or within the vicinity of the power station in order to avoid the transportation of these materials, that is, for the purpose of reducing CO_2 emission during transportation as well as for the purpose of reducing the overall cost.

Thus, the emission reduction by a bio-fuel business during a given year is:

$$CR = NFFG - NBFG. \tag{7.1}$$

where CR is the CO_2 reduction (in metric tonnes), NFFG is the number of fossil fuel power generators, and NBFG is the number of biomass fuel power generators. Equation (7.1) should include the CO_2 that is emitted during the transportation of biomass material to the power generating company.

Emission during soil cultivation and emission during transportation of biomass have been calculated by Thornley (2008), as shown in Table 7.2.

Reduction of emissions,* for example, within the United Kingdom, would benefit the power generating companies when the ROC applied the amount of biomass being used to generate electricity. The emission factor, from a commercial point of view, will benefit the power generating companies in the long term, as experience will be gained during the application of biomass materials. This gained experience can lead to a faster expansion into a new field of clean energy production. The quicker the energy production businesses move into this field, while taking advantages of various governmental incentives and grants available presently (e.g., in the United Kingdom), the more secure the future of a power generating company (or companies) will be (Box 7.1).

Table 7.2 Emission during soil cultivation and emission during transportation of biomass materials (Thornley, 2008)

	Emissions during Soil Cultivation ($g\,kWh^{-1}$)			Emissions during Transportation ($g\,kWh^{-1}$)		
Engine rating (kW)	22–75	75–130	>130	22–75	75–130	>130
CO	0.68	0.68	0.68	1.13	1.13	1.13
NOx	7	6	6	7	6	6
Particulates	0.25	0.25	0.2	0.34	0.3	0.2
Hydrocarbons	0.3	0.3	0.3	0.46	0.46	0.46

* Reduction in emission means a reduction in direct and indirect emission. Indirect emission, according to IPCC website (2013), refers to the formation of greenhouse gases displaced in time and space from the activities that are their ultimate cause.

Box 7.1

EMISSION FACTOR

Emission estimates are usually calculated by applying an emission factor to an appropriate activity statistic. That is: Emission = Factor x Activity.

Emission factors are generally derived from measurements made on a number of sources representative of a particular emission sector.

Total emissions are dominated by the public electricity and heat production sector and other emissions from the combustion of fuel (transport, domestic and industrial sectors). . . . The level of emissions depends highly on the fuel mix and the fuel consumption data.

National Atmospheric Emissions Inventory (2013)

Table 7.3 BF priority listing table

BF	Listing
Business viability	1
Approach	2
Emission	3
Baseline methodology	4
System	5
Applicability	6
Supply	7
Quality	8
Land and water issues	9

Source: Author.

7.4 BF DATA

7.4.1 Introduction

The factor weight (sometimes called the percentage listing or weighing factor) is a term used to describe the importance of a factor or subfactor to the rest of the business. These levels of importance have been represented in the form of percentage values for the factor or subfactor. Table 7.3 refers to these percentage values under the title "factor weight." The value of these factors in the example of Table 7.3 can range from 1% to 8%.

It is possible to say that a new business experiences the values and final results related to business factors (which can be described as a *chain of events*) made up from "within and without" during the first few weeks or months of operation. This means that the creation of a business depends heavily on internal and external factors, which in most cases, are connected. To obtain the scoring for any unpredictable business situation (or probably volatile, as in the case of the present situation with increasing global demand for energy), someone may argue that some of the BF values, if not all of them, are merely "subjective" valuations. One might say that these are subjective valuations if the prediction of an event affects certain factors, such as factors within the current market. Therefore, it is not a specific event, and it is difficult to give it an exact value. The question is: what is the "current market" and how can the various subfactors be evaluated? Furthermore, how might these subfactors affect the final result within the BF scoring system?

The current market and subfactors relating to it are a reflection of the present prices of biomass energy products, that is, the raw materials used to produce fuel. The buying and selling of shares related to the biomass energy business and related areas and other additional factors such as an international political situation may directly or indirectly impact upon the market. There are other influences, such as natural disasters and artificial market manipulation. Each of the "possibilities," if we refer to what might happen in the future regarding the biomass energy market, may have a "subjective" valuation based on a prediction from experience of current *events* (Capoor and Ambrosi, 2006).

What the BF section in the methodology aims for is the removal, or at least the reduction, of these kinds of "subjective predictions." The question here is how is it possible to make successful predictions and on what basis should they be made? Before going into the full details, clarification must be made concerning an important fact about BF. The fact is that only a small part of the BF can be related to the market. In fact, only 6% of the BF are connected to such data. The rest of the factor's data can be obtained easily and accurately without the need for a prediction or a subjective conclusion. That means 94% of BF data stability and accuracy is based on the data obtained from the "original source." For example, if data are related to "land issues," then we can easily obtain all the relevant data about "land availability" and "water supply," for a particular biomass project or business in any part of the United Kingdom. This can be done safely in any politically stable country in the world for that matter. Other factors and subfactors, such as "list of risks," "adjustments," "existing systems," and "knowledge," can be obtained through recently available documented data (e.g., data about present existing systems, and knowledge about the present biomass fuel available on the market) and/or from data related to the actual building process of the business/project during and after the completion of every stage. This means that if analysis needs to be made concerning the "list of risks" before the actual business/project starts, then after compiling a list of various risks which the business/project may face, then the fact that a list is already made (regardless of how high or low the "percentage" value of the accuracy on the list is) is enough for BF to remain as a real factor (or factors) that in itself can potentially be measured by experts in this field. This signifies confirmation that there are genuine factors that have direct or indirect effect (on a long and short term basis) on the biomass business and, given the values these factors may carry with them, for the success or failure of the business itself.

Market factor data can be obtained by dividing the market into smaller sections using a "market segmentation" method (Tonks, 2004). In this way, the subfactors related to the "current market" and the "emerging market" and their input data can be easily minimized into the smallest sections possible. Accurate data regarding these and similar factors and subfactors can be obtained using the same method.

Important note regarding commodity prices as reflected in the international market and how these prices establish themselves should be mentioned here. These prices may take the following procedure (FAO, 2003): from Food and Agriculture Organization Corporate Statistical Database (FAOSTAT) and International Monetary Fund (IMF), commodity monthly prices are collected (in both nominal and real terms). The nominal prices are converted into real ones by the World Bank Manufactures Unit Value (MUV) (FAO, 2003). These commodity prices are adjusted according to their related marketing season.

Examples of energy crops price fluctuations in both nominal and real prices are shown in Figure 7.3 (from January 1970 to December 2000).

7.4.2 The Priority List

During the research on this book, taking the form of a survey method (Chapter 8) and conclusions made during the work on biomass and fossil fuel samples (within and outside the laboratory work) for BF and S&T, it became clear that a priority listing table can be made. Unlike the result obtained for the S&T factors by the survey method, the majority of people questioned disagreed on which factor should be put as number 1 in the listing table for BF. Consequently, by not being able to vote clearly on which factor should be at the top of the list, the importance of other factors is disputed as well.

A minority (20%) of the people asked during the survey did agree on a listing priority table for BF. Here, the factor of "business viability" was the number one priority in the majority of cases (Table 7.3).

Out of the 20% who agreed to BF factors listing, most of them agreed that the value percentage they may give to the BF factors and subfactors can be made as follows (Fig. 7.4):

A. Systems, 10%
   Existing systems, 1%
   Emerging systems, 4%
   Technology issues, 2%
   By-products, 3%.
B. Approach, 18%
   Present prices, 7%
   Prices' tendency, 7%

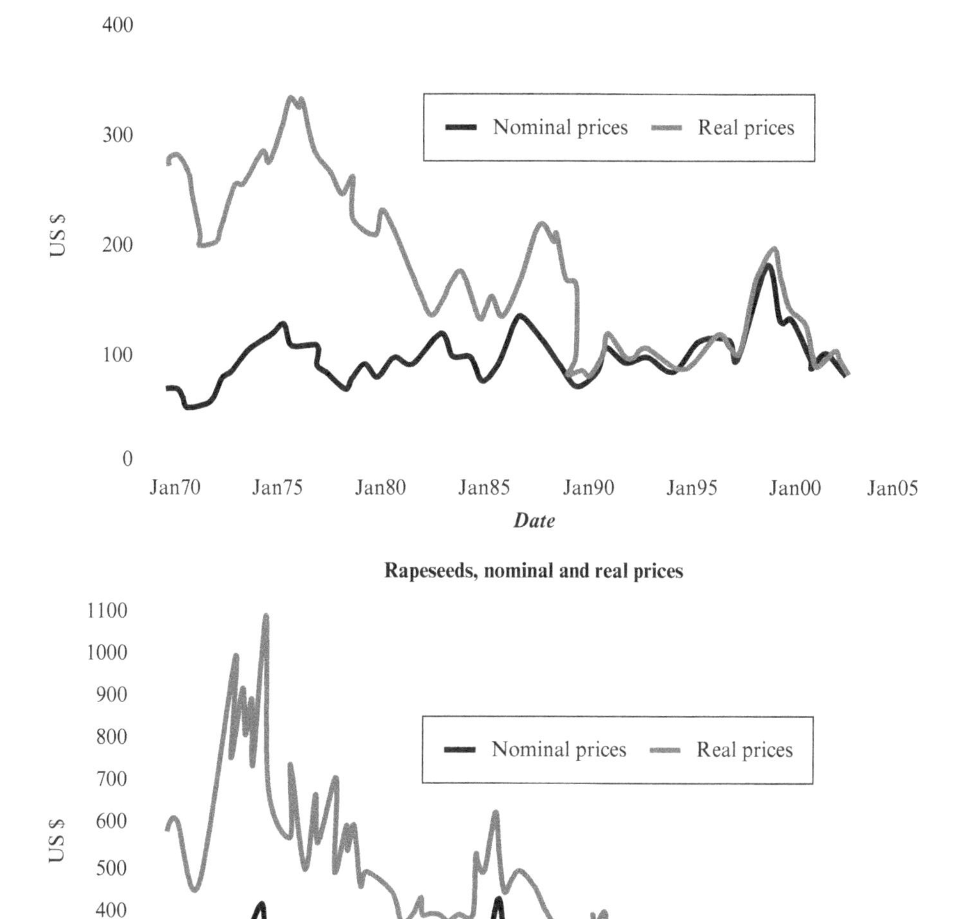

Figure 7.3 Selected energy crop price fluctuations on the international market (for both nominal and real prices). *Source*: Adapted from FAO (2003).

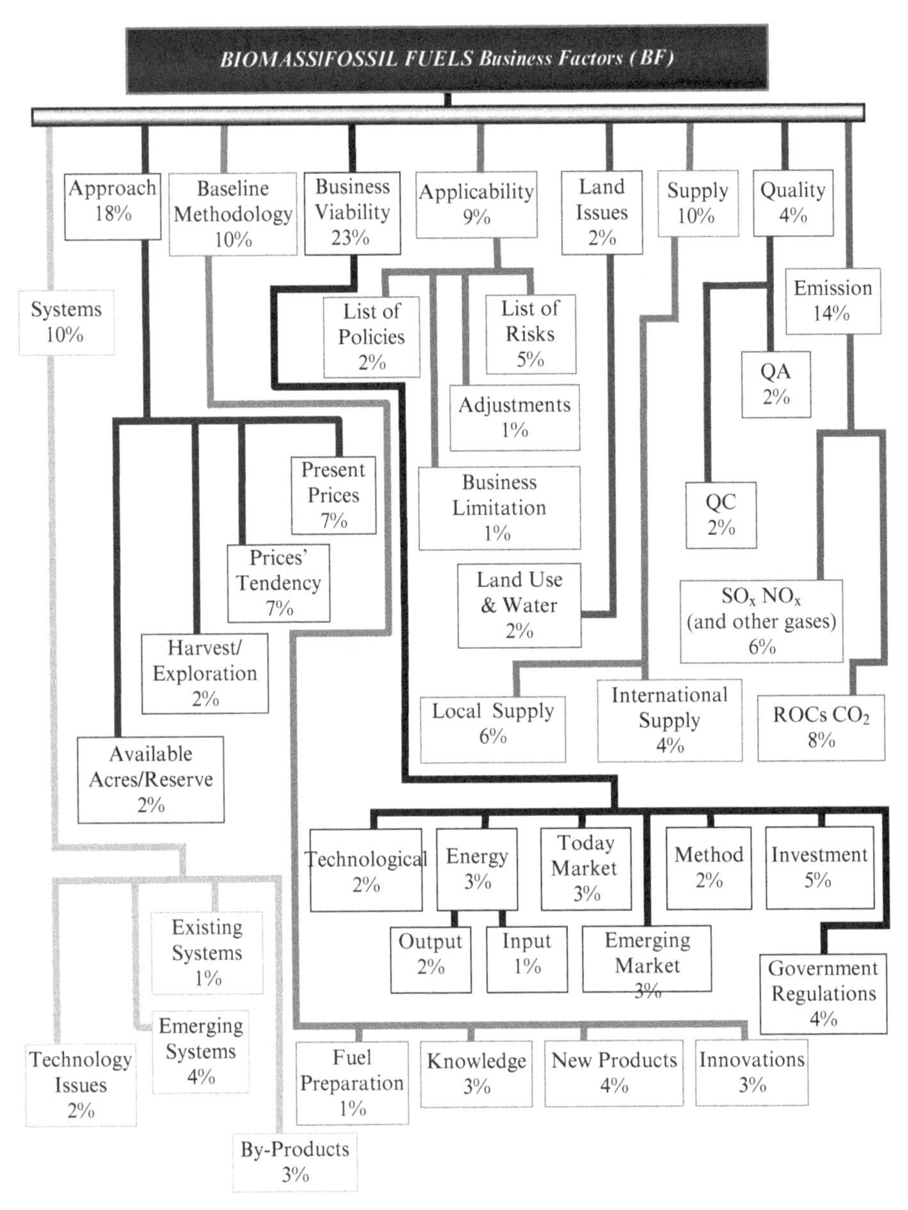

Figure 7.4 Main factors and subfactors for the BF methodology with an example of percentage values. *Source*: Author.

Harvest/reserve, 2%
Available acres/mining, 2%.

C. Baseline methodology, 11%
Innovations, 3%
Knowledge, 3%
New products, 4%
Fuel preparation, 1%.

D. Business viability, 23%
Government regulations, 4%
Investment, 6%
Method, 2%
Today's market, 3%
Emerging market, 3%
Energy (input, 1%; output, 2%)
Technological, 2%.

E. Applicability 9%
List of policies, 2%
List of risks, 5%
Adjustments, 1%
Business limitation, 1%.

F. Land Issues
Land use and water, 2%.

G. Supply, 9%
Local supply, 4%
International supply, 5%.

H. Quality, 4%
QC, 2% and QA, 2%.

I. Emission, 14%
CO_2 emission, 8%
Other gases emission, 6%.

All these factors have been given scoring percentage priority according the data (or information) available at the time of compiling this section of the book (Table 7.4). The data, in many cases, are mostly related to biomass materials, biomass energy, technical factors, regulations, land and water, transportation and storage, commercial businesses and markets.

The BF may change from time to time, and the scoring consequently may change accordingly. The reason why some factors are given higher percentages than others is simply related to the final result of how much influence these factors have on the success of the business itself. Also, the degree to which these

Table 7.4 BF data factors using REA1 methodology applying 75%[a] of the total value of REA

Straw Pellets			
Methodology	Factor Name	Factor Weight (%)	Value
Applicability	Adjustments	1	–1
	Business limitation	1	–1
	List of policies	2	1
	List of risks	5	–1
Approach	Available acres/mining	2	1
	Harvest/mining/reserve	2	1
	Present prices	7	1
	Prices tendency	7	1
Baseline methodology	Fuel preparation	1	0
	Innovations	3	0
	Knowledge	3	1
	New products	4	0
Business viability	Emerging market	3	1
	Energy input	1	1
	Energy output	2	–1
	Government regulations	4	1
	Investment	6	0
	Method	2	0
	Technological	2	0
	Today market	3	0
Emission	CO_2 emission	8	1
	Other gases	6	–1
Land issues	Land use and water	2	1
Quality	Quality assurance	2	0
	Quality control	2	0
Supply	International supply	5	1
	Local supply	4	1
Systems	By-products	3	–1
	Emerging systems	4	–1
	Existing systems	1	0
	Technology issues	2	1
	Business factor total fitness:		42.900%

[a]The remaining value of 25% is left out, which belongs to S&T factors, BF market values date: August 8, 2008.

Source: Author.

factors contribute to the overall picture of the final effort needed to establish the business—plus the contribution which can be made by the local and national governments—enable the business to continue to function successfully.

To summarize, the listing table for BF as well as the percentage values for them, are the outcome of the survey method for this book (Table 7.5). However, each business has its own approach, its own priority listing factors, and their own percentage values (Fig. 7.5, Box 7.2, and Box 7.3).

Table 7.5 BF percentage values

BF	%
Business viability	23
Approach	18
Emission	14
Baseline methodology	11
System	10
Applicability	9
Supply	9
Quality	4
Land and water issues	2

Source: Author.

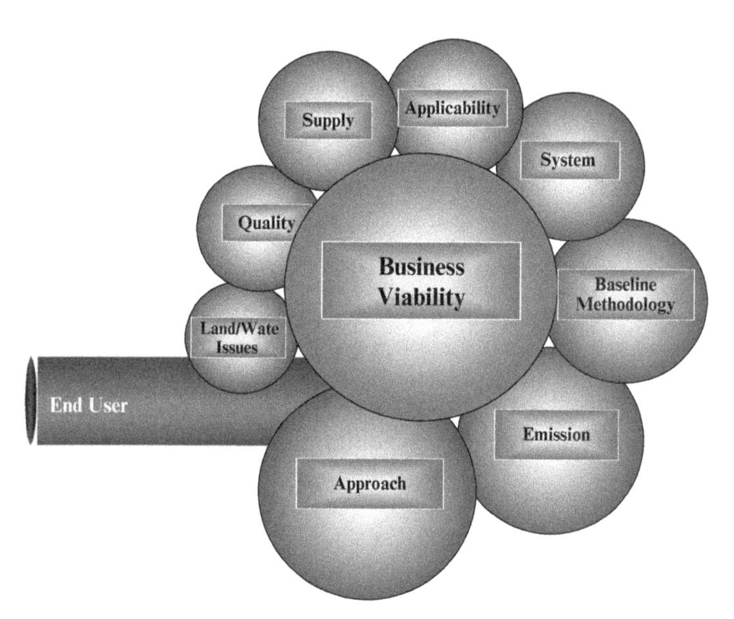

Figure 7.5 Main business factors (BF), their connections to other factors, and the "end user."
Source: Author.

7.5 CONCLUSION

Important factors need to be examined and applied to the selection of a commercial biomass fuel, using nine main business factors and their subfactors.

By obtaining the correct values for BF and applying them through the REA1 methodology, a clear picture emerges, indicating how a biomass energy business can be established and run successfully. The business factors are vital

Box 7.2

BIOMASS ELECTRIC GENERATION SYSTEM

The basic main components for a steam cycle biomass electric generation system (thermal generating plant) can be summarized as follows:

1. Boiler 2. Condenser 3. Fuel Storage System 4. Pumps 5. Furnace 6. Turbine 7. Generator 8. System Controls 9. Fans 10. Cooling Tower 11. Emission Controls

Box 7.3

EUROPE AND WOOD

Which source of renewable energy is most important to the European Union? Solar power, perhaps? (Europe has three-quarters of the world's total installed capacity of solar photovoltaic energy.) Or wind? (Germany trebled its wind-power capacity in the past decade.) The answer is neither. By far the largest so-called renewable fuel used in Europe is wood.

The Economist, 2013

and an important key as to whether or not a selected type of biomass material should be used in a certain business scenario. These factors are connected and related to each other in various ways. Long-term planning, combined with practical application, is an essential ingredient for the success of any biomass energy enterprise (Dubois, 2008). In addition, the following points are vital for success:

1. The market factors should be investigated in the light of success or failure of similar projects/businesses undertaken in various parts of the world. It must also be considered that every biomass project/business is unique, despite similarities which exist among biomass businesses.
2. Analyzing the biomass market is an essential part of using the market segmentation method whenever required, for example,, during the planning, as well as during the concluding stages, and after.
3. Government and local authority legislations should be utilized and discussed with the lawmakers in order to obtain additional support for a variety of schemes and projects related to the biomass businesses and future development.

One of the most important factors within BF is the "business viability" factor. This importance is concluded partly as a result of expert opinion in the

Box 7.4

BIOMASS ENERGY

Proposed Business

A. Location
B. Business size
C. Method(s) of controlling harmful emissions
D. Plant power capacity
E. Viability and effect of the business on local community/local economy
F. Type of raw materials that will be used to generate energy
G. Transport, processing and storage of biomass materials
H. The original source (supplier[s]) of biomass materials
I. Physical distance between power station and supplier(s)
J. The amount of financial investments required
K. Possible reduction of CO_2 emission on a local and national level
L. National/local government and local community support
M. Environmental issues
N. Profitability
O. Future plans/expansion.

field of biomass energy (survey method, Chapter 8), and partly from the author's own research during this particular stage of the book. Without this factor, a profitable long-term commercial biomass energy business would be an impossible task. Therefore, the purpose is to secure long-lasting commercial success in the field of biomass energy production. For this reason, this section of the biomass methodology has been given additional attention during the design, selection, and the implementation of the BF compared with other sections of the methodology (Box 7.4).

REFERENCES

Agnolucci P (2005) Factors influencing the likelihood of regulatory change in renewable electricity markets. Policy studies institute, 5th BIIE academic conference, St John's College, Oxford.

Andresen PD, Chrsitensen J, Kossmann J (undated) Emerging and future bioenergy technologies. Risø energy report 2, Risø National Laboratory, Emmanuel Kouklos, National Technical University of Athens, Greece, 18–22.

Andrew SS (2006) Crop residue removal for biomass energy production: effects on soils and recommendations. USDA-Natural Resource Conservation Service. http://soils.usda.gov/sqi/management/files/agforum_residue_white_paper.pdf (last accessed August 14, 2009).

Business Insights (2007) Energy crops, the environment effect of biomass power generation, Chapter 5. The future of global biomass power generation, the technology, economic and impact of biomass power generation.

Capoor K, Ambrosi P (2006) State and trends of the carbon market 2006. The World Bank, IETA.

Department of Health and Ageing and enHealth Council (June, 2002) Environmental health risk assessment: guidelines for assessing human health risks from environmental hazards. Australia. May 17, 2009.

Dubois O (2008) How good enough biofuel governance can help rural livelihoods: making sure that biofuel development works for small farmers and communities. FAO. http://www.fao.org/forestry/media/15346/0/0/ (last accessed September 12, 2008).

The Economist (2013) Wood: the fuel of the future. http://www.economist.com/news/business/21575771-environmental-lunacy-europe-fuel-future (last accessed December 4, 2013).

FAO (1999) Land rehabilitation, food production and biodiversity. The multifunctional character of agriculture and land: the energy function. Cultivating Our Futures: background papers. http://www.fao.org/docrep/x2775e/X2775E04.htm (last accessed December 4, 2013).

FAO (2003) Consultation on agricultural commodity price problems. Appendix I, charts of time series of selected commodity prices. Corporate Document Repository. http://www.fao.org/DOCREP/006/Y4344E/y4344e0h.htm#TopOfPage (last accessed December 4, 2013).

Field CB, Campbell JE, Lobell DB (2008) Biomass energy: the scale of the potential resource. Trends in Ecology and Evolution 23(2):65–72.

National Atmospheric Emission Inventory (2013) Emission factors. http://naei.defra.gov.uk/data/emission-factors (last accessed October 17, 2013).

Thornley P (2008) Airborne emissions from biomass based power generation systems. Environmental Research Letters 3 (January–March).

Tonks GD (2004) Identifying market segments in consumer markets: variable selection and data interpretation. Track, working paper—department of marketing, the Management School, University of Lancaster.

Turnbull HJ (1994) Developing sustainable integrated biomass systems. Electric Power Research Institute, Woody Crops, Mechanization in short rotation, Intensive culture (SRIC) Forestry. http://www.woodycrops.org/mechconf/turnbull.html (last accessed March 8, 2009).

8

RESULTS: PART 1

8.1 STATISTICAL DATA AND ERRORS

8.1.1 Introduction

Statistical data have been collected, which through analysis, have been processed into facts. These facts are the foundation for a "knowledge" base (Arsham, 2009), which is the main objective for original research and applications in any statistical approach (Fig. 8.1). For this reason, a process is needed to develop business (BF) and scientific and technical (S&T) factors through the application of a methodology. The methodology employed is REA1, which covers a wide field of research and data, such as examining governmental regulations, estimating commercial viability, incorporating facts from laboratory tests, and business-related calculations leading to the overall final scoring. In all of these processes, there is the possibility of error occurring before, during, or after each step. Examination of error is usually referred to as an elimination process, that is, it is part of the mathematical theory of errors. This theory assumes that by performing a number of *repeated* tests on the same sample, results *supposedly* are expected to be produced that are close to each other in value, if not identical. When all possible sources of error have been taken into account, the source of any other additional unexpected error could be made up from a large number of other different methodology factors, which may not have been considered to have an effect in the first instance. However,

The Selection Process of Biomass Materials for the Production of Bio-fuels and Co-firing, First Edition. Najib Altawell.

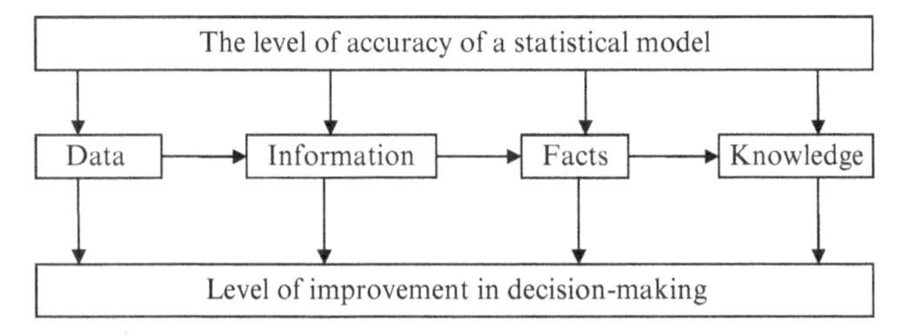

Figure 8.1 The relationship between the accuracy of statistical models and the improvement of decision-making as data progresses toward the final objective, that is, knowledge. *Source*: Adapted from Arsham (2009).

during the construction of the methodology, there was consistent effort to pinpoint the source of possible error during each stage. This kind of approach led to the consistent analysis of various parts of the methodology in order to make sure that possible sources of error were eliminated and, consequently, discrepancies reduced.

The two sources of possible errors are *precision* (which is related to the random error distribution of an experiment, such as the case of a number of parameters which may have wider variations between different samples) and *accuracy* (which is connected to an existence of error [or errors] within the system being used, such as an inaccuracy in calibration) (Department of Physics, Ryerson University, 2009). Both of these factors have been examined and then listed as a possible source of error. For this reason, the REA1 methodology is divided into three different areas where sources of errors are possible. These areas contain parameters, which can be designated as a high, medium, and low source of errors. Thus, the division of these areas is represented by the following:

1. BF (high)
2. Technical (medium)
3. Scientific (low).

The scientific and technical errors within the methodology have been combined together. Consequently, the term used (S&T) represents the outcome of possible errors from both.

There are important factors that may increase or decrease the error percentage in the S&T section of the methodology (Chapter 6, Section 6.3) if not observed correctly. These factors can be summarized under the following points:

1. Regular availability of reliable testing equipment.
2. Relying on a local permanent biomass materials producer rather than on imported and/or temporary suppliers.
3. For energy crops, deciding on the plantation season and harvesting time.

4. Transportation and storage should be made under similar conditions whenever this is possible.
5. Age of the biomass materials should be considered, in particular, if laboratory testing is needed after (or before) the mixing of these materials.
6. Laboratory testing should be conducted immediately, that is, prior to the use of the biomass materials for the purpose of generating energy.
7. Certain standard conditions should be followed when it comes to milling, determent of moisture content, and the mixing of biomass materials (see Chapter 3).
8. Finally, different farming methods and prevailing weather conditions can have an effect on the outcome of the S&T factors.

Another important area of possible error is the boundary level scoring value. This is because a 100% absolute value in this part of the methodology is impossible to obtain. For this reason, the allocation of the 3% methodology level value, as the test showed, was the nearest possible figure that allows the scoring to ascend or descend to a different scoring area, that is, level. Further details on this aspect are in the following section.

8.2 METHODOLOGY LEVEL VALUE (BOUNDARY LEVEL SCORING VALUE)

> Everybody believes in the exponential law of errors: the experimenters, because they think it can be proved by mathematics; and the mathematicians, because they believe it has been established by observation.
>
> Gabriel Lippmann (quoted in Whittaker and Robinson, 1967)

There is a boundary, which takes the scoring from one level to another within the methodology scoring mechanism. As an example for the samples used in this book, the value of this "stepping" boundary is equal to 3%. This figure partly depends on the type of samples and the degree of similarities/dissimilarities among them. The 3% allocated for the calculation of values separating the three scoring levels (+1, 0, and −1) is an important factor within the scoring mechanism of the methodology, as it is part of the *deciding mechanism* in the selection process. A question may be asked as to why 3% instead of a higher or a lower value?

As mentioned previously, the 3% value is used here as an "example" and, therefore, the percentage value can increase or decrease according to the requirement of the energy business applying this type of methodology. This means that the power generating companies should choose their own boundary value, relevant to their own needs. Their boundary value is connected to the experience/knowledge that comes from running an energy business and is also related to "day-to-day" business operations. This means that the percentage value should be in accordance with the economic, scientific, technical, and other factors connected to the business in addition to the type of biomass materials regularly used to generate energy.

The value chosen is a good example of a boundary level for this work. This knowledge comes from various tests performed during the construction and application of REA1 methodology. These tests have shown that when a lower value is used, it has an insufficient influence on the scoring. However, using a value higher than 3% has shown a significant effect on the scoring level. By manipulating the boundary's percentage value, that is, by using lower or higher value than 3%, the final value is decided upon where "noticeable changes" start to take place. Usually, large differences begin to occur as the tests move away from this value, that is, higher than 3%. For this reason, the decision was made that a "safe" boundary level (or limit) can occur "only" after the figure of 3% is reached, that is, above it is too large and below it is too small. However, there are two tables (fossil and biomass fuels, Chapter 5, Table 5.2 and Table 5.3) designed specifically for this book to define the limits (or boundaries) within the scoring level for S&T section; which, consequently, there may not be the need to refer to the value of 3% boundary for the above section—if and when you are using these two tables (further details are in Chapter 5).

8.3 CALCULATING STANDARD DEVIATION AND RELATIVE ERROR

The standard deviations (SDs) of the BF and S&T factors distribution are necessary parameters. The SD can help to establish how the methodological factors voting results are related to each other, as well as their importance within the biomass energy businesses and industries. Using the two statistical formulas, the SD (and population SD) have been calculated as shown in the following equations:

$$\mathrm{SD} = \sqrt{\frac{\sum (X - M)^2}{n-1}} \tag{8.1}$$

$$(\mathrm{Population})\mathrm{SD} = \sqrt{\frac{\sum (X - M)^2}{n}}. \tag{8.2}$$

The BF and S&T factors' SDs, population SD, and the coefficient of variation (relative error), have all been listed in the following sections. In addition, the number of votes for each factor and the percentage value of the voting itself have also been added (Box 8.1).

Box 8.1

ABSOLUTE ERROR AND RELATIVE ERROR

The amount of physical error, that is, the accepted value of the error (AV) during measurement, is usually referred to as absolute error (AE):

$$\text{AE} = \pm\text{Error value.}$$

Relative to the size of the object being measured, that is, measured value (MV) is what is usually referred to as relative error (RE):

$$(\text{RE}) = \frac{\text{MV} - \text{AV}}{\text{AV}}.$$

8.3.1 S&T Factors

1. Energy Factor (EF)

Number of Votes	Energy % Value Voted for X	% Value of the Voting	Mean M Vote	SD $\text{SD} = \sqrt{\frac{\sum (X-M)^2}{n-1}}$	Population SD	Coefficient of Variation (Relative Error)
11	40	5.31	28.75	$7203.42/206 = 34.96$		
110	30	53.14	28.75	$\sqrt{34.96} = 5.91$	$7203.42/207 = 34.79 = \sqrt{34.79} = 5.89$	$5.89/33 = 0.17 \times 100 = 17\%$
57	25	27.53	28.75			
22	20	10.62	28.75			
7	50	3.38	28.75			
Total: 207	Average: 33%					

2. Combustion (Combustion Index Factor, CIF)

Number of Votes	Combustion % Value Voted for X	% Value of the Voting	Mean M Vote	SD $\text{SD} = \sqrt{\frac{\sum (X-M)^2}{n-1}}$	Population SD	Coefficient of Variation (Relative Error)
26	18	32.09	15.41	$805.44/80 = 10.06$		
21	14	25.92	15.41	$\sqrt{10.06} = 3.17$	$805.44/81 = 9.94 = \sqrt{9.94} = 3.15$	$3.17/15 = 0.21 \times 100 = 21\%$

(*Continued*)

Number of Votes	Combustion % Value Voted for X	% Value of the Voting	Mean M Vote	SD $SD = \sqrt{\frac{\sum (X-M)^2}{n-1}}$	Population SD	Coefficient of Variation (Relative Error)
19	15	23.45	15.41			
8	8	9.87	15.41			
7	20	8.64	15.41			
Total: 81	Average: 15%					

3. Volatile Matter Factor (VMF)

Number of Votes	VM % Value Voted for X	% Value of the Voting	Mean M Vote	SD $SD = \sqrt{\frac{\sum (X-M)^2}{n-1}}$	Population SD	Coefficient of Variation (Relative Error)
23	15	33.84	14.62	268.62/64 = 4.19		
17	16	23.07	14.62	$\sqrt{4.19} = 2.04$	268.62/65 = 4.13 = $\sqrt{4.13}$ = 2.03	2.04/14 = 0.14 × 100 = 14%
11	12	21.53	14.62			
9	17	13.84	14.62			
5	10	7.69	14.62			
Total: 65	Average: 14%					

4. Moisture Factor (MF)

Number of Votes	Moisture % Value Voted for X	% Value of the Voting	Mean M Vote	SD $SD = \sqrt{\frac{\sum (X-M)^2}{n-1}}$	Population SD	Coefficient of Variation (Relative Error)
19	15	5.27	12.64	239.52/53 = 4.51		
14	12	3.11	12.64	$\sqrt{4.51} = 2.12$	239.52/54 = 4.43 = $\sqrt{4.43}$ = 2.10	2.12/11.6 = 0.18 × 100 = 18%
9	13	2.16	12.64			
9	10	1.66	12.64			
3	8	0.44	12.64			
Total: 54	Average: 11.6%					

5. Ash Factor (AF)

Number of Votes	Ash % Value Voted for X	% Value of the Voting	Mean M Vote	SD $SD=\sqrt{\frac{\sum (X-M)^2}{n-1}}$	Population SD	Coefficient of Variation (Relative Error)
11	12	34.37	10.89	$72.67/31=2.34$		
8	10	25	10.89	$\sqrt{2.34}=1.53$	$72.67/32$ $=2.27$ $=\sqrt{2.27}$ $=1.50$	1.53/11 $=0.13\times 100$ $=13\%$
7	11	21.87	10.89			
4	8	12.90	10.89			
2	14	6.25	10.89			
Total: 32	Average: 11%					

6. Density Factor (DF)

Number of Votes	Density % Value Voted for X	% Value of the Voting	Mean M Vote	SD $SD=\sqrt{\frac{\sum (X-M)^2}{n-1}}$	Population SD	Coefficient of Variation (Relative Error)
7	10	35	9.3	$129.07/19=6.82$ $\sqrt{6.82}=2.61$	$129.07/20$ $=6.45$ $=\sqrt{6.45}$ $=2.53$	**2.53/10.6** $=0.23$ 0.23×100 $=23\%$
5	8	25	9.3			
5	7	25	9.3			
2	13	10	9.3			
1	15	5	9.3			
Total:" 20	Average: 10.6					

7. Nitrogen Emission Factor (NEF)

Number of Votes	NEF % Value Voted for X	% Value of the Voting	Mean M Vote	SD $SD=\sqrt{\frac{\sum (X-M)^2}{n-1}}$	Population SD	Coefficient of Variation (Relative Error)
5	6	31.25	7.17	$283.3/15=18.88$		
5	4	31.25	7.17	$\sqrt{18.88}=4.34$		
4	12	25	7.17		.3/16 $=14.89$ $=\sqrt{14.89}$ $=3.85$	$4/8=0.5$ 0.5×100 $=50\%$
1	2	6.25	7.17			
1	15	6.25	7.17			
Total: 16	Average: 8%					

Factor (S&T)	Standard Deviation
Energy	5.91
Combustion (combustion index)	3.17
Volatile matter	4.19
Moisture	2.12
Ash	1.53
Density	2.61
Nitrogen emission	4.34
S&T average error: 2.70	

8.3.2 Business Factors (BF)

1. Business Viability (BV)

Number of Votes	BV % Value Voted for X	% Value of the Voting	Mean M Vote	SD $SD=\sqrt{\frac{\sum(X-M)^2}{n-1}}$	Population SD	Coefficient of Variation (Relative Error)
28	26	29.78	23.75	3746.14/93 = 40.28	3746.14/94	6.34/25.6
27	25	28.72	23.75	$\sqrt{40.28}$ = 6.34	= 39.85	= 0.24
18	20	19.14	23.75		= $\sqrt{39.85}$	0.24 × 100
16	17	17.02	23.75		= 6.31	= 24%
5	40	5.31	23.75			
Total: 94	Average: 25.6					

2. Approach (Business Risks, BR)

Number of Votes	BR % Value Voted for X	% Value of the Voting	Mean M Vote	SD $SD=\sqrt{\frac{\sum(X-M)^2}{n-1}}$	Population SD	Coefficient of Variation (Relative Error)
14	20	45.16	18.50	263.75/30 = 8.79		
8	18	25.80	18.50	$\sqrt{8.79}$ = 2.96	263.75/31 = 8.50 = $\sqrt{8.50}$ = 2.91	2.96/17.6 = 0.16 0.16 × 100 = 16%
6	15	19.35	18.50			
2	25	6.45	18.50			
1	10	3.22	18.50			
Total: 31	Average: 17.6%					

3. Emission

Number of Votes	Emission % Value Voted for X	% Value of the Voting	Mean M Vote	SD $SD = \sqrt{\frac{\sum (X-M)^2}{n-1}}$	Population SD	Coefficient of Variation (Relative Error)
5	20	23.80	14.45	259.2/20 = 12.96		
4	15	19.04	14.45	$\sqrt{12.96} = 3.6$	259.2/21 = 12.34 = $\sqrt{12.34}$ = 3.51	3.6/14.2 = 0.25 0.25 × 100 = 25%
4	12	19.04	14.45			
4	10	19.04	14.45			
4	14	19.04	14.45			
Total: 21	Average: 14.2%					

4. Baseline Methodology (BM) (Fuel Preparation, FP)

Number of Votes	FP % Value Voted for X	% Value of the Voting	Mean M Vote	SD $SD = \sqrt{\frac{\sum (X-M)^2}{n-1}}$	Population SD	Coefficient of Variation (Relative Error)
25	14	27.17	10.86	548.12/91 = 6.02		
22	12	20.24	10.86	$\sqrt{6.02} = 2.45$	548.12/92 = 5.95 = $\sqrt{5.95}$ = 2.44	2.45/10.4 = 0.23 0.23 × 100 = 23%
17	10	18.47	10.86			
17	7	18.47	10.86			
11	9	11.95	10.86			
Total: 92	Average: 10.4%					

5. System(s)

Number of Votes	System % Value Voted for X	% Value of the Voting	Mean M Vote	SD $SD = \sqrt{\frac{\sum (X-M)^2}{n-1}}$	Population SD	Coefficient of Variation (Relative Error)
30	14	32.96	10.48	792.81/90 =		
20	11	21.97	10.48	$\sqrt{8.80} = 2.96$	792.81/91 = 8.71 = $\sqrt{8.71}$ = 2.95	2.96/9.4 = 0.31 0.31 × 100 = 31%
19	9	20.87	10.48			
12	8	13.18	10.48			
10	5	10.98	10.48			
Total: 91	Average: 9.4%					

6. Applicability (Production and Market, P&M)

Number of Votes	P&M % Value Voted for X	% Value of the Voting	Mean M Vote	SD $SD=\sqrt{\frac{\sum(X-M)^2}{n-1}}$	Population SD	Coefficient of Variation (Relative Error)
15	10	42.85	9.6	$234.2/34=6.88$		
10	8	28.57	9.6	$\sqrt{6.88}=2.62$	$234.2/35$ $=6.69$ $=\sqrt{6.69}$ $=2.58$	$2.62/11$ $=0.23$ 0.23×100 $=23\%$
6	12	17.14	9.6			
3	5	8.57	9.6			
1	20	2.85	9.6			
Total: 35	Average: 11					

7. Supply

Number of Votes	Supply % Value Voted for X	% Value of the Voting	Mean M Vote	SD $SD=\sqrt{\frac{\sum(X-M)^2}{n-1}}$	Population SD	Coefficient of Variation (Relative Error)
28	10	37.33	9.42	$639.87/74=8.64$		
19	9	25.33	9.42	$\sqrt{8.64}=2.94$	$639.87/75$ $=8.53$ $=\sqrt{8.53}$ $=2.92$	$2.94/9$ $=0.32$ 0.32×100 $=32\%$
11	15	14.66	9.42			
11	5	14.66	9.42			
6	6	8	9.42			
Total: 75	Average: 9%					

8. Quality

Number of Votes	Quality % Value Voted for X	% Value of the Voting	Mean M Vote	SD $SD=\sqrt{\frac{\sum(X-M)^2}{n-1}}$	Population SD	Coefficient of Variation (Relative Error)
8	5	40%	4.35	$52.5/19=2.76$		
6	4	30%	4.35	$\sqrt{2.76}=1.66$	$52.5/20$ $=2.62$ $=\sqrt{2.62}$ $=1.62$	$1.66/4.8$ $=0.34$ 0.34×100 $=34\%$
3	3	15%	4.35			
2	2	10%	4.35			
1	10	5%	4.35			
Total: 20	Average: 4.8%					

9. Land and Water

Number of Votes	Quality % Value Voted for X	% Value of the Voting	Mean M Vote	SD $SD = \sqrt{\frac{\sum (X-M)^2}{n-1}}$	Population SD	Coefficient of Variation (Relative Error)
5	3%	31.25	2.31	$19.72/15 = 1.31$		
5	2%	31.25	2.31	$\sqrt{1.31} = 1.14$	$19.72/16$ $= 1.23$ $= \sqrt{1.23}$ $= 1.10$	$1.14/3$ $= 0.38$ 0.38×100 $= 38\%$
4	1%	25	2.31			
1	4%	6.25	2.31			
1	5%	6.25	2.31			
Total: 16	Average: 3%					

Factor (BF)	Standard Deviation
Business viability	6.34
Approach (business risks, "BR")	2.96
Emission	3.6
Baseline methodology (BM) (or fuel preparation, "FP")	2.45
System(s)	2.96
Applicability (production and market, "P&M")	2.62
Supply	2.94
Quality	1.66
Land and water	1.14
BF average error: 2.96	

8.3.3 Methodology Standard Deviation for S&T

Refer to Box 8.2 for a discussion of error theory.

S&T Factors	SD	$(SD)^2$	Sum of $(SD)^2$	Overall SD $SD = \sqrt{\frac{\sum (X-M)^2}{n-1}}$	Population SD	Coefficient of Variation (Relative Error)
Energy	5.91	34.92	81.59	$81.59/6 = 13.59$		
Combustion	3.17	10.04		$\sqrt{13.59} = 3.68$		
VM	2.04	4.16			$81.59/7$ $= 11.65$ $= \sqrt{11.65}$ $= 3.41$	$3.68/30.35$ $= 0.12$ 0.12×100 $= 12\%$
Moisture	2.12	4.49				
Ash	1.53	2.34				
Density	2.61	6.81				
N. emission	4.34	18.83				

Box 8.2

WHAT IS "ERROR THEORY"?

It is not possible to obtain a 100% accurate measurement for any physical quantity. This is because there are an infinite number of direct and indirect factors that influence the final result, and consequently, deviates it from its "true" value. However, the word "true" can exist only in theoretical terms, and therefore it is meaningless in practical applications.

There are two types of errors: random errors (precision) and systemic errors (bias).

Random errors are usually difficult to control and/or obtain their actual values. These can be related to human errors, the state of the object being measured/tested and the environment surrounding it (e.g., temperature change, the state of the air, molecule structure, and earth movement), and accidental errors. Systematic errors can be less difficult to control as these errors mostly originate from the instruments/equipment being used. With systematic errors, it is possible to reduce the percentage of errors via the use and application of viable and accurate hardware.

8.3.4 Methodology Standard Deviation for BF

BF	SD	$(SD)^2$	Sum of $(SD)^2$	Overall SD $SD = \sqrt{\frac{\sum (X-M)^2}{n-1}}$	Population SD	Coefficient of Variation (Relative Error)
Business viability	6.34	40.19	96.11	$96.11/8 = 12.01$ $\sqrt{12.01} = 3.46$		
Approach (business risks, "BR")	2.96	8.76				
Emission	3.6	12.96			$96.11/9 = 10.67 = \sqrt{10.67} = 3.26$	$3.46/31.98 = 0.10$ $0.10 \times 100 = 10\%$
Baseline methodology (BM) (or fuel preparation, "FP")	2.45	6.00				
System(s)	2.96	8.76				
Applicability (production and market, "P&M")	2.62	6.86				
Supply	2.94	8.64				
Quality	1.66	2.65				
Land and water	1.14	1.29				

8.3.5 Methodology Standard Deviation

Methodology Sections	SD	$(SD)^2$	Sum of $(SD)^2$	Overall SD $SD=\sqrt{\frac{\sum(X-M)^2}{n-1}}$	Population SD	Coefficient of Variation (Relative Error)
S&T	3.68	13.54	25.51	25.51/1 = 25.51	25.51/2	5.05/31.16
BF	3.46	11.97		$\sqrt{25.51}$ = 5.05	= 12.75	= 0.16
					= $\sqrt{12.75}$	0.16 × 100
					= 3.57	= 16%

8.4 ANALYSIS

In order to establish certain facts, a research survey has been carried out for the sake of obtaining fresh data concerning biomass energy aspects. These facts have shed light on the importance of BF and S&T factors in the field of biomass energy, including the co-firing aspects and the production of bio-fuels. These factors play an important role in refining and reducing the overall cost of producing the fuel. Some of these factors can be listed under the following titles:

1. The importance of S&T to the business
2. The importance of BF to the business
3. The priorities of both BF and S&T to each business.
4. The percentage values (weighting factor) that each business believes to be their "right" values.

The analytical technique, in essence, is simple. Basic written questionnaires have been used in a field of research and business that is still relatively new. Most of the answers have been chosen from the knowledge, experience, and background of the people who took part in this survey. Obviously, limitations can occur if there is insufficient data (which is not the case in this survey) or if the respondents gave inaccurate information, as indicated in Chapter 5. The response to the S&T factors (Fig. 8.2) came close to 80% of the respondents, by putting the energy factor at the top of their list. Other S&T factors, already discussed in detail (Chapters 6 and 7), ranged from the top of the list to the bottom of it, but only voting related to nitrogen emission was located in a way that was erratic in choosing their preferred position in the survey form. However, the number of respondents who voted for this factor was the least, in comparison with the votes obtained for the rest of S&T factors.

The value of SD for the energy factor, compared with the rest of the S&T factors, is relatively high at around 6% (Fig. 8.3). This indicates that a large number of voters (207 votes) gave similar types of responses when it comes to the priority, in general, and the weighing factor, in particular. The SD for

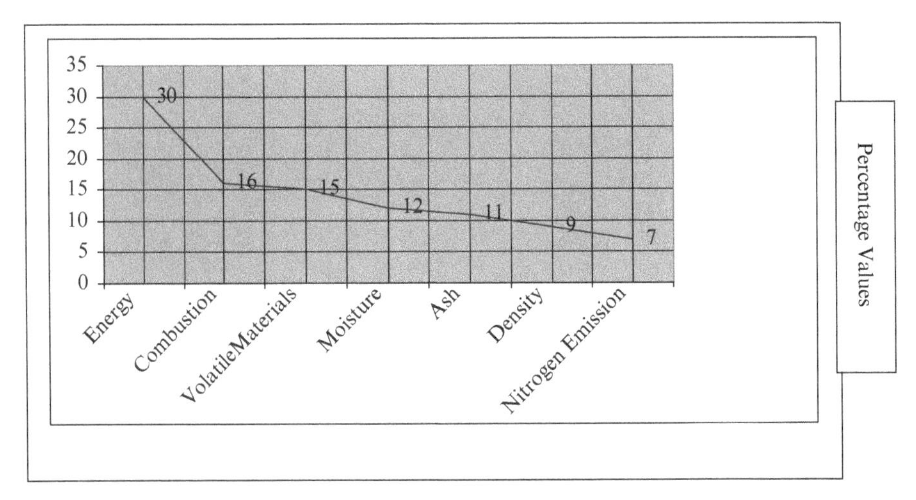

Figure 8.2 Average scoring votes for the S&T factors. *Source*: Author.

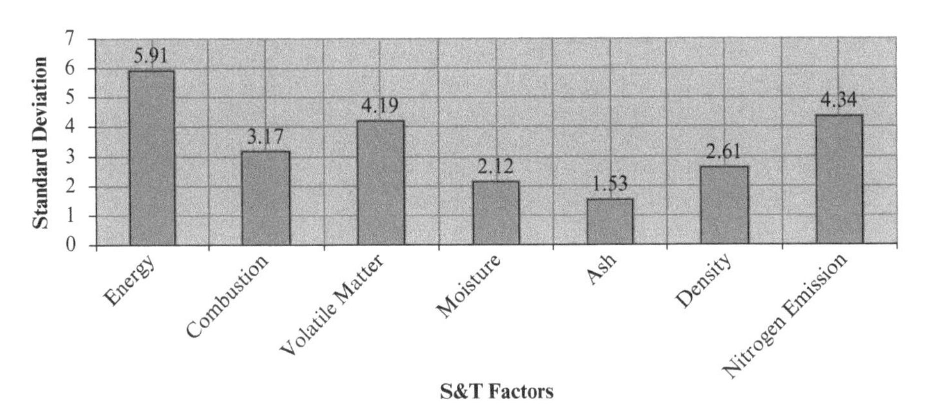

Figure 8.3 Comparison of the standard deviation values for S&T factors. *Source*: Author.

VMF and NEF are very close to each other, that is, with values of 4.19 and 4.34, respectively. Apart from the CIF, the closeness of the SDs for the rest of the factors reflects the smaller spreads of their voting values.

A plot for the overall SD of S&T factors produces a shallow rising graph with a final steep increase due mainly to the high score from the energy factor voting. This also indicates consistent results which do not deviate much from the mean value.

A. Bell Curve

Briefly, in a statistical term, a "normal curve" or "bell curve," is a representation of the function of what is known as a "normal probability distribution." In general, the word "normal" refers to "Gaussian distribution," that is, the normal way we would expect errors to be distributed (Calkins, 2005). Figure 8.4 provides the value of SD for S&T factors using Bell Curve representation.

B. The Average Score

Regarding the survey, the average score has been taken from the high and low preferences marked by those who took part. The average value percentage for the energy factor ranged from 20% up to 50%, which, if the average is taken, comes close to a third of 100%. This is around the percentage value given in the examples of the S&T factors (see Chapters 7 and 9). In the same manner, the rest of the S&T factors have been calculated according to the average votes that were given to them in the survey (Fig. 8.5).

C. BF Voting

Concerning BF voting, the top five were in the following order:

1. Business viability
2. Approach (business risks)

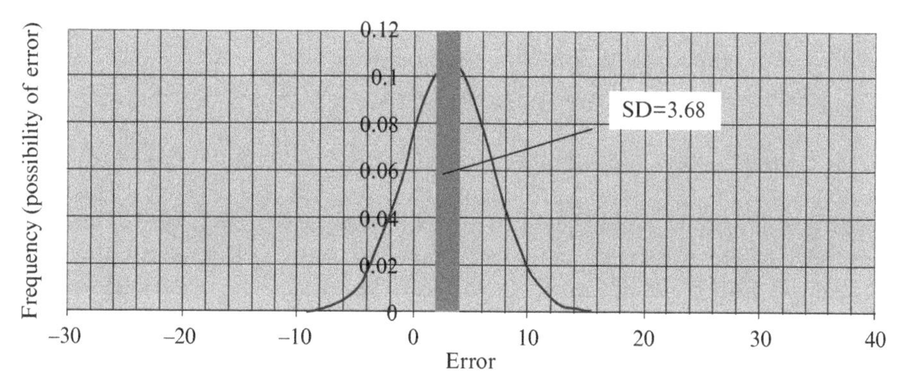

Figure 8.4 Bell curve representation of S&T factors with an SD of 3.68. *Source*: Author.

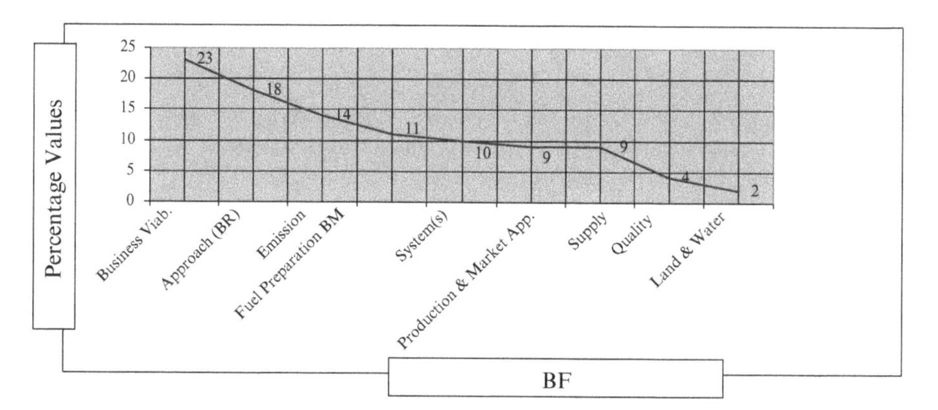

Figure 8.5 Average scoring votes for the BF. *Source*: Author.

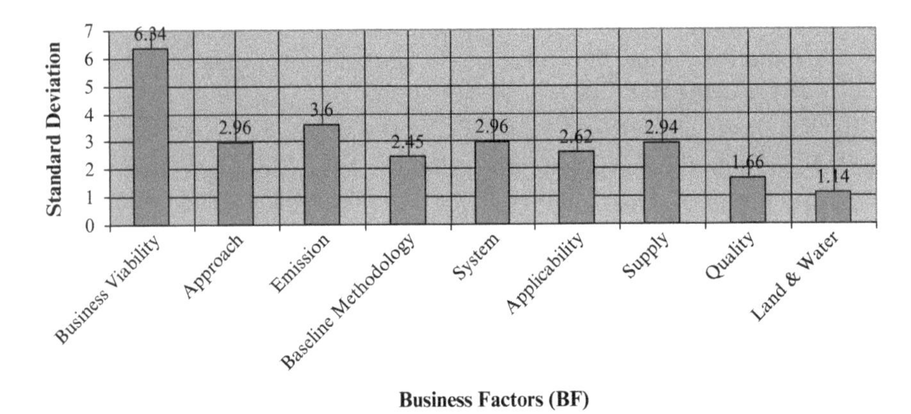

Figure 8.6 Comparison of the standard deviation values for BF. *Source*: Author.

3. System(s)
4. Fuel preparation (baseline methodology)
5. Production and market (applicability).

Apart from a business viability factor (BVF) with an SD score of 6.34, the closeness in the SD values for the majority is around 2.92 (Fig. 8.6). This indicates a similarity in voting with relation to these factors, but with a different approach to the priority given by the voters. This is especially the case because the Supply Factor was not within the top five. The SD for the overall BF section shows that the top of the graph is slightly away from the base 0 (similar to S&T, Fig. 8.4), but still steep. Results again are regularly reproduced and do not deviate much from the mean value. This is reflects the high precision and accuracy/closeness of the data. For example, the SD for S&T is 3.68 while for BF is 3.26, which are close to each other. The result of 5.05 is the overall SD of the methodology, which is still a small figure when it comes to the possible overall error.

The coefficient of variation (relative error) for each factor within the S&T section of the methodology showed for all, apart from NEF, a close result with an average value of 0.17. On the other hand, the coefficient of variation (relative error) for BF—even though the variations are slightly higher than S&T—there are no any other factors that differ more than the rest as was the case with the energy factor within the S&T section. The average value for the coefficient of variation (relative error) of BF is 0.27, which is higher than that of the S&T factors. Calculating the overall coefficient of variation for the S&T factors, a figure of 0.12 is obtained, that is, relative error would be 12%, while the overall coefficient of variation for BF is 0.10, that is, the relative error is 10%.

Finally, the overall methodology coefficient of variation is 0.16, that is, the relative error is 16%. It can be concluded that the factors for the entire business section methodology have closer relationships than the actual data obtained from laboratory tests. The procedure is very similar to that for S&T; however, here, there are different interpretations for the word "system," in particular, between those respondents who filled in the questionnaire in the United States and those who completed it in the United Kingdom. Fortunately, the majority of the BF forms have been completed with direct discussion about the meaning of the above word, that is, with each individual who volunteered to complete the questionnaire. The small difference in voting between the three business factors provided, on average, an example for a priority listing table. Regarding the percentage value, a substantial number of people (close to 50%) either declined to give the above value or only partially managed to answer this particular section. Written comments at the bottom of the questionnaires from participants ranged from the quality/value of using biomass materials for the purpose of generating energy, to the final economic value of the energy produced. Various other comments, mostly verbal, were mainly criticisms of world politicians and the lack of, or slow pace, of international cooperation involving this field.

8.5 CONCLUSION

When designing any type of methodology, there will always be the possibility of errors connected to the final result. Some of these errors can be random and some systematic. Classification of errors depends mostly on the parameters within the methodology sections. In addition, tools from the methodology can be employed to check for errors within the whole system.

The conclusion, therefore, can be summarized in the following points:

1. Some errors can be traced by checking each variable and the environment it originated from and/or was influenced by it. However, it is not always true that if there were practical methodologies, almost free from errors, they would provide the user with the correct answer. Methodologies with some shortcomings may provide relatively correct (or close to correct) results. This is because either the error is small or the shortcoming may not affect or be related to the aspect of the biomass selection process.
2. From the analysis provided in this chapter, it has been concluded that the S&T section of the methodology may contain the lowest percentage of errors. These errors, when they do occur, can be easily minimized by following the procedures outlined previously. In contrast, the BF section is more complex when it comes to the exact percentage of error and their probable origin. This is because BF cannot be measured within a laboratory environment, as is the case with S&T factors.

3. The methodology value, as a possible source of error, is a good example where tools can be used for the purpose of reducing error. This can be done when approaching the stage related to the scoring value for each factor in a particular biomass sample.
4. The possibility for imprecision/precision (or reproducibility) in the survey (i.e., relative error) is 12% for S&T. This is surprisingly two points higher than the BF section of the methodology. The error can be reduced within the S&T section if and when certain criteria are followed when dealing with and processing biomass materials (Section 8.1).
5. The REA1 methodology error is relatively small and falls within an expected standard error in any scientific/methodological research or corrective procedure. The SD figure of 5% (with a population SD of 3.5%) means that the relative error of the REA1 methodology is close to 16%. These figures are expected and are acceptable, especially when all the factors have been taken into account, covering the survey methods, laboratory tests, and researched business/commercial aspects.
6. As mentioned in point 3, the survey itself produced a higher possibility of error related to the S&T section than the BF section of the methodology. This is mainly because of then uncertainty noticed among a number of people when trying to vote (in order of preference) for S&T factors. On the other hand, the BF has a much more straightforward approach to errors. The range of error for both is estimated according to the percentage of responses, as well as from those who partially completed the questionnaires.
7. The final result of the methodology CV (relative error) of 16% is valid, considering that voting took place in three different parts of the world. An anomaly among the voting is related to the nitrogen emission factor. The widely different percentage valuation given to this and the reluctance (or not knowing how) to give this a priority value, produced a wide variation in the NEF end result (Box 8.3).

Box 8.3

LIQUID FUELS: DISTRIBUTION COSTS

Distribution costs for liquid fuels are lowest for pipeline transport and highest for truck transport with rail transport falling somewhere in between. Based on industry sources, the current cost of distributing liquid fuel over 1000 miles is approximately $0.03/gal via pipeline, $0.16/gal via rail, and more than $0.40/gal via truck. Because ethanol is currently delivered mainly by rail and truck, delivery costs are higher than delivery costs for petroleum fuels which utilize pipeline infrastructure.

USDE (2011)

REFERENCES

Arsham H (2009) Topics in statistical data analysis: revealing facts from data. Merrick School of Business, University of Baltimore. http://home.ubalt.edu/ntsbarsh/stat-data/Topics.htm (last accessed April 26, 2009).

Calkins GK (2005) An introduction to statistics: lesson 6, the bell-shaped, normal, Gaussian distribution. Berrien County Math & Science Center, Andrews University, Berrien Springs, MI. http://www.andrews.edu/~calkins/math/webtexts/stat06.htm (last accessed May 7, 2009).

Department of Physics, Ryerson University (2009) Introduction to errors and error analysis. http://www.ryerson.ca/physics/current/lab_information/experiments/IntroToErrorsFinal.pdf (last accessed August 8, 2009).

USDE (2011) Biomass Multi-Year Program Plan. Energy Efficiency and Renewable Energy. http://www1.eere.energy.gov/biomass/pdfs/mypp_april_2011.pdf (last accessed March 22, 2013).

Whittaker ET, Robinson G (1967) Normal frequency distribution. In The Calculus of Observations: A Treatise on Numerical Mathematics, 4th ed. New York: Dover, pp. 164–208, chapter 8.

9

RESULTS: PART 2

9.1 DATA AND METHODOLOGY APPLICATION

9.1.1 Introduction

Biomass data obtained through the application of the methodology showed potential for a newly designed selection process for the most suitable (best) biomass samples. The listing of these samples takes place according to their scoring. In this way, both the scientific/technical and commercial viabilities become part of this selection. The data provide a model of possible future production and usage for the top four (or five) selected energy crops. Bio-fuels and power generating companies have the advantage of providing facilities to deal with (on a wider scale) the selected biomass materials by following similar procedures outlined in the application of the Renewable Energy Analyser One (REA1) methodology.

Results sources:

1. Data from the survey
2. Data from laboratory tests
3. Data from the commercial environment.

Most of the current commercial biomass energy statistics and data are either out of date and/or kept secret, mainly for commercial reasons. This is

The Selection Process of Biomass Materials for the Production of Bio-fuels and Co-firing, First Edition. Najib Altawell.

why data and results, which deal with the biomass energy business case, in addition to the scientific and technical (S&T) case (such as those provided in this chapter), are urgently needed as they are so vital in advancing the biomass energy production and development worldwide.

Calculating potential production scenarios using specifically designed commercial models is vital in every proposed energy business case. This is because these models are required to validate data (such as S&T) at the same level.

This chapter, therefore, focuses on data and results obtained by the application of the methodology which has been examined in the previous chapters. When they are truthful and have been applied correctly, the data and results are the only reflection of a correct methodology being applied.

Concerning samples, each sample of S&T data has been listed separately with the values of their factors. Every sample has been compared with the standard sample of "coal." The selected top four or five samples, in relation to S&T factors, have been plotted against coal for the sake of a final comparison. Similar procedures have been followed regarding biomass BF and their comparisons with coal BF values. Various lists are provided for all the samples with their own data and reports in connection with S&T. In addition, a number of BF examples have been added to show the scoring mechanism of this section of the methodology. The top five biomass samples for S&T and BF are listed independently of each other. Combining both factors together, that is, BF and S&T, using 50% from each (or 25% from S&T and 75% from the BF), the result for the best biomass sample in comparison with the standard sample has been obtained.

Finally, this part of the book will examine the heating values using results obtained via the application of REA1 methodology.

9.2 TESTS

9.2.1 Experimental Tests

9.2.1.1 Aim of the Tests. The aim of these tests is to obtain data related to the values of each S&T factor for every sample. The purpose of obtaining this data is for comparison with the reference sample (in order to aid the selection process), and in order to assist in the makeup of a new type of hybrid bio-fuel.

9.2.1.2 Testing Devices. A number of devices were used to test the samples for various values and characteristics. These devices were located within the Department of Chemical and Environmental Engineering and the Department of Materials Engineering at the University of Nottingham, UK.

1. For energy content, an IKA calorimeter system was used to obtain the number of joules per gram in each sample (Fig. 9.1).

Figure 9.1 IKA Calorimeter System Model C 5000. *Source*: Author.

2. For combustion index, volatile matter, ash elements, and moisture, a Pyris 1 thermogravimetric analyzer was used (Fig. 9.2).
3. To find out the percentage of various gaseous emissions and other elements, a Thermo Electron Flash EA was used (Fig. 9.3).
4. For absolute density the tests were conducted using a gas pycnometer device (Fig. 9.4).
5. For elemental composition, particles size, and internal volume (space), the data and images were obtained by employing a scanning electron microscope (SMS, Fig. 9.5).

The majority of the tests were performed on each sample for a minimum of three times, either manually or automatically by the device itself. Factor values tested during laboratory analysis concentrated on the following values:

1. Energy
2. Ignition
3. Burning period
4. Volatile matter
5. Absolute density
6. Packing density
7. Moisture

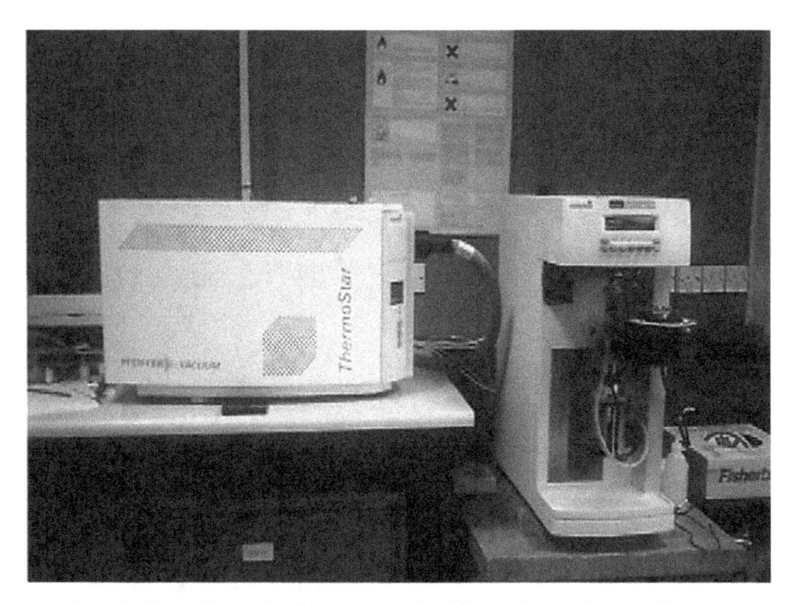

Figure 9.2 Pyris 1 TGA Thermograimetric Analyzer. *Source*: Author.

Figure 9.3 Thermo Electron Flash EA 1112 series 3. *Source*: Author.

Figure 9.4 Gas Pycnometer (AccuPcyc 1330). *Source*: Author.

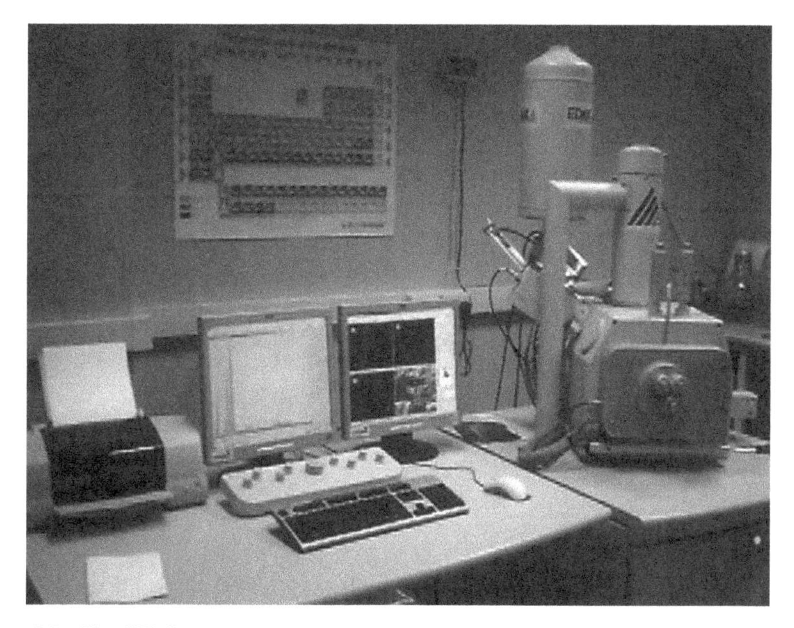

Figure 9.5 The FEI Quanta 600 scanning electron microscope is one of the devices used to asses a variety of S&T factors from biomass and fossil fuel samples. *Source*: Author.

8. Ash quality (PNaCaK and Mg)
9. Ash quantity
10. Nitrogen emission
11. Sulfur emission
12. Carbon dioxide emission
13. Hydrogen emission
14. Internal void space (internal biomass/fossil fuel vacuum).

9.2.1.3 Basic Characterizations. Specific parameters are needed to develop the standard procedures and protocols for the quantitative characterization of biomass samples and their standard samples, the fossil fuels. These parameters should provide reproducible measurements if and when these tests are conducted again under similar conditions. Fossil fuels and biomass materials have been used during various approaches and experiments in order to determine the values of BF and S&T factors. The materials used were in the form of powders, small crushed pieces, pellets, or in original plant produce form, for example, seeds. As has been mentioned at the beginning of this book, part of these biomass samples have been provided by one of the power generating companies in the United Kingdom. Other biomass samples have been selected at random but according to published reports concerning positive scoring of various S&T and BF factors related to these samples.

9.2.1.4 S&T and BF Factors Data. Basic methods are used for calculating some of the biomass S&T data. The standard sample, that is, the South African bituminous coal (Kleinkopje) various data (Chapter 9, Section 9.3.1.1, Coal Data,) are obtained in a similar way. There were simple calculations performed based on the following factors:

1. Moisture
2. Ash
3. Packing density
4. Ignition
5. Burning period
6. Ash quantity.

Other calculations, such as energy, absolute density, ash quality, space, VM, N, C, H, and S have all been carried out automatically, using the devices mentioned in Section 9.2.1.2. With regard to BF, the value of these should be obtained during the calculation time. As has been mentioned previously, this is because there is no standard value for BF, as in the case with S&T. BF values should always be obtained at the time they are needed. This means that much of the BF data have a very short life cycle and therefore, cannot be saved for future use. The changes to some of the BF in day-to-day situations have already been explained in Chapter 7. Applying the data can be done manually, following the steps in the methodology REA1. Obviously, the calculation and comparison should provide an accurate result based on the accuracy of the input data. The meaning of 100% S&T, 75% S&T, 50% S&T, and 25% S&T—as well as 100% BF, 75% BF, 50% BF, and 25% BF—is that an increase (or decrease) of percentage value of either S&T and BF when they are combined and calculated together to obtain the final result for a particular sample. The final value does not depend on the number of S&T or BF factors, rather the final value depends on the percentage allocation from each side, that is, the percentage of BF and the percentage of S&T together. Clearly, if the percentages are equal to each other, that is, 50% BF and 50% S&T, then the value or the influence from each section is equal.

A formula has been applied in order to calculate the scoring for each sample. A software programmer has coded the formula as follows:

$$\begin{aligned}\text{Score} = {} & \text{SUM(NonLab Factor} * \text{Non Lab Factor Weight)} \\ & * \text{NonLabFactorPercentage} + \text{SUM(Lab Factor} * \text{Lab Factor Weight)} \\ & * (1 - \text{NonLabFactorPercenage})\end{aligned}$$

where Non Lab Factor = BF Lab Factor = S&T

The weight (percentage) can be found in the trees: Figure 6.12 and Figure 7.4.

Finally, the coal REA1 report (see Section 9.3.1.2, Coal S&T Report) provides an example for the S&T value range of between 100% S&T and 25% S&T. Examples of biomass S&T methodology and reports related to coal and the 15-biomass samples are illustrated in the following pages.

9.3 S&T SAMPLES DATA AND REPORTS (RESULTS)

9.3.1 Fossil Fuel

9.3.1.1 Coal Data (Standard Sample)

Energy (H) J/g	Mos %	Abs Den g/cm^3	Pac Den g/cm^3	Ash Qun %	Ign T °C	VM %	N Em %	C Em %	H Em %	S Em %	Space %	Ash Qua % P, Na, Ca, K, Mg	Burning Period °C
26,819	1.6	1.5034	0.9618	15.4	432	23	1.65	67.86	3.67	3.36	35	11.54	250

Mos, moisture; Abs Den, absolute density; Pac Den, packing density; Ash Qun, ash quantity; Ign, ignition; VM, volatile materials; Ash Qua, ash quality; Em, emission.

9.3.1.2 *Coal S&T Report.* South African bituminous coal (Kleinkopje), in the form of a powder (this sample was donated by a power generating company in the United Kingdom). For further details, see Chapter 4.

Factor Name	Factor Weight	Value	Final Value
Moisture	4	1.6	1
Ash quality	10	11.54	−1
Ash quantity	5	15.4	0
Ignition	8	432	1
Absolute density	6	1.5034	1
Packing density	5	0.9618	1
Volatile matter	15	23	0
Nitrogen emission	5	1.65	0
Energy content	30	26,819	1
Burning period	8	250	1

S&T total fitness:* 61.25 (100% S&T), 45.94 (75% S&T), 30.62 (50% S&T), and 15.31 (25% S&T).

9.3.2 Biomass Materials

9.3.2.1 Apple Pruning Data

Energy (H) J/g	Mos %	Abs Den g/cm^3	Pac Den g/cm^3	Ash Qun %	Ign T °C	VM %	N Em %	C Em %	H Em %	S Em %	Space %	Ash Qua % P, Na, Ca, K, Mg	Burning Period °C
16,971	7	1.4236	0.3352	2.4	256	68	2.82	48.98	5.54	~0	39	92	180

Mos, moisture; Abs Den, absolute density; Pac Den, packing density; Ash Qun, ash quantity; Ign, ignition; VM, volatile materials; Ash Qua, ash quality; Em, emission.

9.3.2.2 *Apple Pruning S&T Report.* Crushed apple tree wood (this sample was donated by a power generating company in the United Kingdom). For further details, see Box 4.11.

* As a fuel without negative aspects.

Factor Name	Factor Weight	Value	Final Value
Ash quality	10	92	1
Ash quantity	5	2.4	0
Ignition	8	256	1
Absolute density	6	1.4236	1
Packing density	5	0.3352	0
Volatile matter	15	68	1
Nitrogen emission	5	2.82	0
Energy content	30	16,971	0
Burning period	8	180	0

S&T total fitness: 43.53 (100% S&T), 32.65 (75% S&T), 21.76 (50% S&T), and 10.88 (25% S&T).

9.3.2.3 *Corn Data*

Energy (H) J/g	Mos %	Abs Den g/cm^3	Pac Den g/cm^3	Ash Qun %	Ign T °C	VM %	N Em %	C Em %	H Em %	S Em %	Space %	Ash Qua % P, Na, Ca, K, Mg	Burning Period °C
17,334	12.5	1.3911	0.7157	2.2	280	78	3.76	44.83	6.39	~0	48	100	160

9.3.2.4 ***Corn S&T Report.*** Crushed into low grade powder (this sample was selected for analysis in this book). For further details, see Chapter 4.

Factor Name	Factor Weight	Value	Final Value
Ash quality	10	100	1
Ash quantity	5	2.2	0
Ignition	8	280	1
Absolute density	6	1.3911	1
Packing density	5	0.7157	0
Volatile matter	15	78	0
Nitrogen emission	5	3.76	−1
Energy content	30	17,334	0
Burning period	8	160	0

S&T total fitness: 24.67 (100% S&T), 18.5 (75% S&T), 12.34 (50% S&T), and 6.17 (25% S&T).

9.3.2.5 *Distillers Dried Corn Data*

Energy (H) J/g	Mos %	Abs Den g/cm^3	Pac Den g/cm^3	Ash Qun %	Ign T. °C	VM %	N Em %	C Em. %	H Em %	S Em %	Space %	Ash Qua % P, Na, Ca, K, Mg	Burning Period °C
18,680	12.4	1.3565	0.4629	4.5	230	73	6.60	49.92	6.38	~0	66	87.23	140

9.3.2.6 *Distillers Dried Corn S&T Report.* Crushed pellets (this sample was donated by a power generating company in the United Kingdom). For further details, see Box 6.2.

Factor Name	Factor Weight	Value	Final Value
Moisture	4	12.4	0
Ash quality	10	87.23	1
Ash quantity	5	4.5	0
Ignition	8	230	0
Absolute density	6	1.3565	1
Packing density	5	0.4629	0
Volatile matter	15	73	0
Nitrogen emission	5	6.6	−1
Energy content	30	18,680	0
Burning period	8	140	−1

S&T total fitness: 16.67 (100% S&T), 12.5 (75% S&T), 8.34 (50% S&T), and 4.17 (25% S&T).

9.3.2.7 *Miscanthus Data*

Energy (H) J/g	Mos %	Abs Den g/cm^3	Pac Den g/cm^3	Ash Qun %	Ign T °C	VM %	N Em %	C Em %	H Em %	S Em %	Space %	Ash Qua % P, Na, Ca, K, Mg	Burning Period °C
16,847	9.6	1.4511	0.2989	2	253	72	2.86	46.24	5.45	~0	79	48.77	130

9.3.2.8 *Miscanthus S&T Report.* Dried crushed form (this sample was donated by a power generating company in the United Kingdom). For further details, see Chapter 4.

Factor Name	Factor Weight	Value	Final Value
Moisture	4	9.6	1
Ash quality	10	48.77	0
Ash quantity	5	2	1
Ignition	8	253	1
Absolute density	6	1.4511	1
Packing density	5	0.2989	0
Volatile matter	15	72	0
Nitrogen emission	5	2.86	0
Energy content	30	16,847	0
Burning period	8	130	−1

S&T total fitness: 23.65 (100% S&T), 17.74 (75% S&T), 11.82 (50% S&T), and 5.91 (25% S&T).

9.3.2.9 *Niger Seed Data*

Energy (H) J/g	Mos %	Abs Den g/cm³	Pac Den g/cm³	Ash Qun %	Ign T °C	VM %	N Em %	C Em %	H Em %	S Em %	Space %	Ash Qua % P, Na, Ca, K, Mg	Burning Period °C
25,918	7	1.1273	0.6812	4	268	82	6.04	58.67	8.06	~0	35	100	160

9.3.2.10 *Niger Seed S&T Report.* Original seeds used in the tests (this sample was selected for analysis in this book). For further details, see Chapter 4.

Factor Name	Factor Weight	Value	Final Value
Moisture	4	7	1
Ash quality	10	100	−1
Ash quantity	5	4	0
Ignition	8	268	1
Absolute density	6	1.1273	−1
Packing density	5	0.6812	0
Volatile matter	15	82	−1
Nitrogen emission	5	6.04	−1
Energy content	30	25,918	1
Burning period	8	160	0

S&T total fitness: 42.18 (100% S&T), 31.64 (75% S&T), 21.09 (50% S&T), and 10.54 (25% S&T).

9.3.2.11 *Pot Barley Data*

Energy (H) J/g	Mos %	Abs Den g/cm^3	Pac Den g/cm^3	Ash Qun %	Ign T °C	VM %	N Em %	C Em %	H Em %	S Em %	Space %	Ash Qua % P, Na, Ca, K, Mg	Burning Period °C
15,740	11	1.4186	0.8637	1	286	75	2.59	41.81	6.37	~0	39	96.4	210

9.3.2.12 ***Pot Barley S&T Report.*** Original seeds used in the tests (this sample was selected for analysis in this book). For further details, see Chapter 4.

Factor Name	Factor Weight	Value	Final Value
Moisture	4	11	0
Ash quality	10	96.4	−1
Ash quantity	5	1	1
Ignition	8	286	1
Absolute density	6	1.4186	1
Packing density	5	0.8637	0
Volatile matter	15	75	0
Nitrogen emission	5	2.59	0
Energy content	30	15,740	−1
Burning period	8	210	1

S&T total fitness: 27.29 (100% S&T), 20.47 (75% S&T), 13.64 (50% S&T), and 6.82 (25% S&T).

9.3.2.13 *Rapeseed Data*

Energy (H) J/g	Mos %	Abs Den g/cm^3	Pac Den g/cm^3	Ash Qun %	Ign T C°	VM %	N Em %	C Em %	H Em %	S Em %	Space %	Ash Qua % P, Na, Ca, K, Mg	Burning Period C°
26,387	8.3	1.0978	0.7467	5.2	261	83	4.16	53.28	8.02	~0	32	29.14	180

9.3.2.14 *Rapeseed S&T Report.* Original seeds used in the tests (this sample was donated by a power generating company in the United Kingdom). For further details, see Chapter 4.

Factor Name	Factor Weight	Value	Final Value
Moisture	4	8.3	1
Ash quality	10	29.14	1
Ash quantity	5	5.2	−1
Ignition	8	261	1
Absolute density	6	1.0978	−1
Packing density	5	0.7467	0
Volatile matter	15	83	−1
Nitrogen emission	5	4.16	−1
Energy content	30	26,387	1
Burning period	8	180	0

S&T total fitness: 52.13 (100% S&T), 39.1 (75% S&T), 26.06 (50% S&T), and 13.03 (25% S&T).

9.3.2.15 *Rapeseed Meal Data*

Energy (H) J/g	Mos %	Abs Den g/cm^3	Pac Den g/cm^3	Ash Qun %	Ign T °C	VM %	N Em %	C Em %	H Em %	S Em %	Space %	Ash Qua % P, Na, Ca, K, Mg	Burning Period °C
17,943	11	1.3530	0.6203	6	221	70	7.41	43.34	5.94	~0	54	91.32	190

9.3.2.16 *Rapeseed Meal S&T Report.* Original dried state used in the tests (this sample was donated by a power generating company in the United Kingdom). For further details, see Box 10.2.

Factor Name	Factor Weight	Value	Final Value
Moisture	4	11	0
Ash quality	10	91.32	−1
Ash quantity	5	6	−1
Ignition	8	221	0
Absolute density	6	1.353	0
Packing density	5	0.6203	0
Volatile matter	15	70	1
Nitrogen emission	5	7.41	−1
Energy content	30	17,943	0
Burning period	8	190	0

S&T total fitness: 15.61 (100% S&T), 11.71 (75% S&T), 7.81 (50% S&T), 3.90 (25% S&T).

9.3.2.17 *Reed Canary Grass Data*

Energy (H) J/g	Mos %	Abs Den g/cm^3	Pac Den g/cm^3	Ash Qu %	Ign T °C	VM %	N Em %	C Em. %	H Em %	S Em %	Space %	Ash Qua % P, Na, Ca, K, Mg	Burning Period °C
17,035	8	1.3166	0.1690	5	266	78	3.36	48.27	5.22	~0	87	18.54	130

9.3.2.18 *Reed Canary Grass S&T Report.* Dried crushed state (this sample was donated by a power generating company in the United Kingdom). For further detail, see Box 3.2.

Factor Name	Factor Weight	Value	Final Value
Moisture	4	8	1
Ash quality	10	18.54	1
Ash quantity	5	5	0
Ignition	8	266	1
Absolute density	6	1.3166	0
Packing density	5	0.169	−1
Volatile matter	15	78	0
Nitrogen emission	5	3.36	−1
Energy content	30	17,035	0
Burning period	8	130	−1

S&T total fitness: 22.56 (100% S&T), 16.92 (75% S&T), 11.28 (50% S&T), and 5.64 (25% S&T).

9.3.2.19 *Rice Data*

Energy (H) J/g	Mos %	Abs Den g/cm^3	Pac Den g/cm^3	Ash Qun %	Ign T °C	VM %	N Em %	C Em %	H Em %	S Em %	Space %	Ash Qua % P, Na, Ca, K, Mg	Burning Period °C
15,188	13	1.4690	0.8966	0.4	289	78	2.97	39.45	5.78	~0	39	4.21	195

9.3.2.20 *Rice S&T Report.* Milled into powder (this sample was selected for analysis in this book). For further details, see Chapter 4.

Factor Name	Factor Weight	Value	Final Value
Moisture	4	13	0
Ash quality	10	4.21	1
Ash quantity	5	0.4	1
Ignition	8	289	1
Absolute density	6	1.469	1
Packing density	5	0.8966	0
Volatile matter	15	78	0
Nitrogen emission	5	2.97	0
Energy content	30	15,188	−1
Burning period	8	195	0

S&T total fitness: 29.37 (100% S&T), 22.03 (75% S&T), 14.68 (50% S&T), and 7.34 (25% S&T).

9.3.2.21 *Straw Pellets Data*

Energy (H) J/g	Mos %	Abs Den g/cm³	Pac Den g/cm³	Ash Qun %	Ign T °C	VM %	N Em %	C Em %	H Em %	S Em %	Space %	Ash Qua % P, Na, Ca, K, Mg	Burning Period °C
16,465	9.6	1.4825	0.5315	6.4	257	70	3.08	43.34	5.90	~0	64	31.48	175

9.3.2.22 *Straw Pellets S&T Report.* Dried crushed state (this sample was donated by a power generating company in the United Kingdom). For further details, see Box 2.9.

Factor Name	Factor Weight	Value	Final Value
Moisture	4	9.6	1
Ash quality	10	31.48	0
Ash quantity	5	6.4	−1
Ignition	8	257	1
Absolute density	6	1.4825	1
Packing density	5	0.5315	0
Volatile matter	15	70	1
Nitrogen emission	5	3.08	−1
Energy content	30	16,465	0
Burning period	8	175	0

S&T total fitness: 33.53 (100% S&T), 25.15 (75% S&T), 16.76 (50% S&T), and 8.38 (25% of S&T).

9.3.2.23 *Black Sunflower Seed Data*

Energy (H) J/g	Mos %	Abs Den g/cm³	Pac Den g/cm³	Ash Qun %	Ign T °C	VM %	N Em %	C Em %	H Em %	S Em %	Space %	Ash Qua % P, Na, Ca, K, Mg	Burning Period °C
24,711	6.4	1.0848	0.4813	2.3	269	88	5.60	64.68	9.33	~0	56	98.83	160

9.3.2.24 ***Black Sunflower Seed S&T Report.*** Dried seeds used in the test (this sample was selected for analysis in this book). For further details, see Chapter 4.

Factor Name	Factor Weight	Value	Final Value
Moisture	4	6.4	1
Ash quality	10	98.83	−1
Ash quantity	5	2.3	0
Ignition	8	269	1
Absolute density	6	1.0848	−1
Packing density	5	0.4813	−1
Volatile matter	15	88	−1
Nitrogen emission	5	5.6	−1
Energy content	30	24,711	1
Burning period	8	160	0

S&T total fitness: 42.13 (100% S&T), 31.6 (75% S&T), 21.06 (50% S&T), and 10.53 (25% S&T).

9.3.2.25 *Striped Sunflower Seed Data*

Energy (H) J/g	Mos %	Abs Den g/cm³	Pac Den g/cm³	Ash Qun %	Ign T C°	VM %	N Em %	C Em %	H Em %	S Em %	Space %	Ash Qua % P, Na, Ca, K, Mg	Burning Period C°
27,099	7	1.0826	0.4857	2.3	269	88	6.80	67.31	9.48	~0	55	98.71	160

9.3.2.26 *Striped Sunflower Seed S&T Report.* Dried complete seeds used in the test (this sample was selected for analysis in this book). For further detail see Chapter 4.

Factor Name	Factor Weight	Value	Final Value
Moisture	4	7	1
Ash quality	10	98.71	−1
Ash quantity	5	2.3	0
Ignition	8	269	1
Absolute density	6	1.0826	−1
Packing density	5	0.4857	−1
Volatile matter	15	88	−1
Nitrogen emission	5	6.8	−1
Energy content	30	27,099	1
Burning period	8	160	0

S&T total fitness: 42.13 (100% S&T), 31.60 (75% S&T), 21.06 (50% S&T), and 10.53 (25% S&T).

9.3.2.27 *Switch Grass Data*

Energy (H) J/g	Mos %	Abs Den g/cm^3	Pac Den g/cm^3	Ash Qun %	Ign T C°	VM %	N Em %	C Em %	H Em %	S Em %	Space %	Ash Qua % P, Na, Ca, K, Mg	Burning Period C°
17,138	7.8	1.3317	0.1397	3.3	271	91	2.91	48.80	5.27	~0	89	32.25	120

9.3.2.28 *Switch Grass S&T Report.* Dried crushed state (this sample was donated by a power generating company in the United Kingdom). For further details, see Box 9.1.

Factor Name	Factor Weight	Value	Final Value
Moisture	4	7.8	1
Ash quality	10	32.25	0
Ash quantity	5	3.3	0
Ignition	8	271	1
Absolute density	6	1.3317	0
Packing density	5	0.1397	−1
Volatile matter	15	91	−1
Nitrogen emission	5	2.91	0
Energy content	30	17,138	0
Burning period	8	120	−1

Box 9.1

"SPATIALLY EXPLICIT ANALYSIS"

The availability of agricultural land for energy crop production is estimated by taking into account the use of land for the production of food and other purposes, using scenario analysis that take into account agricultural policies, technological development, population growth, income growth, and so forth. A type of land that has received special attention is degraded and marginal land because this type of land is partially or entirely unsuitable for conventional agriculture. So the use of these types of areas does not lead to competition with food. The same approach is applied when estimating the potential of forestry and forestry residues, as well as agricultural residues and organic waste. However, the difference with statistical analysis is somewhat arbitrary because spatially explicit datasets are usually calibrated with statistical data.

Biomass Energy Europe (2010)

S&T total fitness: 12.56 (100% S&T), 9.42 (75%), 6.28 (50% S&T), and 3.14 (25% S&T).

9.3.2.29 *Wheat Data*

Energy (H) J/g	Mos %	Abs Den g/cm^3	Pac Den g/cm^3	Ash Qun %	Ign T C°	VM %	N Em %	C Em %	H Em %	S Em %	Space %	Ash Qua % P, Na, Ca, K, Mg	Burning Period C°
15,128	14	1.4712	0.8338	1	283	80	1.46	39.67	6.20	~0	43	97.41	212

9.3.2.30 *Wheat S&T Report.* Dried complete seeds used in the test (this sample was selected for analysis in this book). For further details, see Chapter 4.

Factor Name	Factor Weight	Value	Final Value
Moisture	4	14	0
Ash quality	10	97.41	−1
Ash quantity	5	1	1
Ignition	8	283	1
Absolute density	6	1.4172	1
Packing density	5	0.8338	0
Volatile matter	15	80	0
Nitrogen emission	5	1.46	0
Energy content	30	15,128	−1
Burning period	8	212	1

S&T total fitness: 27.29 (100% S&T), 20.47 (75% S&T), 13.64 (S&T 50%), and 6.82 (25% S&T).

9.4 BF SAMPLES REPORTS EXAMPLES (RESULTS)

9.4.1 Coal BF Data (Altawell, GSTF, 2012)

Factor Name	Factor Weight	Value
Existing systems	1	1
Emerging systems	4	1
Technology issues	2	1
By-products	3	–1
Present prices	7	0
Prices tendency	7	0
Harvest/mining/reserve	2	1
Available acres/mining	2	1
Innovations	3	1
Knowledge	3	1
New products	4	1
Fuel preparation	1	1
Government regulations	4	1
Investment	6	1
Method	2	1
Today's market	3	1
Emerging market	3	1
Energy input	1	1
Energy output	2	1
Technological	2	1
List of policies	2	0
List of risks	5	0
Adjustments	1	1
Business limitation	1	1
Land use and water	2	1
Local supply	4	1
International supply	5	1
Quality control	2	1
Quality assurance	2	1
CO_2 emission	8	–1
Other gases	6	–1

Total business fitness: 62.21 (100% BF), 46.66 (75% BF), 31.1 (50% BF), and 15.55 (25% BF) (observing market and business conditions during April 2009).

9.4.2 Rapeseed BF Report

Factor Name	Factor Weight	Value
Existing systems	1	0
Emerging systems	4	1
Technology issues	2	–1
By-products	3	–1
Present prices	7	1
Prices tendency	7	0
Harvest/mining/reserve	2	1
Available acres/mining	2	1
Innovations	3	1
Knowledge	3	0
New products	4	0
Fuel preparation	1	0
Government regulations	4	1
Investment	6	1
Method	2	0
Today's market	3	1
Emerging market	3	1
Energy input	1	0
Energy output	2	0
Technological	2	0
List of policies	2	0
List of risks	5	0
Adjustments	1	0
Business limitation	1	0
Land use and water	2	1
Local supply	4	1
International supply	5	1
Quality control	2	0
Quality assurance	2	0
CO_2 emission	8	1
Other gases	6	0

Total business fitness: 53.42 (100% BF), 40.07 (75% BF), 26.71 (50% BF), and 13.36 (25% BF) (observing market and business conditions during April 2009).

9.4.3 Black Sunflower Seed BF Report

Factor Name	Factor Weight	Value
Existing systems	1	0
Emerging systems	4	1
Technology issues	2	0
By-products	3	–1

Factor Name	Factor Weight	Value
Present prices	7	1
Prices tendency	7	0
Harvest/mining/reserve	2	1
Available acres/mining	2	1
Innovations	3	0
Knowledge	3	0
New products	4	0
Fuel preparation	1	0
Government regulations	4	1
Investment	6	0
Method	2	0
Today's market	3	1
Emerging market	3	1
Energy input	1	0
Energy output	2	0
Technological	2	0
List of policies	2	0
List of risks	5	0
Adjustments	1	–1
Business limitation	1	0
Land use and water	2	1
Local supply	4	0
International supply	5	1
Quality control	2	0
Quality assurance	2	0
CO_2 emission	8	1
Other gases	6	–1

Total business fitness: 40.50 (100% BF), 30.38 (75% BF), 20.25 (50% BF), and 10.12 (25% BF) (observing market and business conditions during April 2009).

9.4.4 Niger Seed BF Report

Factor Name	Factor Weight	Value
Existing systems	1	0
Emerging systems	4	1
Technology issues	2	0
By-products	3	–1
Present prices	7	1
Prices tendency	7	0
Harvest/mining/reserve	2	1
Available acres/mining	2	1
Innovations	3	0
Knowledge	3	0

(Continued)

Factor Name	Factor Weight	Value
New products	4	0
Fuel preparation	1	0
Government regulations	4	1
Investment	6	0
Method	2	0
Today's market	3	0
Emerging market	3	1
Energy input	1	0
Energy output	2	0
Technological	2	0
List of policies	2	0
List of risks	5	0
Adjustments	1	0
Business limitation	1	0
Land use and water	2	1
Local supply	4	0
International supply	5	1
Quality control	2	0
Quality assurance	2	0
CO_2 emission	8	1
Other gases	6	–1

Total business fitness: 37.54 (100% BF), 28.16 (75% BF), 18.77 (50% BF), and 9.38 (25% BF) (observing market and business conditions during April 2009).

9.4.5 Apple Pruning BF Report

Factor Name	Factor Weight	Value
Existing systems	1	0
Emerging systems	4	1
Technology issues	2	0
By-products	3	–1
Present prices	7	1
Prices tendency	7	0
Harvest/mining/reserve	2	0
Available acres/mining	2	0
Innovations	3	0
Knowledge	3	1
New products	4	–1
Fuel preparation	1	1
Government regulations	4	1
Investment	6	0
Method	2	0
Today's market	3	0

Factor Name	Factor Weight	Value
Emerging market	3	0
Energy input	1	0
Energy output	2	0
Technological	2	0
List of policies	2	1
List of risks	5	–1
Adjustments	1	0
Business limitation	1	0
Land use and water	2	0
Local supply	4	1
International supply	5	0
Quality control	2	0
Quality assurance	2	0
CO_2 emission	8	1
Other gases	6	–1

Total business fitness: 33.49 (100% BF), 25.12 (75% BF), 16.74 (50% BF), and 8.37 (25% BF) (observing market and business conditions during April 2009).

9.4.6 Striped Sunflower Seed BF Report

Factor Name	Factor Weight	Value
Existing systems	1	0
Emerging systems	4	1
Technology issues	2	0
By-products	3	–1
Present prices	7	–1
Prices tendency	7	0
Harvest/mining/reserve	2	1
Available acres/mining	2	1
Innovations	3	0
Knowledge	3	0
New products	4	0
Fuel preparation	1	0
Government regulations	4	1
Investment	6	0
Method	2	0
Today's market	3	0
Emerging market	3	1
Energy input	1	0
Energy output	2	0
Technological	2	0
List of policies	2	0
List of risks	5	0

(*Continued*)

Factor Name	Factor Weight	Value
Adjustments	1	0
Business limitation	1	0
Land use and water	2	1
Local supply	4	0
International supply	5	1
Quality control	2	0
Quality assurance	2	0
CO_2 emission	8	1
Other gases	6	–1

Total business fitness: 30.54 (100% BF), 22.9 (75% BF), 15.27 (50% BF), and 7.63 (25% BF) (observing market and business conditions during April 2009).

9.5 THE FINAL BIOMASS SAMPLES

9.5.1 S&T Results

According to the results obtained by the application of the methodology REA1 (Table 9.1), the top most suitable biomass samples according to S&T (excluding the business influence, i.e., 100% S&T) are the following:

1. Rapeseed
2. Apple pruning
3. Niger seed
4. Black sunflower seeds } Same level
5. Striped sunflower seeds. }

All the 15 biomass samples (Fig. 9.6) and the top five of these samples (Fig. 9.7) tested during the work on this book have certain qualities and characters which makes some of them suitable for the production of bio-fuels. However,

Table 9.1 Top S&T five biomass samples with coal (100% S&T)

Number	Biomass Sample	S&T Percentage of Fitness
1	Rapeseed	52.13
2	Apple pruning	43.53
3	Niger seed	42.18
4	Black sunflower seeds	42.13
5	Striped sunflower seeds	42.13
6	**Coal**	**61.25**

Source: Author.

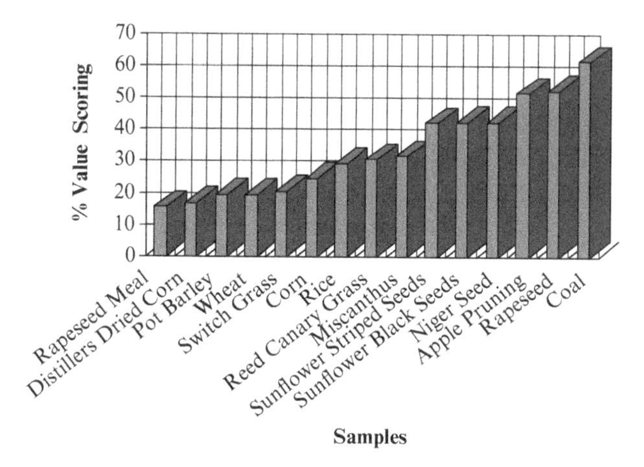

Figure 9.6 Fitness for all 15 biomass samples applying only S&T (compared with coal) *Source*: Author.

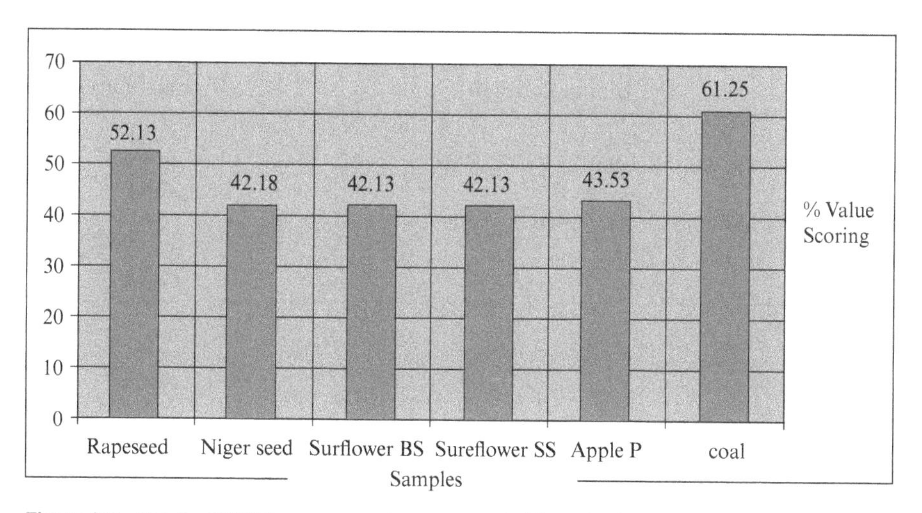

Figure 9.7 Top five S&T biomass samples compared with coal using 100% S&T. *Source*: Author.

the demand for high quality fuel that has fewer side effects on the hardware, low cost, and the ability to be stored for a long period of time without degradation are all vital but basic determining factors.

The processes related to the testing and calculating of the values of the various factors within the REA1 methodology have been made according to the main principles related to: energy content, combustion index, volatile matter, moisture, ash, density, and nitrogen emission. All of these factors have been discussed in detail in the previous sections and chapters.

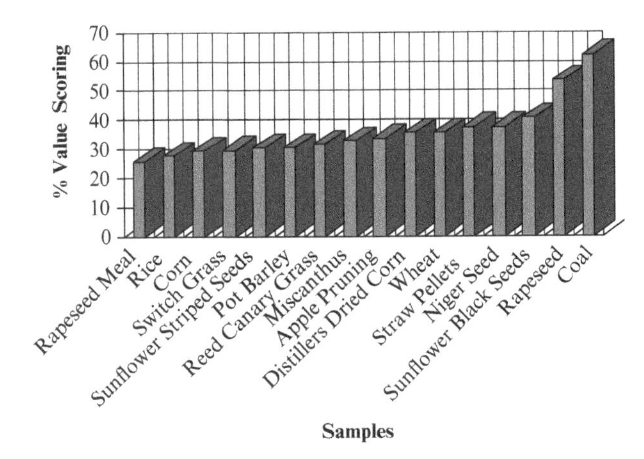

Figure 9.8 All samples' BF value percentages applying BF scores only. Observing market and business conditions during April 2009. *Source*: Author.

Table 9.2 The top five BF biomass samples with coal (100% BF), observing market and business conditions during April 2009

Number	Biomass Sample	BF Percentage of Fitness
1	Rapeseed	53.42
2	Sunflower black seeds	40.50
3	Niger seed	37.54
4	Straw pellets	37.51
5	Wheat	35.56
6	**Coal**	**62.21**

Source: Author.

9.5.2 BF Results

Even though there are strong connections between BF and S&T factors in deciding on the outcome, the BF follow a different route in the selection process.

Looking at the overall BF (with the exclusion of the S&T factors), the potential trends of the business is in the shape of a graph (Fig. 9.8), from one biomass sample to another, providing a much more gradual/smoother approach than the graph obtained for the S&T (Fig. 9.6). The reason for this is that laboratory tests on the samples showed bigger differences among the samples than those exhibited in the business section. This is one of the reasons that not all the top five in the S&T analysis appeared at the business top of the final business selection.

All the 15 samples for BF (Fig. 9.8) and the top five BF (Table 9.2) have two different biomass samples to those in the top five for the S&T (Table

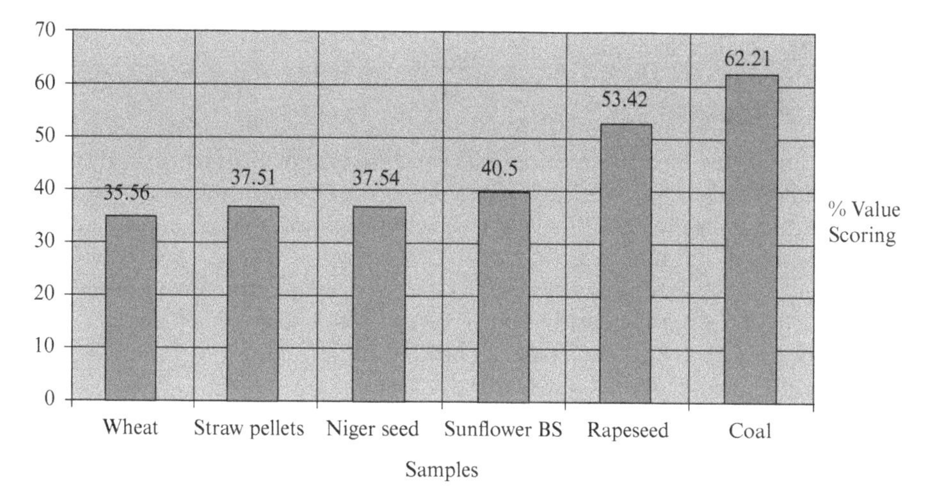

Figure 9.9 Top five BF biomass samples with coal using 100% BF scoring, observing market and business conditions during April 2009. *Source*: Author.

9.1 and Fig. 9.7). The reason for this is that S&T good quality in one sample may not have the market and business quality in the same sample—at the same time. This is why there are two different classifications (i.e., S&T only and BF alone), which have been created to highlight these differences before combining the two sections of the methodology (see Chapters 6 and 7) to attain the final biomass samples.

The striped sunflower seeds, apple pruning (Table 9.1), straw pellets plus wheat (Table 9.2) have certain differences that are the main cause of having two different samples within the two final selection of the S&T and BF tables. According to today's market prices, that is, at the time of compiling this section of the book, striped sunflower seeds cost more to purchase than black sunflower seeds. From the BF perspective (Fig. 9.9), black sunflower seeds scored more positively than the other type. Straw pellets, according to the same market, are also cheaper to purchase than apple wood, while the constant availability of wheat on the international market provides wheat with higher score in this respect. Rapeseed, similar to black sunflower seeds, occupied both tables for having better scoring in both BF and S&T factors than the other samples.

9.6 SAMPLES FINAL FITNESS

The following are percentage values of biomass samples compared with coal in a ratio of 50% BF and 50% S&T. These figures have been

Box 9.2

SWITCH GRASS (SG)

SG is native to North America and used as rangeland forage and hay. Sometime SG is referred to as "tall panic grass." Other common names for SG include wobsqua grass, lowland switch grass, blackbent, tall prairie grass, wild redtop, and thatch grass

SG is one of the dominant species of the tall grass prairie. A warm-season plant, it can be seen occasionally along roadsides. Growth begins in late April or mid May. SG height can be up to 7 ft.

The chemical elements of SG are the following:

Carbon	48.80%
Hydrogen	5.27%
Nitrogen	2.91%
Oxygen	43.00%
Sulfur	0.02%

Source: Author.

generated by REA1 using the methodology principles discussed in Chapters 5, 6, and 7.

Note that the order of fitness will change as and when the percentage value of BF and/or S&T changes, accordingly. Market and business conditions play an important role as changes occur on a daily basis. The market and business aspects observed during the month of April 2009—which produced the following results—may not be applied at a later date, and consequently the order of the following list may change as well. The following pages provide a summary list of "sample fittest" (Box 9.2).

Coal

Summary of Sample Fitness:

S&T factors fitness 30.62%

BF fitness 31.10%

Total Fitness: 61.73%

1. Rapeseed
Summary of Sample Fitness:
S&T factors fitness........26.06%
BF fitness........26.71%
Total Fitness: 52.78%

2. Black Sunflower Seeds
Summary of Sample Fitness:
S&T factors fitness........21.06%
BF fitness........20.25%
Total Fitness: 41.32%

3. Niger Seed
Summary of Sample Fitness:
S&T factors fitness........21.09%
BF fitness........18.77%
Total Fitness: 39.86%

4. Apple Pruning
Summary of Sample Fitness:
S&T factors fitness........21.76%
BF fitness........16.74%
Total Fitness: 38.51%

5. Striped Sunflower Seeds
Summary of Sample Fitness:
S&T factors fitness........21.06%
BF fitness........15.27%
Total Fitness: 36.33%

Top five biomass samples as 50% BF and 50% S&T were applied via REA1 methodology

6. Straw Pellets
Summary of Sample Fitness:
S&T factors fitness........16.76%
BF fitness........18.76%
Total Fitness: 35.52%

7. Wheat
Summary of Sample Fitness:
S&T factors fitness........13.64%
BF fitness........17.78%
Total Fitness: 31.42%

8. Pot Barley
Summary of Sample Fitness:
S&T factors fitness........13.64%
BF fitness........15.33%
Total Fitness: 28.98%

9. Rice
Summary of Sample Fitness:
S&T factors fitness 14.68%
BF fitness 13.85%
Total Fitness: 28.53%

10. Miscanthus
Summary of Sample Fitness:
S&T factors fitness 11.82%
BF fitness 16.33%
Total Fitness: 28.15%

11. Corn
Summary of Sample Fitness:
S&T factors fitness 12.34%
BF fitness 14.77%
Total Fitness: 27.11%

12. Reed Canary Grass
Summary of Sample Fitness:
S&T factors fitness 11.28%
BF fitness 15.83%
Total Fitness: 27.11%

13. Distillers Dried Corn
Summary of Sample Fitness:
S&T factors fitness 08.34%
BF fitness 17.78%
Total Fitness: 26.12%

14. Switch Grass
Summary of Sample Fitness:
S&T factors fitness 06.28%
BF fitness 14.84%
Total Fitness: 21.12%

15. Rapeseed Meal
Summary of Sample Fitness:
S&T factors fitness 07.81%
BF fitness 12.86%
Total Fitness: 20.66%

When applying a 50% ratio of importance to both BF and S&T factors, the rapeseed sample came at the top of the list (Fig. 9.10). Rapeseed is also at the top of the list when other ratios are used during a number of tests under

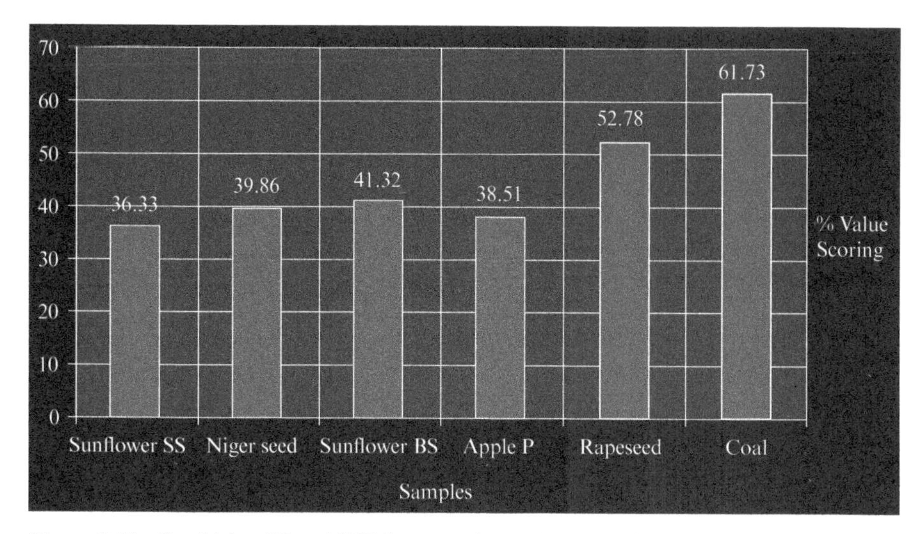

Figure 9.10 Combining BF and S&T factors using 50% scoring for both of them, the final top five are compared with a coal sample, observing market and business conditions during April 2009. *Source*: Author.

Table 9.3 Final top five biomass samples using the influence of both BF and S&T factors on an equal basis, observing market and business conditions during April 2009

Number	Biomass Sample	BF and S&T Percentage of Fitness (50–50%)
1	Rapeseed	52.78
2	Sunflower BS	41.32
3	Niger seed	39.86
4	Apple P	38.51
5	Sunflower SS	36.33
6	**Coal**	**61.73**

Source: Author.

the application of REA1 methodology. Other biomass samples followed behind rapeseed in the scoring, as shown in Table 9.3.

9.7 DISCUSSION AND ANALYSIS

Researching 15 biomass samples has provided a considerable amount of data. At the same time, the selection of different results (according to the percentage value applied to the two sections of the methodology, i.e., BF and S&T; see Chapters 6 and 7) also added additional materials to the data. For this reason, the process followed during the work on each sample is by checking a variety of possibilities connected to the commercial production of biomass energy (Fig. 9.11).

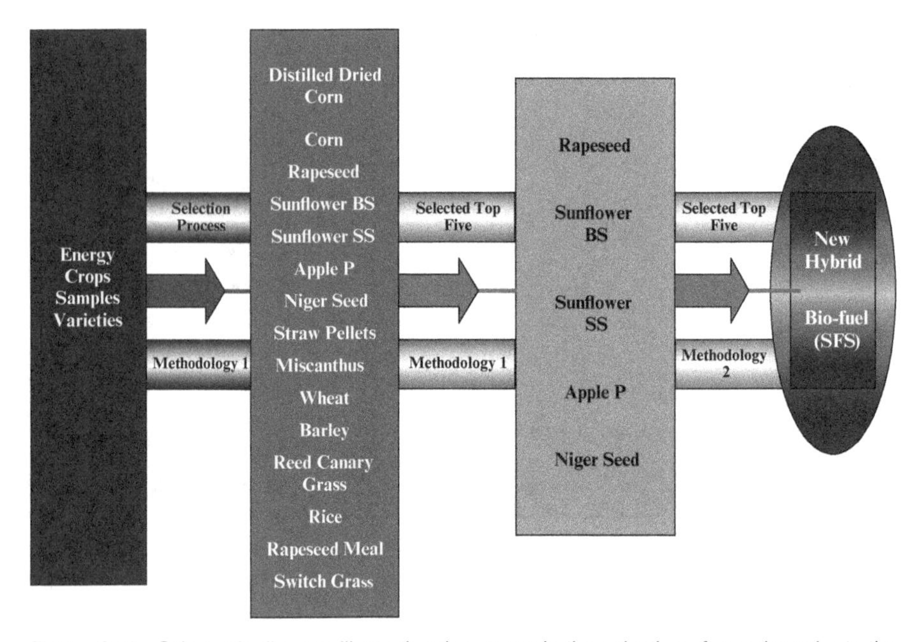

Figure 9.11 Schematic diagram illustrating the stages in the selection of samples prior to the final results, which lead toward the development of a hybrid (SFS) bio-fuel. *Source*: Author.

Two of the standard samples, such as natural gas and crude oil derivatives, have not been included in this chapter as standard/reference examples. The "standard" sample of coal, however, has been used regularly and listed at the beginning of BF and S&T data and their reports.

Coal is still one of the largest fossil fuels used by power stations around the world to generate electricity. For example, half of the electricity generated in the United States is from coal (American Coal Foundation, 2009). Reportedly, to use coal as a standard, when dealing with the aspects of generating electricity, that is when electricity is generated fully or partly from biomass materials—many regarded this procedure to be one of the best options available now.

The energy from coal matches the energy obtained from some of the biomass samples already examined in this book (see Section 9.3.2.26 Sunflower SS). An energy content of the same level as coal, or higher than coal, is one of the positive aspects attributed to some of the biomass materials. On the other hand, many of the biomass materials exhibit a negative tendency in that they absorb moisture. Coal's comparative moisture factor is very much in its favor. In comparison with the rest of the biomass samples, the corn sample recorded the highest percentage of moisture.

The other negative aspect (but it can be also a positive one) of biomass compared with coal is the high VM each sample has shown during the laboratory tests.

In regard to absolute and packing density, coal came at the top, while switch grass (for packing density) and sunflower SS (for absolute density) were tanked the lowest. Having said that, as the sunflower samples used in this test were in seed form rather than a powder form (as in the case with coal and other samples), then sunflower in fact could be high on the list when it comes to absolute density and packing density. In comparison with other biomass samples, coal has the highest ignition temperature at 432°C, while rapeseed meal has the lowest ignition temperature (221°C). In other important factors, such as the amount of ash left over, biomass samples scored positively in this respect.

In regard to the emission factor among biomass materials, the results are in favor of the biomass materials used, that is, apart from low CO_2 emissions, energy crops are also able to neutralize CO_2 by replanting what has been already burned. In addition, the emission of sulfur from the biomass samples used is close to zero, while in coal, there is a much higher percentage of emission compared with the samples, that is, around 3.36%. However, tests on the biomass samples have shown that they contain a higher level of nitrogen than the standard sample, that is, the average percentage value for the 15 biomass samples is 4.16%.

These are some basic facts concerning biomass materials in comparison with coal. These, of course, are already well-known facts that have been looked at and followed from the start of the research on this book. Consequently, characteristics of this nature do not always provide a guideline or a straightforward answer regarding the selection and refining process of biomass samples, nor for the final production of a new type of bio-fuel. Despite this, negative characteristics of biomass materials can be improved by searching for a counterpart solution within the same sample or a "group of samples." This means that biomass materials, unlike fossil fuels, can be cheaply adjusted to a particular type of environmentally friendly fuel without any reduction in their energy output or damaging any other positive aspects. There is already a method to help in achieving these; however, discussing the details of this method would take this research beyond the scope of this book.

Looking at the result of each sample, the single section (100%) is given in order to show the result as it stands, that is, without the effect/influence of the business or the S&T factors—on either side. These kinds of single factor results are useful, if there is no business influence involved and/or no scientific/technical side. They are also useful to consult before trying to obtain the final combined result of the two factors, that is, knowing the percentage value of each section (BF or S&T) may influence, in a constructive manner, the calculation of the BF. The BF needs to assess the effect of certain information/data on the business side, on both a short and a long-term basis. For example, with regard to the amount of energy obtained from certain types of biomass material, would it be better, from a commercial point of view, to choose energy crop(s) with the highest energy content, but which require additional work (additional cost), or would it be better to choose a crop with a lower amount

Table 9.4 Listing of the final results for biomass samples with their average fitness value, observing market and business conditions during April 2009

Rapeseed	53%
Sunflower BS	41%
Niger seed	39%
Apple P	38%
Sunflower SS	36%
Straw pellets	35%
Wheat	31%
Pot Barley	29%
Rice	28%
Miscanthus	28%
Corn	27%
Reed canary grass	27%
Distillers dried corn	26%
Switch grass	21%
Rapeseed meal	21%
Average fitness top 4 biomass samples	**43%**
Average fitness top 5 biomass samples	**42%**
Average fitness for 15 biomass samples	**32%**

Source: Author.

of energy but involving less work (costing less)? The answer to a question such as this depends entirely on the energy business itself.

Changing the percentage value between BF and S&T for the purpose of obtaining the full final results, as indicated earlier, can in itself change the final scoring value. This is because the scoring mechanism of 1, 0, −1 has an effect on the calculation itself. Zero scoring, even though it is neutral in its nature, in the methodology has a very small positive value of around 0.05 (Table 9.4).

Therefore, a neutral factor is in fact a very small element with a positive nature. This means that if the scoring in the BF is mostly zeros, small accumulations of positive scoring will occur. This option has been taken from the early stages of the construction of the methodology for a good reason. No negative impact is in fact a positive in its nature. However, the option is left open, that is, if the energy business wishes to change the neutral value to an actual zero, then this can be done easily within the methodology system. The total fitness, therefore, is purely an individual business's choice. What is presented in this chapter are merely examples of what is calculated and researched to be as close to the reality of a certain situation at a certain time. This is because BF are constantly changing and even the present scientific facts obtained from biomass laboratory analysis can, sometimes, slightly change. The season/time of the energy crops plantation, location, soil fertility, amount of water, amount of sunlight, method of harvesting, and moisture contents all may differ from time to time (Biomass Energy Centre, 2008). All these factors have a certain influence on the chemical make-up of the biomass materials obtained and that in itself opens a path for changes to the overall results. The BF can show a higher percentage of errors than the laboratory tested S&T values, however, result from the survey method indicated the opposite, that is S&T is slightly higher in relative error value than BF (see Chapter 8).

The decision of what samples to choose from the final selected methodology list should be made on the basis of the strength of both sections of the methodology, that is, BF and S&T. This means power generating companies and bio-fuel manufacturers should consider which section of the methodology has a higher priority to them, that is, what percentage value should be given to each section of the methodology. In this chapter, equal importance is given to BF and S&T, that is, 50% for each section. This approach is used here as an example to illustrate the final result by considering that both sections of the methodology are equally important. Having decided on the percentage value for the two sections of the methodology, the results obtained, such as from those of the 15-biomass samples used in this book, should be divided into areas of strength and weakness. This can be done from the figures obtained from each section of the methodology and/or from the results of both.

Rapeseed achieved 53% with BF and S&T, that is achieving approximately 26% for each. However, Switch Grass and Rapeseed Meal came at the lowest on the list, both achieving around 21%. This is mainly to do with a higher BF score for Switch Grass than Rapeseed Meal (15% and 13% respectively), while the opposite is true for the S&T section for the two samples, that is 6% and 8% respectively.

The results obtained for all of the 15 samples should be looked at in a similar manner but with emphasis on the margin of difference between one sample and another. Those samples with a lower scoring and close to other samples in their values will not be considered as a viable option in the mixing process for the Super Fuel sample (SFS) hybrid fuel. Those samples at the top of the list will be considered, not just for their higher value, but also for scoring related to BF. The reason BF is important for the top five samples is mainly because they all obtained 21% for the S&T (apart from the rapeseed sample), that is what make one sample better than another is how the business and commercial performance is rated to it.

Finally, the choice for the bio-fuel mixture can be achieved by mixing the top four or five biomass samples. This will require a new research method (i.e., developing a new methodology) to calculate the exact percentage and the actual makeup for each sample prior to the manufacture of the hybrid bio-fuel. An optimization method will be needed to speed up the process. This kind of optimization can be achieved with the development of the present REA1 methodology. Further details regarding the new bio-fuel (SFS) can be found in the following chapter (Box 9.3).

Box 9.3

WHICH FUEL WILL WE BE USING IN THE FUTURE?

The following basic factors will be needed:
Environmentally friendly
Secured and reliable (supply)
Economical—affordable
Available on demand
(International, national, and local—cities, towns, and villages)
Compatible with a wide range of technologies
Grades—produced in a range of qualities and forms
Storage and lifecycle
Present and future developments are possible.

9.8 CONCLUSION

The selected 15-biomass samples, together with their standard fossil fuel counterpart, have undergone laboratory testing and were then categorized accordingly through the application of the methodology.

The data and reports for BF and S&T provided a selection of results. Result selection may depend on the variation of the percentage value of each section of the methodology, that is, when both factors BF and S&T are combined together in order to provide the final result, changes of their shares (percentage value) can change the value of the overall result.

To conclude the final findings in this chapter, the following points have been noted:

1. Displaying full reports for BF and S&T for all the samples has provided a wider choice for the user. This may meet the biomass selection needs for those who are involved in the biomass energy business.
2. Data and results suggest that the selection process is simple and therefore can be less prone to errors. Factual results can be obtained by using the approach outlined in the two sections of the methodology. This approach can be used by power generating companies. It will be useful during the early stages of the use of biomass materials and during their development for the commercial generation of electricity.
3. Out of the 15 selected biomass samples, the conclusion borne from applying REA1 methodology is that the energy crop "rapeseed" is the most suitable source of energy. Combining "rapeseed" with the other four or five energy crops at the top of the list, the scenario of producing new type of biomass fuel is possible;
4. Based on the results and the way these have been obtained, there does not appear to be any pronounced errors in the approach mechanism used to identify the required biomass materials.

5. The difference between one sample and another in percentage value (weighting factor) is greatly affected in the way the scoring is administered via the REA1 methodology. This change can be particularly noticed when inputting the scoring of BF. For example, the changes in market conditions, business situations, and local/national/international regulations could change a positive score into a neutral or even into a negative one. The difference in percentage value does exist (as illustrated in Chapter 8), but these kind of differences can be easily reduced by manipulating the input scoring values. Therefore, a relative error of 15–16% would be noticed if the exact percentage valuations, that is, as those inputted during the survey method, produced larger differences between one sample and another. This possibility would rarely happen, as each business and every situation within the business itself is different, if not unique.
6. The percentage value for BF and S&T is attributed by the power generating companies, that is, which one has a higher priority than the other. The final values for the top five samples will be decided only by the BF scoring, rather than via both sections. This is because the top five samples (apart from Rapeseed) scored the same value in the S&T section.
7. This chapter has provided the basis for the development of a hybrid biomass fuel (SFS) as the biomass samples have been ranked in order of their usability for bio-fuels production and electricity generation.

In the future, when data of different sample(s) characteristics have been optimized to be as close as possible to the standard sample(s), an additional new method might be needed (Box 9.4).

Box 9.4

CO_2

So far, land plants and the ocean have taken up about 55 percent of the extra carbon people have put into the atmosphere while about 45 percent has stayed in the atmosphere. Eventually, the land and oceans will take up most of the extra carbon dioxide, but as much as 20 percent may remain in the atmosphere for many thousands of years.

The changes in the carbon cycle impact each reservoir. Excess carbon in the atmosphere warms the planet and helps plants on land grow more. Excess carbon in the ocean makes the water more acidic, putting marine life in danger.

Without greenhouse gases, Earth would be a frozen −18 degrees Celsius (0 degrees Fahrenheit). With too many greenhouse gases, Earth would be like Venus, where the greenhouse atmosphere keeps temperatures around 400 degrees Celsius.

NASA (2013)

REFERENCES

Altawell N (June 2012) Energy crops optimisation selection process for the commercial production of bio-fuels. GSTF Journal of Engineering Technology 1(1):25–30.

American Coal Foundation (2009) Fast facts about coal. http://www.teachcoal.org/aboutcoal/articles/fastfacts.html (last accessed April 8, 2009).

Biomass Energy Centre (2008) Effect of moisture content in biomass material. U.K. Forestry Commission, Forest Research, (defra). http://www.biomassenergycentre.org.uk/portal/page?_pageid=75,17656&_dad=portal&_schema=PORTAL (last accessed May 19, 2008).

Biomass Energy Europe (2010) Methods & Data Sources for Biomass Resource Assessments for Energy. Version 3. Issue/Rev.: 1.

NASA (2013) Effects of changing the carbon cycle. Earth observatory. http://earthobservatory.nasa.gov/Features/CarbonCycle/page5.php (last accessed June 19, 2013).

10

ECONOMIC FACTORS

10.1 BIOMASS FUEL ECONOMIC FACTORS AND SFS

10.1.1 Introduction

Various international, national, and local laws and agreements have been passed concerning two important issues: climate change and finding an alternative source of energy. The imposition of these legislations and regulation, accompanied by higher prices for fossil fuels, has pushed forward new ideas, new research, and new enterprises. The goal is to solve what is commonly termed "the energy crisis" and issues related to "global warming," at the same time. However, without consideration for commercial drivers, the above may not produce a viable and economically stable structure for many renewable energy enterprises. For this reason, the methodology proposed 75% of the scoring should be allocated to the business section.

Incentives and encouragement to produce and use fuels from biomass materials are both vital; publicized in the form of marketing and advertising by national and local governments and, therefore, should be introduced on a regular basis at various levels. The business factors (BF) section of the methodology (Chapter 7) has identified and examined a number of incentives which can be made. These may be made in the form of exemption from taxation, the offering of full or partial grants and/or free-interest loans and/or feed-in tariffs. A number of these incentives already exist in some countries.

The Selection Process of Biomass Materials for the Production of Bio-fuels and Co-firing, First Edition. Najib Altawell.

The social aspect is another important factor for the success of any business, especially for one that is an entirely new venture such as a commercial business in the field of renewable energy (Graham and Walsh, 1995). However, when it comes to the establishment of large new renewable energy projects and businesses, some investors are hesitant due to the uncertainty of profitable return (Murphy, 2001).

The social aspect has been discussed in detail during the survey conducted with a number of business people with commercial biomass energy experiences. The outcome of this discussion—which drives some of these businessmen and women to invest in commercial biomass energy—can be summarized in the following two points:

1. Rising prices of fossil fuels. This new incentive for the renewable energy field will only be a realistic venture if prices continue to rise further in the near future (all economic, scientific, and political indications are pointing toward this direction), then it will be unrealistic and unprofitable to continue to use fuels, such as crude oil, for the purpose of generating energy.
2. The prospect for biomass energy as a business on a local, national, and global scale is a very promising venture indeed. This was established according to the results from 17 studies examined and evaluated for the potential of producing energy from biomass materials by Berndes more than 10 years ago (Berndes et al., 2003).

Finally, the economic potential of biomass depends largely on the productivity of the land where it is grown. This can be one factor in reducing the overall cost. The other factor is technological advancement and mechanization before, during, and after each step within the process of obtaining energy from biomass, mainly for the purpose of saving money and time.

10.2 ECONOMIC FACTORS

The open discussion about the commercial viability of biomass energy during the BF section of the survey highlighted some common factors. These factors have been summarized in the following list:

1. Biomass fuels are not competitive with fossil fuels.
2. Biomass energy sources are perceived as beneficial to society by emphasizing that it will protect the environment, reduce global warming, result in a reduction in gas emissions, and will create new industries—which in turn will create additional jobs (to offset their economic disadvantages);
3. Balance benefit and cost. There is a huge difference when it comes to the amount of investment needed and the beneficial return. This does not

mean biomass energy is not economically viable, but rather it is within the time scale that this kind of industry needs before it can establish itself and mature.

4. It is difficult to give market values to goods and services not traded in any market. This can dissuade many potential investors, and consequently, delay the progression of these kinds of services and industries.
5. Environmental impacts are still not well identified. Since biomass is new as a method of producing energy on a large commercial scale, the environmental impacts are still not fully covered. This may change the attitude of politicians and businessmen, if, later on, new restrictions and costly regulations introduced.
6. Costs will differ during each step of manufacturing. This means we may find that obtaining the original biomass materials is very cheap indeed. However, trying to take these materials into the following stages, that is, taking these biomass materials from the source (e.g., farms) to the final stage as a useable energy fuel, involves many processes, some of them with high economical costs.
7. There are four basic steps for the establishment of a biomass energy business. Consideration of this kind of "chain," where various steps take place before and after the final biomass energy production, has been examined briefly in Section 10.3.

In many cases, there is financial support to offset negative aspects for the present economic situation within biomass energy sectors. However, to get financial support from a national or local government, in most cases, a number of applications will have to be made. Conditions attached to eligibility are not always straightforward. For example, in the European Union, energy plants taking part in R.D&D programs are eligible for funding from their national governments and from the European Union, which is nothing but good news for those who are planning to launch a business in this field. However, in some cases, there are complicated procedures that need to be followed. These kinds of procedures can be the cause of lengthy delays in obtaining finance. In some cases, this is urgently needed and may lead to a loss of support from investors.

There is a positive picture emerging in relation to investments. According to the Global Status Report (REN21, 2008), worldwide investment in renewable energy is growing fast. The report cites that the year 2007 showed a higher percentage of investments than at any other time, supported with a number of examples. The top three countries, according to the above report, were: Germany, with more than $14 billion, China with around $12 billion, and the United States with $10 billion. Around $4 billion was invested in plants and equipment for bio-fuel production and $16 billion of public and private funds were invested in research and development (REN21, 2008). In regard to recent investments (from 2011 to 2013), the investment in renewable energy worldwide is still rising. This is the case even though there was bad news during 2011 and 2012 concerning the solar energy field. The bankruptcy of Solyndra and the damaging competition between the United States and China (tariff wars) meant there was a 66% and 35% loss during 2011 and 2012, respectively, in regard to Ardour Solar Energy Index (SOLRX).

10.3 BIOMASS BUSINESS

How can biomass energy be an active trading partner in the international market? The following basic factors play an important role (Heinimö et al., 2007):

1. Politics connected to biomass energy
2. Market strategies and investment
3. Biomass research activities.

The countries that have signed the Kyoto Protocol are in a position to provide extra momentum for the renewable energy industries, in particular the countries where biomass energy is already part of their energy production. The ability to see what kind of future the world is heading toward in terms of energy supply, production, prices, and environmental issues (such as "climate change") are all part of the same momentum that can drive additional investments in this field. A model examining various possibilities would help to shape a better future for the biomass industries. For example, the Intergovernmental Panel on Climate Change (IPCC) provides a scenario in a similar way, foreseeing the future development of greenhouse emissions. Other research and studies related to future development should concentrate more on the energy aspect and related environmental issues. These examples of possible future outcomes connected to the energy field, in particular, and the environment, in general, would help in speeding up the development of clean energy to be traded on the international market.

A brief outline of major factors to encourage the growth of biomass businesses have been examined in the following sections. In order to understand the principles of the establishment and the growth of a biomass energy business, there are a number of basic steps which should be examined and considered. These steps have been explained in the next sections.

10.3.1 Step 1

In this step, which some researchers refer to as "level 1" or "the farm scale" (Graham and Walsh, 1995), economic characteristics of this level can be obtained through a local business (crop farm) analysis. In order for a complete

Table 10.1 Farm switch grass projected production characteristic: as stated by McLaughlin et al., 2002: "Influence of costs (−) and benefits (+) of switch grass production on costs of energy embodied in switch grass as a renewable energy feedstock"

Factors	Production Characteristics		
Farm gate price ($/Mg)	$30.31	$44.0	$52.37
Hectares planted to switch grass (millions)	3.1 ha	16.8 ha	21.3 ha
Yield (Mg/ha)	11.1 Mg/ha	9.4 Mg/ha	9.0 MG/ha

Source: Adapted from McLaughlin et al. (2002).

analysis to be achieved, the details concerning the production and input and output prices for all crops produced must be known:

1. Crop prices (production cost)
2. Volume of production
3. Expected profit.

One of the main problems for many farmers is the price fluctuation of their harvests, for example, cash grain prices received by farmers verses cost fluctuation of grains, such as in the case of soybean (for biodiesel) and corn (Manella, 2006). These changes (e.g., lower prices for their crops) have a negative effect on newly established farms who converted their main harvests to be sold as a source for energy generation.

An example of a farmed energy crop, such as Switch Grass (Table 10.1) may benefit a local farm. This is summarized in the following points (McLaughlin et al., 2002):

1. Direct benefits to the farm economy
2. Indirect economic benefits related to crop pricing
3. Subsidy farm payments
4. Soil and water improvement
5. Reduction in greenhouse emission.

10.3.2 Step 2

The economic study at this step (community or local society level) (Graham and Walsh, 1995) examines and investigates various factors that can affect the local economy directly and indirectly. This happens as a result of changes in local farming practice during the production of energy crops. Farming crops to be sold *solely* as a source for generating energy bring with it factors that can vary between local environments. However, in many cases, it is difficult to quantify these factors as it is difficult to measure all of them, that is, both before

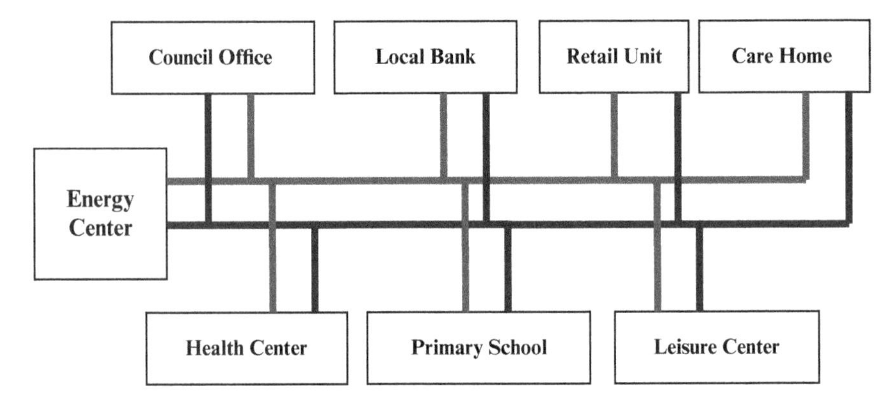

Figure 10.1 Energy Center system in a Scottish town providing heat/electricity to a number of local locations. *Source*: Adapted from Mcilwraith (2006).

biomass energy production is proposed and shortly after production has started. A network of a simple rural system is an essential part in the development of biomass energy at the community/village level. The main factors fall into the following categories:

1. Number and types of jobs gained and lost
2. Changes in local and national government law and regulations
3. New local infrastructure
4. Benefits and disadvantages for local business
5. Changes in prices for locally produced crops.

The system of biomass energy in a town or in a small city is sometimes referred to as "biomass community energy system" (Mcilwraith, 2006). The system supplies energy to the local population in the form of heating and/or electricity. Figure 10.1 provides an example illustrating the distribution of locally produced energy to main locations within a community in a Scottish town.

10.3.3 Step 3

This may involve a national economic study, which can be carried out over a much longer term for research and analysis. Clearly, the picture here and the consequent effects cover a wider area than the previous two levels. The national economy may depend partly on energy produced from crops* specifically farmed for this purpose. This kind of change can steer the economy in a new direction, simply by creating a self-dependency in the field of energy

* Farming output national statistics for individual countries. Energy crop production listed annually.

production. The change of direction may increase demands for land and water along with other requirements associated with the production of energy crops.

A national economic analysis may include the following main factors (Graham and Walsh, 1995):

1. Total national gross domestic product, that is, the positive and negative effects on the activity of the national economy
2. National job creation and job losses
3. National trade balance
4. Positive and negative impact on the national government expenditure
5. How it may affect national security with regards to energy supply
6. Possible changes in the national environmental regulations, their cost, and how these regulations can be applied
7. National crop prices.

Support from national governments promoting the development and strengthening of sustainable biomass energy businesses at this level is vital. The lack of regulations, laws and financial incentives to support growth related to this energy industry would make it very difficult, if not impossible, for businesses to establish themselves. Fortunately, various governments around the world, and the United States/Europe in particular, are providing *some* of the incentives and regulations biomass energy businesses require (Fig. 10.2).

Governmental support may include some of the following guidelines (Rösch and Kaltschmitt, 1999):

1. The handling of the support regulations should be simple and easy
2. Guaranteed funding applications within a short and defined timescale
3. Better coordination among national and local funding organizations.

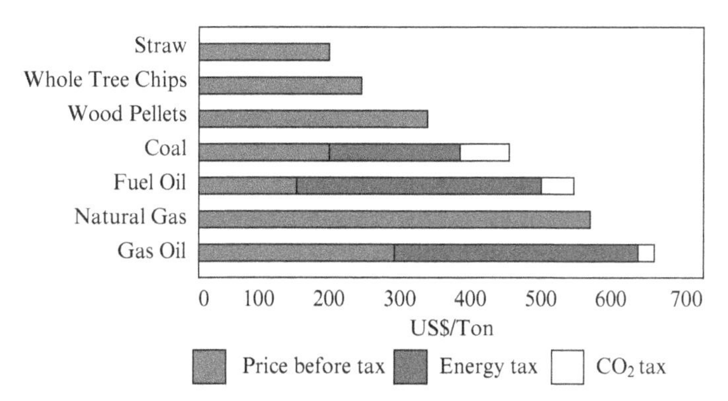

Figure 10.2 Example of national bio-energy support (Denmark national fuel types pricing differentiation for the purpose of promoting the development of biomass energy). *Source*: Adapted from Biomass Energy in ASEAN Member Countries (undated).

10.3.4 Step 4

The final step is related to the international level. An international economic analysis cannot be achieved without proper cooperation taking place in the field of energy crops—informed by well researched data.* This newly generated data should be exchanged freely, possibly through an international institute, such as the United Nations, with sufficient financial support alongside it.

The economic analysis would be on the work carried out individually by each country. The data collected would be compared with that from other countries. The pool of international data will eventually form the backbone for economic analysis, providing possible prospects and short and long term implications, that is, positive as well as negative effects on various parts of the world. Consequently, this data will give the world's economic output, based on biomass energy supply and its benefits and/or drawbacks.

The previous three steps are a very important part of the international level, and therefore, whatever happens to these levels may affect the results on the international level, directly or indirectly, especially if the changes are beyond the national government's control (such as in the case of natural disasters, unexpected crop failures, or as a result of political or armed activities). An international economic analysis is, in some ways, similar to analysis at a national level, except that security and supply are mostly considered for each individual country, as a whole.

Considerations for an international economic analysis are the following:

1. International economy gross product
2. National development based on energy production activity
3. International trade balance
4. The impact on total world expenditure
5. International environmental regulations, their cost and how these regulations can be applied in each individual country
6. Energy crop prices on the international market.

There are a number of other economic issues which should be looked at to ascertain whether any of them could be an obstacle. These issues can be related to *investments*, *technological issues*, and *methods of practice*.

Investment, whether using governmental or private funds, should not matter at the early stages of biomass business establishment. This is because financial investments are a lifeline and vital for the purpose of generating income and development. What matters is that the funding on a long-term basis should be secured and supported by the public along with governmental and nongovernmental organizations. Shares in the biomass industries and businesses should be encouraged and floated on the international market, similar to the

* Data related to a variety of crop production can be obtained from FAO.

present situation with fossil fuels. *Technological issues* can be dealt with through the present established hardware manufacturing industries, regardless of whether these industries are specialized in supplying power generating companies or not. Emphasis should be on specifications, requirements, new designs, and long term costs of maintenance for all types of energy production businesses. *Methods of practice* are the accepted procedures followed worldwide concerning the mechanisms of getting energy from fossil fuels. The idea here is to use the knowledge and expertise gained from the economic analysis of fossil fuels and apply it in a similar way to the economic aspects of a biomass energy business, with some modifications. Methods of practice can also mean the way that an energy business is being operated individually, whether in accordance to local and/or national requirements or not. Each business has its own procedure and methods, (mostly kept secret) on how to deal with certain aspects of production, business management, and marketing style. These individual methods of practice are unique and are constantly being researched and updated. They are in effect the actual reasons behind the success of *any* business, let alone energy business.

10.4 BIOMASS FUEL SUPPLY CHAIN

Developing highly efficient supply chains for biomass fuel is vital for the success of a biomass energy business. Biomass fuel, in comparison with fossil fuel, is a new type of commercial energy with which the market, in general, is still not fully familiar.

Table 10.2 and Table 10.3 give an example of three types of biomass transportation costs and estimated U.K. fuel prices. There are a number of important steps that need to be undertaken before the final biomass fuel is ready for use by the end user. These steps may include farming, harvesting, transport, processing, energy conversion, and supply to the end user. In each step, there will be the need for direct and/or indirect monetary investment. These types of financial injections, together with the appropriate expertise and technology, are necessary until the biomass energy business is able to generate its own profits.

The supply chain, whether for the purposes of obtaining raw materials and/or processing them (or simply to transport the fuel products to the end user), is not yet fully developed in most countries across the world.

Table 10.2 Estimated transportation cost of biomass fuel

Distance/Fuel	Logs (45% MC[a])	Chips (45% MC)	Pellets (10% MC)
50 km	7.9	12.4	4.1
200 km	20.8	24.1	11.1

[a]MC, moisture content (% weight basis).

Source: Adapted from Suurs (2002).

Table 10.3 Estimated U.K. fuel prices (excluding VAT)

Fuel	Price/Unit	kWh/Unit	Pence/kWh
Wood chips (30% MC)	£80 per ton	3500 kWh/t	2.3p/kWh
Wood pellets	£185 per ton	4800 kWh/t	3.9p/kWh
Natural gas	4.1p/kWh	1	4.1p/kWh
Heating oil	33p per liter	10 kWh/L	3.3p/kWh
LPG (bulk)	34p per liter	6.6 kWh/L	5.2p/kWh
Electricity	14.0p/kWh	1	14.0p/kWh

Source: Adapted from Biomass Energy Centre (2009).

To understand the mechanisms, and most importantly the costs involved, the regional supply chain from the original sources of the raw materials to the final destination must be examined. The main points of investigation can be summarized as follows:

1. The total cost of the first supply point in the chain plus the cost from the commercial farming site.
2. Knowing and calculating the biomass materials needed in the form of supply and demand for the region. This may help to establish a regular supply route that can provide the business with a lower cost within the supply chain.
3. Type of system being used and related cost issues.
4. Identifying other types of cost which may occur within the supply chain.

10.5 THE DEMAND FOR A NEW BIOMASS FUEL

The biomass energy market is slowly growing and trying to establish itself in competition with other sources of energy. Recent statistics indicate that the market for biomass energy is pointing upward, particularly in the West (Frankfurt School, 2012). For example, around 63% of renewable energy during 2012 is sourced from biomass within the European Union. However, the cost for this type of fuel is still high compared with fossil fuels.

Product demand is a yardstick for the success or failure of the biomass fuel production industry and for its future implications. Without regular demand from the commercial market (international, national, and local), as well as from individual customers requiring fuels produced from biomass sources, long-term success is nothing but a mirage. These demands are vital, as without them no progress can be made.

The production of a new type of biomass fuel is a necessary step to fill the gap in a highly demanding energy market. For this reason, the characteristics of a new biomass fuel should satisfy the requirements and needs of the market/end users (Table 10.4). These demands are set within the areas of the fuel

Table 10.4 SFS compared with coal

Fuels	SFS	Coal
Type	Renewable	Nonrenewable
Source	Crops—unlimited	Mining—limited
CO_2 Emission	Negative	Positive
Fuel quality	High	High
Prices	Lower than present coal prices as indicated by the original cost of raw materials and the cost of processing	High
Fuel commercial transportation use	Yes	No
Fuel commercial systems heating use	Yes	Yes
Fuel commercial electricity generating use	Yes	Yes
Fuel storage	Dry environment	Common present storage facilities at power stations
Fuel transportation	Similar to coal	Common present transportation facilities
Market	Dependent on market demand	High demand
Availability	Dependent on investment and marketing	Positive

Source: Author.

quality as well as market availability and price. In addition, it should be possible for the new fuel to be used within present combustion systems with minor adjustments if necessary.

The research related to the current biomass materials formed during the compilation of this book has looked at the above factors and implemented them through the selection process discussed in the previous chapters.

10.6 THE SFS ECONOMIC VALUE SCENARIO

There are a number of economic advantages when using Super Fuel sample (SFS) as a biomass fuel:

1. Low market cost as the fuel is made up from four/five different crop sources, that is, the market value of the main elements is divided by four or five. This in turn should not destabilize the market prices for the crops, as there may be an increase in demand of 20% in total and not 100% as in the case of one single biomass crop or raw biomass material presently being used in the production of biomass fuel.

2. The four/five crops that makeup the SFS are available in most local markets in various parts of the world, including the United Kingdom. This means that there will be no need to import any of the original crops and that in itself will save energy and may reduce cost, as well as avoid the emission of CO_2.
3. The mixing of the various biomass crop components to produce SFS can vary and may change according to the required grade of the fuel and its use. This kind of flexibility in production and the ability to create different varieties and grades of SFS make this fuel much more desirable on the local and international market. Different grades can be used for a selection of different applications with accordingly low or high prices.
4. Storage and lifecycle times of the SFS can be sustained as long as any of the fossil fuels available presently on the market. This is because the SFS ingredients (and the proportions being used in the mixing recipe) are measured and mixed according to certain factors, such as the length of the storage period required, where the fuel needs to be used, the grade type, and any other special specifications required.

Thus, cheaper types of SFS (which need to be used within a short period of time) only require a short storage time in their life cycle. This means reducing the processing of the original ingredients, which will reduce the cost further. Storing biomass materials for longer periods can be achieved both successfully and economically by employing ancient techniques, that is, silage methods.* This can also done by preventing microbial growth taking place on the ingredients before and after mixing. Fuel used for power generating companies will be mostly in the form of powder or pellets. Tests have shown within the food and feed industries that by removing the air completely from container or large storage systems, the biomass materials will stay in the original form for a very long time. On the other hand, if the SFS is in a liquid form, then a percentage of preservative, that is, alcohol (which can be formed as a by-product from the process of making the SFS fuel), will preserve the liquid biomass fuel for a longer period.

10.7 DISCUSSION

As mentioned on a number of occasions, the final biomass fuel source (SFS) is madeup from four or five selected biomass samples. Using the methodology REA1, the selection process to help in the making of the new fuel is designed with great emphasis on the business factors (BF). For this reason,

* An ancient commonly used method to preserve cattle and sheep feed (with a high percentage of moisture) via controlled fermentation.

a large number of factors related (in one way or another) to the commercial side of power generating have been included. This is in addition to the scientific and technical (S&T) factors, which can cover many aspects related to the *quality* of the fuels presently being used worldwide. The fitness data for the top four or five biomass samples (considering only business and economical factors), has been calculated by the methodology employed within the REA1 system (see Chapter 9). The S&T testing factors section of the methodology showed that sunflower and rapeseed were also well suited as a biomass energy product. switch grass, straw pellets, and apple prunings were included in the top seven samples. Concerning the new bio-fuel price, SFS should be lower than the price of coal fuel, as indicated in the present market prices of the five crop products. There is no point in introducing a new type of fuel if the price is higher than the fossil fuel, even if the new fuel is 100% environmentally friendly. Therefore, how can a newly introduced fuel be cheaper than present fossil fuels? Apart from the original principle, which has already been explained, that is, the four/five crops that may influence the final price of SFS, the price is predicted to be low simply because the demand on the market for the biomass materials will be divided by four or five. There are two additional factors that can reduce the price of SFS. First, the simple production mechanism means that there are no huge costs during the mining and refining process, unlike fossil fuels such as crude oil. Second, the nonexistent cost of long-distance transportation, as mentioned earlier. SFS will be produced locally in each country, that is, the fuel transportation from one country to another does not exist, and this in itself will further reduce the cost to the end user. SFS can be used for transportation, for heating/cooling systems, and for generating electricity. Initially, there will be three main basic types of SFS fuel, solid, liquid, and gas fuels. The solid type, in the form of pellets, powder, or briquettes, will be used by the power generating companies. The liquid and gas fuels can be used for transportation and heating/cooling systems. The storage of SFS (solid fuel) is similar to the present facilities for storing coal, but with emphasis on a dry environment (for a short period of time). This means there is no need to build or design new storage facilities, specifically for the purpose of storing the new fuel. The present local transport system for coal, that is, for short distances only, can be applied in transporting SFS fuels whenever this is necessary. Transportation using pipes can be used for the liquid and gas type of SFS. Market demand for SFS depends mostly on the availability of the fuel on the local and international markets, as well as the price when compared with other fossil fuel products (Willis et al., 2008).

Since the quality of SFS will be close to the fossil fuel quality, the market demand is likely to be high, especially if the prices remain low or close to the present fossil fuel prices. Solutions for various negative characteristics (Table 10.5) (e.g., energy content, ignition quality, and lifecycle) of any biomass fuel can be found and applied according to the type of use for the fuel, that is, the grade required and the acceptable market price worldwide.

Table 10.5 Negative characteristics of biomass fuel

Characteristics	Problem	Solution
Low pH (acidic)	Corrosion	Balancing materials composition, neutralization of acid
High viscosity	Handling	Add liquid (e.g., water)
	Pumping	Add solvent
Instability and temperature sensitivity	Storage	Avoid contact with hot surfaces
	Phase Separation	
	Decomposition and Gum formation. Viscosity increase.	Stabilisation or refining through catalytic treatment Add dilutants or water
Char and solid contents	Combustion problems. Equipment blockage. Erosion.	Filtration Hot gas filtration
Alkali metals	Deposition of solids in boilers, engines and turbines	Biomass pretreatment Hot gas filtration Catalytic upgrading
Water content	Complex effect on problem recognition heating value, viscosity, pH, homogeneity, and other characteristics	Problem recognition Optimization and control of water content according to application

Source: Adapted from Song and Elliott (2007).

10.8 CONCLUSION

There are several barriers to the adoption of renewable energy technologies but opportunities exist to overcome them, in particular when fossil fuel prices are constantly rising on the international market. These constraints are mostly financial. As the technologies are new to the capital markets, there is more risk in investing in renewable energies, including biomass energy, than for using established technologies. The higher the perceived risk, the higher the required rate of return demanded on capital. The perceived length and difficulty of the permitting process is an additional determinant of risk. Other possible market risks for biomass energy businesses are mainly due to future changes in the energy market.

1. Due to changes in the energy market, some analysts believe there will be uncertainties concerning demand for heat or electricity (or both).
2. Some of the guaranteed reimbursements/support concerning electricity and/or heat produced from biomass may change due to changes in environmental and national energy regulations.

3. The future will be different for biomass energy production in Europe as the European Union has proposed changes in the Common Agriculture Policy (CAP), which may change land regulations for energy crop production.
4. When planning and starting a new biomass energy business, a basic but important four steps (Section 10.3) should be considered and implemented whenever the factor/case is relevant. The "biomass fuel supply chain" is concluded to be one of the most important factors for the success of the biomass energy market to the general public.

Incentives, such as grants and/or low interest loans or loan guarantees, might serve to reduce perceived investor risk. Tax credits for renewable energy technology production through the early high-risk years of a project may provide another mechanism for biomass energy development when producing a new type of biomass-derived environmentally friendly fuel (SFS). Keeping the prices and the availability of the fuel in a stable form, that is at an affordable price for the individual user and easily obtainable at numerous locations, are important to the commercialization process. This kind of stability will reduce the pressure on the market for the original sources of the materials of which the SFS is composed (Box 10.1 and Box 10.2).

Box 10.1

RENEWABLE TRANSPORT FUEL OBLIGATION (RTFO)

The RTFO obliges fossil fuel suppliers to produce evidence that a specified percentage of their fuels for road transport in the UK comes from renewable sources, including biomethane. This can include the use of biomethane as a road transport fuel.

Defra (2011)

Box 10.2

RAPESEED MEAL (RM)

Rapeseed meal is a by-product obtained from crushing rapeseed in order to obtain oil. It can be obtained during anytime of the year. RM has the tendency to draw moisture easily; therefore, when it comes to storage, certain conditions have to be followed. These include the dryness of the environment in the place of storage, the temperature (low temperature). If the conditions of RM are required to stay the same before being burned, then RM has only a limited period of storage (approximately 2 months).

The market for RM is within the following areas:

a. Co-firing for power stations
b. Gasification
c. Pyrolysis
d. Animal feed
e. Industrial uses for proteins in the meal (e.g., bioplastics for biodegradability and coatings for water resistance)

The chemical elements of rapeseed meal are the following:

Carbon	43.34%
Hydrogen	5.94%
Nitrogen	7.41%
Oxygen	43.27%
Sulphur	~0.04%

Source: Author.

A sample of rapeseed meal.

REFERENCES

Berndes G, Hoogwijk M et al. (2003) The contribution of biomass in the future global energy supply: a review of 17 studies. Biomass and Bioenergy 23:1–28.

Biomass Energy Centre (2009) Facts and figures, fuel cost per kWh. http://www.biomassenergycentre.org.uk/portal/page?_pageid=75,59188&_dad=portal&_schema=PORTAL (last accessed July 8, 2009).

Biomass Energy in ASEAN Member Countries (undated) Biomass: more than a traditional form of energy. FAO regional Wood Energy Development Programme in Asia in cooperation with the ASEAN-EC Energy Management Training Centre and the EC-ASEAN COGEN Programme. http://144.16.93.203/energy/HC270799/RWEDP/acrobat/asean.pdf (last accessed April 19, 2009).

Defra (2011) Anaerobic Digestion Strategy and Action Plan. http://www.defra.gov.uk/publications/files/anaerobic-digestion-strat-action-plan.pdf (last accessed August 11, 2012).

Frankfurt School (2012) Global trends in renewable energy investment 2012. http://fs-unep-centre.org/sites/default/files/publications/globaltrendsreport2012final.pdf (last accessed April 5, 2013).

Graham RL, Walsh ME (1995) Evaluating the economic costs, benefits and tradeoffs of dedicated biomass energy systems: the importance of scale. From the Proceedings, Second Biomass Conference of the Americas: Energy, Environment, Agriculture and Industry; pages 207–215. Meeting held August 21–24, Portland, Oregon; published by National Renewable Energy Laboratory, Golden, Colorado. http://bioenergy.ornl.gov/papers/bioam95/graham1.html (last accessed July 11, 2008).

Heinimö J, Pakarinen V, Ojanen V, Kässi T (2007) International bioenergy trade—scenario study on international biomass market in 2020. Lappeenranta University of Technology, Department of Industrial Engineering and Management Research Report 181.

Manella M (2006) The federal biomass R&D technical advisory committee updates its vision for bioenergy and biobased products in the United States. Biomass Initiative newsletter, feature article, October 2006. http://209.85.229.132/search?q=cache:HXwXKcaqgr4J:www.danbio.info/Admin/Public/DWSDownload.aspx%3FFile%3D%252FFiles%252FFiler%252FBiomass%252FBiomass_October_Newsletter.pdf+establishing+biomass+business+graph&cd=8&hl=en&ct=clnk&gl=uk (last accessed April 19, 2009).

Mcilwraith B (2006) Glenshellach biomass district heating project. Slides presentation, Scottish Community Renewables Network, WHHA Glenshellach Biomass Community Energy System.

McLaughlin SB, Ugarte DG, Garten CT, Lynd LR, Sanderson MA, Tolbert VR, Wolf DD (2002) High value renewable energy from prairie Grasses. Environmental Science and Technology 36:2122–2129. http://pubs.acs.org/doi/full/10.1021/es010963d?cookieSet=1 (last accessed March 4, 2009).

Murphy ML (2001) Repowering options: retrofit of coal-fired power boilers using fluidized bed biomass gasification. Energy products of Idaho. http://www.energyproducts.com/documents/Gasifier%20Retrofit%20MLM.PDF (last accessed July 11, 2008).

REN21 (2008) Renewables 2007: global status report. Renewable Energy Policy Network for the 21st Century. http://www.ren21.net/pdf/RE2007_Global_Status_Report.pdf (last accessed May 4, 2009).

Rösch C, Kaltschmitt M (1999) Energy from biomass—do non-technical barriers prevent an increased use? ScienceDirect, Biomass and Bioenergy 16(5):347–356.

Song C, Elliott DC (2007) Thrust 2: utilization of petroleum refinery technology for biofuel production. Slides presentation—Pacific Northwest National Laboratory, Battelle, U.S. Department of Energy. http://www.ecs.umass.edu/biofuels/Presentations/Thrust2-Overview.pdf (last accessed October 8, 2008).

Suurs RA (2002) Long distance bioenergy logistics: an assessment of costs and energy consumption for various biomass energy transport chains' Utrecht University Copernicus Institute.

Willis A, Furey J, Franco E, Hurd J (2008) Biomass vs. fossil fuels. Biomass—website data and information. http://blogs.princeton.edu/chm333/f2006/biomass/comparison/biomass_vs_fossil_fuels/ (last accessed August 31, 2008).

11

CONCLUSION

11.1 GENERAL CONCLUSION

This book has examined in detail the selection process of biomass materials for the purpose of using them as a source of a fuel. In the process, the study has presented an insight into the field of renewable energy in general, and biomass energy in particular. This book also provided a mapping process for a number of biomass samples, examining in four different steps the viability of each biomass sample in the initial stage of the first selection process. In addition to this, the book provided full details regarding the construction process of Renewable Energy Analyser One (REA1) methodology in three successive chapters. The survey results and the results obtained via the application of the methodology have covered 15 different biomass samples, in addition to coal used as the main reference sample. Finally, a general economic study of biomass energy leading toward the development of a new hybrid super fuel sample (SFS) fuel has provided a guideline for a prospective commercial biomass energy enterprise.

The Selection Process of Biomass Materials for the Production of Bio-fuels and Co-firing, First Edition. Najib Altawell.

11.2 METHODOLOGY (REA1) AND APPLICATIONS

Examining the overall subject of renewable energy can certainly provide additional insight into the strengths and weaknesses of this new field of "commercial" energy sources. This book has reviewed a large number of research projects that involved different people with different backgrounds, various institutes, organizations and businesses. It looked at different and sometimes unexpected results, and raised large numbers of questions. Consequently, a general conclusion is borne out here as a result of what was carried out and achieved in this book, that is, a methodology that can work not just for biomass energy sources, but for a variety of renewable sources of energy, as well as for fossil fuels.

The methodology therefore can be applied as follows:

1. For any biomass or fossil fuel business where biomass or fossil fuels selection is vital to their work. The REA1 application can be used either for the selection of fuels on a commercial basis or from a scientific and technical viewpoint, or for both of them at the same time.
2. For the purpose of selection of biomass materials for co-firing purposes.
3. For the purpose of selection for the production of various types of biofuels, including those obtained from pyrolysis and gasification. With minor adjustments, the present methodology can also be used for other types of renewable energy production processes.

11.3 WHY BIOMASS?

What's left is the final closing question, which was frequently asked and discussed during the work on this book, that is, why should it be biomass and not other forms of renewable sources of energy? Again, consider the question: "Why should biomass energy work better than any other source of energy?" The answer is simple. When a methodology "made to measure," so to speak, is applied and correctly calculates various factors associated with scientific, technical, and business sides of biomass materials, then the result, in practical terms, can be achieved by any commercial business dealing with biomass energy alone and/or co-firing it with different types of fuel. How it is possible to conclude that biomass is the answer to the present increasing demand for global energy when the commercial long-term case is still not yet fully proven? At the present time, much work and research across the globe is aimed at just that, that is, building the basis of what will be needed (now or later) in order to provide fast, accurate, and affordable ways of employing the principles of biomass energy on a large commercial and international scale. The study of the basic principles of biomass energy sources and the methods/mechanisms of applying it made it desirable to put a case forward for a simple but comprehensive approach in the form of a new biomass methodology. In any technology, no work is complete without the demonstration of practical application. Thus, Chapter 9 of this book has produced results from the methodology applications. Chapter 10 then describes and examines an economical approach with particular emphasis on the creation of a new biofuel. The SFS is a designed hybrid fuel model added by the author at the end of this chapter (Fig. 11.1).

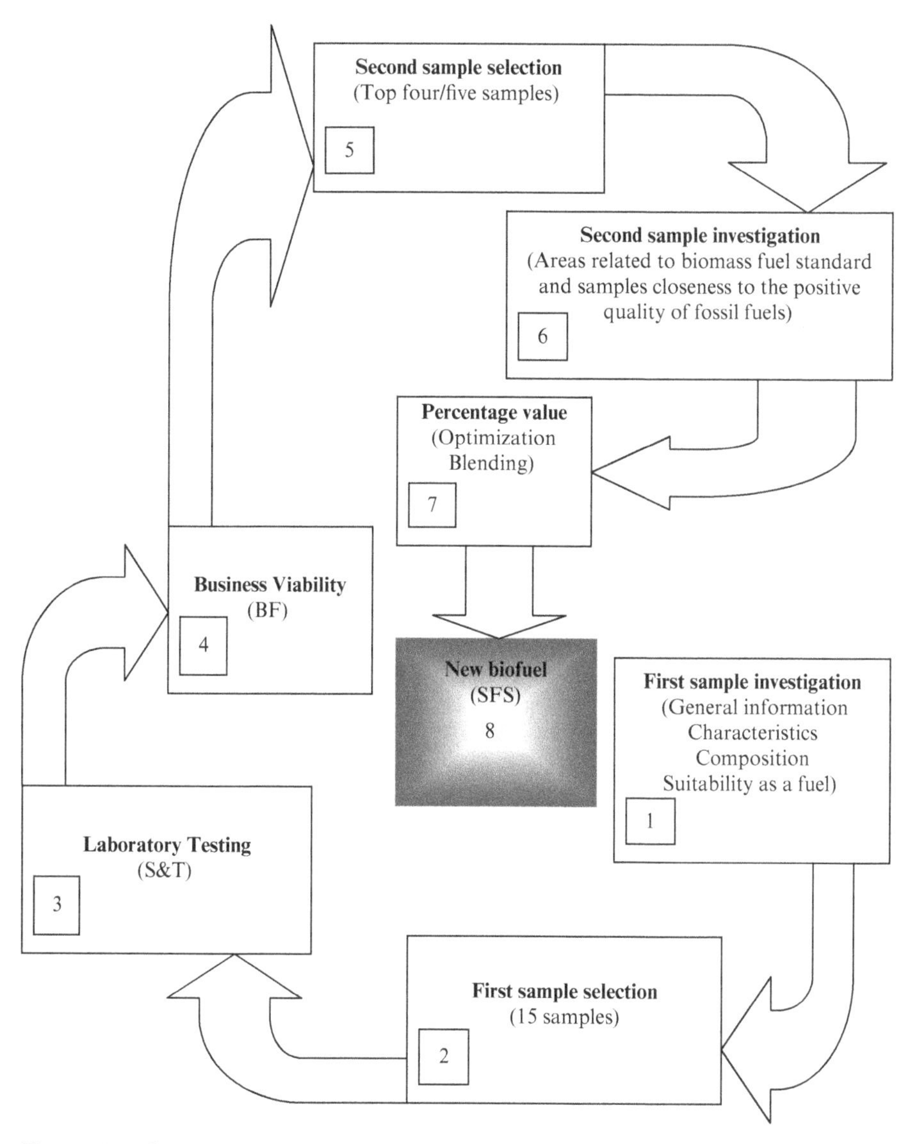

Figure 11.1 Selection model for biomass materials (leading to the production of SFS bio-fuel). *Source*: Author.

11.4 CO-FIRING AND POWER GENERATING

For commercial biomass energy production to establish itself and become an important source of energy, that is, competing at the same level as fossil fuels on the global market, regular awareness programs, and support should be launched by international, national, and local governments. The support programs should include solutions for co-firing problems and costs, as considered within various parts of this book. In a similar way, comprehensive support and publicity for some of the top biomass energy crops would act as another incentive for the farmers and power generating companies to take part in the development. Biomass samples and other types of energy crop have been examined and analyzed in Chapter 4.

The commercial biomass business is still lagging far behind biomass energy research. This is due to a number of reasons, but possibly the most significant is that biomass energy is not yet fully able to pay off large return dividends for potential investors. Yet at the same time, large commercial companies do not want to be left behind when it comes to renewable energy, in particular biomass energy. They were therefore among the first to start their own research in this field.

The survey completed in this book has found out that a number of businessmen, engineers, researchers, and technicians (in the field of biomass energy) still believe that there is not enough support at the present time for co-firing methods and for the bio-fuels industries in general. For this reason, biomass business viability was among one of the most important factors answered and commented upon in the questionnaires. The survey also showed indirectly, that is, via face-to-face questions and answers, that co-firing is still one of the most popular and economical methods used by the power generating companies in introducing biomass materials for the purpose of generating electricity.

At the beginning of researching this book, only literature review materials available were the guideline for the business factor (BF) and scientics and technical (S&T) methodology sections. By completing the survey, a new source of data presented itself as a base for obtaining factual results. It is hoped that this information, which is at the core of providing the final data in Chapters 8 and 9, will play an important role for the power generating companies using the biomass materials selection process to select materials for co-firing. The results can also launch further developments and additional applications within the business, scientific, and technical aspects of biomass, with particular emphasis on bio-fuels and co-firing.

11.5 THE NEW BIOMASS FUEL (SFS)

The goal throughout the past few years while researching this book was new type of biomass fuel should be the end result. This type of fuel should fulfill certain criteria , which the majority of biofuels available on the market cannot provide. The characteristics of SFS bio-fuel have been investigated in Chapter 10. Fulfilment of certain conditions, such as price, quality, and availability, storage, can be partly made from the results of biomass materials added together to form a correct ratio mixture (recipe) needed for the SFS. The number of biomass materials needed for the mix has already been decided upon, that is, between four and five samples. However, details of the percentages for the mix from each sample (methodology 2, i.e., REA2) would take this book beyond the scope of the present research. SFS can easily be blended with other types of fuels, just as in the case of using it for the co-firing. The process works in the same way when using SFS in liquid and gas forms. Fortunately, liquefaction of the SFS can be achieved without the use of additional energy, that is, in the form of heating, such as the method used in pyrolysis.

11.6 THE FUTURE OF CO-FIRING AND BIOMASS ENERGY

The growth of biomass as a source of energy in a number of countries is developing faster than many expected. For example, from 2003 to 2005, the market growth in biomass was 11.8%, and from 2004 to 2005, there was a growth of 16.1% in electricity production (Ballard, 2007). These figures for growth are the result of support by governments, which consequently may not reflect a clear picture of how the growth of biomass energy industries would be if there were no governmental support/intervention.

Depending on the type of technology used and the type of biomass materials, the cost of producing energy from biomass materials can vary. The type and capacity of the power plant also have an important impact on the final cost. Reportedly, co-firing at the present is most cost effective when it uses of biomass materials (U.S. Department of Energy, 2009). Financial costs for projects which deal with co-firing can be small. Co-firing ranges from 1 to 30 MW of bio-power capacity so the cost itself can be recovered within two years when low-cost biomass fuels have been used (U.S. Department of Energy, 2009) which is less than other types of renewable energy.

Regardless of whether co-firing is the best route for the power generating companies, when it comes to using biomass as an alternative to co-firing in terms of cost, the future of biomass energy is still in the hands of the investors and, eventually, the end users. This means that a confident market in the shape of stable investment and growing trading are both vital for the future of biomass energy and co-firing. Alongside them are continuous developments in the scientific and technical aspects that hopefully can lead to a better quality fuel at a lower cost. At the same time, environmental protection should be the number one priority as without this, the whole process of producing a new type of commercial fuel would bring us back to square one.

11.7 FINAL RESULTS AND FINAL CONCLUSION

The overall final conclusion for REA1 methodology factors and biomass samples can be summarized in the following points:

1. The study showed that within each factor of BF and S&T within the survey method, the accuracy was around 97%, that is, an average error of 3%.
2. The study also showed that the average accuracy of BF and S&T sections is 96%, that is, an average error of 4%.
3. For the methodology as a whole, the study showed that the average accuracy is 95%, that is, an average error of 5%.
4. For S&T, the most important factor voted for was "energy." The factor "nitrogen emission" received the lowest number of votes. For BF, "business viability" received the highest number of votes, while the factor "land and water," received the lowest number of vote.
5. The highest total fitness for S&T alone is for rapeseed (26%), while the lowest S&T total fitness is for switch grass (6%). For BF, the lowest is for rice and rapeseed meal (~13%); the highest is for rapeseed (~26%).
6. The methodology results showed that rapeseed has the highest total final fitness (53%), while switch grass and rapeseed meal received the lowest total final fitness (21%).
7. The study has shown that the final top four biomass samples are: rapeseed, sunflower BS, niger seed, and apple tree wood (apple P).
8. From the top four (or five) biomass samples, the book outlined an approach for a new type of hybrid biomass fuel. This type of bio-fuel can be produced without significant impact on the local, national, and international market. The SFS bio-fuel can be optimized at a level close to the positive quality of a fossil fuel, such as coal, that is, the reference sample used in this book.

11.8 POSITIVE OUTLOOK

Visiting biomass businesses and power generating companies, and attending conferences and workshops, the author has found a great deal of support and enthusiasm from researchers, businesses, and power generating companies who deal, in one way or another, in the field of biomass energy. The public support for a sustainable environmentally friendly fuel is already there. What the general public would like to see is better global coordination and more investment in the field of biomass energy.

Some of the institutes and companies visited provided presentations and training on a number of issues related to renewable energy. This in turn has contributed directly and indirectly to the research itself in various positive ways. The next step of continuing the research in this field to a more specialized level, that is, for developing SFS, is vital, rather than allowing it to be left on a shelf, as part of a report accumulating dust.

11.9 WHAT NEXT?

Despite all the difficulties and changes which have taken place during 4 years of research compiling this book, the work has been successfully completed on time, providing a useful contribution in the form of practical results. These results will benefit everyone who took part in this research and those working in the same or related fields.

The details provided in this book should not be forgotten or ignored, especially when the time comes for another decision by the power generating companies to increase (at a higher percentage due to renewable obligation certificate [ROC]) the usage of *selected* biomass materials for the purpose of generating electricity.

This book has provided factual figures indicating that the prospect of biomass materials can be part of a wider commercial energy use. This prospect is possible to achieve worldwide. However, trading biomass energy on "international" level, similar to present day fossil fuel trading, can only happen when additional investments, further research, and international laws and policies in this field are already in place.

REFERENCES

Ballard G (2007) Opportunities for biomass energy programmes: experiences & lessons learned by UNDP in Europe & the CIS. Final report.

U.S. Department of Energy (2009) State energy alternatives: biomass energy. Energy efficiency and renewable energy. http://apps1.eere.energy.gov/states/alternatives/biomass.cfm (last accessed April 26, 2009).

INDEX

Note: Page entries in *italics* indicate figures; tables are noted with *t*.

The Selection Process of Biomass Materials for the Production of Bio-fuels and Co-firing, First Edition. Najib Altawell.

Series Editor: M. E. El-Hawary, Dalhousie University, Halifax, Nova Scotia, Canada

The mission of IEEE Press Series on Power Engineering is to publish leading-edge books that cover the broad spectrum of current and forward-looking technologies in this fast-moving area. The series attracts highly acclaimed authors from industry/academia to provide accessible coverage of current and emerging topics in power engineering and allied fields. Our target audience includes the power engineering professional who is interested in enhancing their knowledge and perspective in their areas of interest.

1. *Principles of Electric Machines with Power Electronic Applications, Second Edition*
M. E. El-Hawary

2. *Pulse Width Modulation for Power Converters: Principles and Practice*
D. Grahame Holmes and Thomas Lipo

3. *Analysis of Electric Machinery and Drive Systems, Second Edition*
Paul C. Krause, Oleg Wasynczuk, and Scott D. Sudhoff

4. *Risk Assessment for Power Systems: Models, Methods, and Applications*
Wenyuan Li

5. *Optimization Principles: Practical Applications to the Operations of Markets of the Electric Power Industry*
Narayan S. Rau

6. *Electric Economics: Regulation and Deregulation*
Geoffrey Rothwell and Tomas Gomez

7. *Electric Power Systems: Analysis and Control*
Fabio Saccomanno

8. *Electrical Insulation for Rotating Machines: Design, Evaluation, Aging, Testing, and Repair*
Greg Stone, Edward A. Boulter, Ian Culbert, and Hussein Dhirani

9. *Signal Processing of Power Quality Disturbances*
Math H. J. Bollen and Irene Y. H. Gu

10. *Instantaneous Power Theory and Applications to Power Conditioning*
Hirofumi Akagi, Edson H. Watanabe, and Mauricio Aredes

11. *Maintaining Mission Critical Systems in a 24/7 Environment*
Peter M. Curtis

12. *Elements of Tidal-Electric Engineering*
Robert H. Clark

13. *Handbook of Large Turbo-Generator Operation and Maintenance, Second Edition*
Geoff Klempner and Isidor Kerszenbaum

14. *Introduction to Electrical Power Systems*
Mohamed E. El-Hawary

15. *Modeling and Control of Fuel Cells: Distributed Generation Applications*
M. Hashem Nehrir and Caisheng Wang

16. *Power Distribution System Reliability: Practical Methods and Applications*
Ali A. Chowdhury and Don O. Koval

17. *Introduction to FACTS Controllers: Theory, Modeling, and Applications*
Kalyan K. Sen and Mey Ling Sen

18. *Economic Market Design and Planning for Electric Power Systems*
James Momoh and Lamine Mili

19. *Operation and Control of Electric Energy Processing Systems*
James Momoh and Lamine Mili

20. *Restructured Electric Power Systems: Analysis of Electricity Markets with Equilibrium Models*
Xiao-Ping Zhang

21. *An Introduction to Wavelet Modulated Inverters*
S.A. Saleh and M.A. Rahman

22. *Control of Electric Machine Drive Systems*
Seung-Ki Sul,

23. *Probabilistic Transmission System Planning*
Wenyuan Li

24. *Electricity Power Generation: The Changing Dimensions*
Digambar M. Tigare

25. *Electric Distribution Systems*
Abdelhay A. Sallam and Om P. Malik

26. *Practical Lighting Design with LEDs*
Ron Lenk and Carol Lenk

27. *High Voltage and Electrical Insulation Engineering*
Ravindra Arora and Wolfgang Mosch

28. *Maintaining Mission Critical Systems in a 24/7 Environment, Second Edition*
Peter Curtis

29. *Power Conversion and Control of Wind Energy Systems*
Bin Wu, Yongqiang Lang, Navid Zargari, and Samir Kouro

30. *Integration of Distributed Generation in the Power System*
Math H. Bollen and Fainan Hassan

31. *Doubly Fed Induction Machine: Modeling and Control for Wind Energy Generation Applications*
G. Abad, J. Lopez, M.A. Rodriguez, L. Marroyo, and G. Iwanski

32. *High Voltage Protection for Telecommunications*
S. Blume

33. *Smart Grid: Fundamentals of Design and Analysis*
James Momoh

34. *Electromechanical Motion Devices, Second Edition*
Paul Krause, Oleg Wasynczuk, and Steven Pekarek

35. *Electrical Energy Conversion and Transport: An Interactive Computer-Based Approach, Second Edition*
George G. Karady and Keith E. Holbert

36. *ARC Flash Hazard and Analysis and Mitigation*
Jay C. Das

37. *Handbook of Electrical Power System Dynamics: Modeling, Stability, and Control*
Mircea Eremia and Mohammad Shahidehpour

38. *Analysis of Electric Machinery and Drive Systems, Third Edition*
Paul Krause, Oleg Wasynczuk, Scott Sudhoff, and Steven Pekarek

39. *Extruded Cables for High Voltage Direct Current Transmission: Advances in Research and Development*
Giovanni Mazzanti and Massimo Marzinotto

40. *Power Magnetic Devices: A Multi-Objective Design Approach*
Scott Sudhoff

41. *Risk Assessment of Power Systems: Models, Methods, and Applications, Second Edition*
Wenyuan Li

42. *Practical Power System Operation*
Ebrahim Vaahedi

43. *The Selection Process of Biomass Materials for the Production of Bio-fuels and Co-firing*
Najib Altawell